Principles of
Statistical Radiophysics 4

S. M. Rytov   Yu. A. Kravtsov
V. I. Tatarskii

Principles of
Statistical Radiophysics 4

Wave Propagation Through Random Media

With 43 Figures

Springer-Verlag Berlin Heidelberg New York
London Paris Tokyo Hong Kong

Professor Sergei M. Rytov

Corresp. Member of the USSR Academy of Sciences,
Department of General Physics and Astronomy, USSR Academy of Sciences,
SU-117901 Moscow, USSR

Professor Yurii A. Kravtsov

General Physics Institute, USSR Academy of Sciences,
SU-117942 Moscow, USSR

Professor Valeryan I. Tatarskii

Corresp. Member of the USSR Academy of Sciences, Inst. of Atmospheric Physics,
USSR Academy of Sciences, Pyzhevsky Per., SU-109017 Moscow, USSR

Translator

Alexander P. Repyev

Samarkandsky Boulevard 13-3-63, SU-109507 Moscow, USSR

Title of the original Russian edition: *Vvedenie v statisticheskuyu radiofiziku II,* sluchainuie polya
2. revised and enlarged edition
© Nauka, Moscow 1978

ISBN-13:978-3-642-72684-2   e-ISBN-13:978-3-642-72682-8
DOI: 10.1007/978-3-642-72682-8

Library of Congress Cataloging-in-Publication Data. (Revised for volume 4) Rytov, S.M., 1908-. Principles of
statistical radiophysics. Translation of: Vvedenie v statisticheskuĭu radiofiziku. 2. rev. and enl. ed. Includes
bibliographies and indexes. Contents: 1. Elements of random process theory–2. Correlation theory of random
processes–v. 4. Wave propagation through random media. 1. Radio waves–Mathematics. 2. Stochastic processes. I. Kravt͡sov, I͡Urii Aleksandrovich. II. Tatarskiĭ, V. I. (Valer'i͡an Il'ich). III. Title. QC661.R9213   1987
537.5'34   87-4644

© Springer-Verlag Berlin Heidelberg 1989
Softcover reprint of the hardcover 1st edition 1989

2157/3150-543210 – Printed on acid-free paper

Foreword

Principles of Statistical Radiophysics is concerned with the theory of random functions (processes and fields) treated in close association with a number of applications in physics. Primarily, the book deals with radiophysics in its broadest sense, i.e., viewed as a general theory of oscillations and waves of any physical nature[1]. This translation is based on the second (two-volume) Russian edition. It appears in four volumes:

1. Elements of Random Process Theory
2. Correlation Theory of Random Processes
3. Elements of Random Fields
4. Wave Propagation Through Random Media.

The four volumes are, naturally, to a large extent conceptually interconnected (being linked, for instance, by cross-references); yet for the advanced reader each of them might be of interest on its own. This motivated the division of the *Principles* into four separate volumes.

The text is designed for graduate and postgraduate students majoring in radiophysics, radio engineering, or other branches of physics and technology dealing with oscillations and waves (e.g., acoustics and optics). As a rule, early in their career these students face problems involving the use of random functions. The book provides a sound basis from which to understand and solve problems at this level. In addition, it paves the way for a more profound study of the mathematical theory, should it be necessary[2]. The reader is assumed to be familiar with probability theory.

In progressing from one volume to the next, the reader will see that the physical problems under consideration become more complex and the mathematical machinery more involved. This results quite naturally from the course-oriented origin of the book, based as it is on summarized lecture notes. Their extensive teaching experience has convinced the authors that such an approach is preferable to a more uniform presentation. Each chapter is followed by a number of problems worked

---

[1] It should be noted that certain questions in statistical radiophysics are not covered in this book. For example, it is neither concerned with quantum radiophysics and quantum electronics, nor with statistical phenomena related to the propagation of waves in nonlinear media, see [1].

[2] Treatments of the mathematical theory are given, for instance, in [2–9], and more specialized applications of the theory to radiophysics and engineering are to be found in [10–19].

out in detail. The problems not only serve as exercises, but in many cases contain additional theoretical material and further literature references.

Moscow, March 1986

S.M. Rytov
Yu.A. Kravtsov
V.I. Tatarskii

Preface

Volume 4 concludes the *Principles of Statistical Radiophysics* and is devoted to specific applications of the general theory of random fields. This is a very extensive field which has to a certain degree been touched upon in Chap. III.4.[1] There we discussed a medium with inhomogeneities that are so small that it is justified to restrict the solution of the wave equation to the first approximation of perturbation theory, i.e., to consider only single scattering events.

Volume 4 treats other formulations of the problem of wave propagation through random media, and is concerned with different approaches to a physical problem mentioned in scheme (2) of the classification introduced in Sect. III.2.1. The only exception is Chap. 5 which deals not with volume, but with surface scattering [scheme (3) of the same classification].

This volume introduces two additional methods of the theory of random functions:

First, the theory of Markov processes (Chap. I.4) is generalized to *Markov fields*. The generalization is nontrivial. In Markov processes we deal with probability distributions for an ensemble of random *variables* [the values of the random function of time t at some successive instants of time $t_1 < t_2 < \ldots$] whereas in the theory of Markov fields we have ensembles of random *functions* of coordinates. The complete stochastic description of a random field is known to be given by a *characteristic functional* that is a probability distribution in an *infinite-dimensional* space (Sect. III.1.7). Therefore it requires more sophisticated mathematical tools. For example, instead of the Einstein-Fokker-Planck equation (a partial differential equation defining the distribution function, see Sect. I.4.5,6) we have now a functional equation defining a functional. To keep the mathematics required to a minimum we confine ourselves in Chap. 2 to Markov fields and use only their statistical moments given by partial differential equations.

Second, Chap. 4 presents the *Feynman diagram technique* that allows to describe graphically the perturbation expansions of averaged characteristics of linear stochastic problems. We use the technique to deal with multiple scattering. But without any modifications it is as well applicable to other problems. Feynman diagrams allow us to obtain integral equations whose solutions are the *sums* of subseries derived from perturbation expansions for statistically *averaged* characteristics of random wave fields. Thus it is possible to find solutions even if the first approximation is insufficient.

---

[1] Labels starting with a Roman number refer to the respective volume of the Principles, e.g., Chap. III.4 stands for Chap. 4 of Vol. 3.

Each chapter examines at length the basic idea of the respective approximation method and the region of validity of the emerging results.

Chaps. 1 and 5 were written by Yu.A. Kravtsov, the others by V.I. Tatarskii. S.M. Rytov acted as the editor of this volume.

The authors and the editor are grateful to professor A.M. Yaglom for his valuable comments and advice.

Moscow,
February 1989

Yu.A. Kravtsov
V.I. Tatarskii

Contents

1. Wave Propagation in Media with Large-Scale Random Inhomogeneities. Geometrical Optics Method

This and two subsequent chapters of this volume are concerned with approximate methods of solving problems involving wave propagation through media containing large-scale random inhomogeneities, whose characteristic dimensions l_ϵ are large as compared to the wavelength λ. In this process, large-angle scattering (and, in particular, backscattering) is negligible. As a result, fluctuations of the wave field are dominated by those inhomogeneities that lie in the path of the wave, i.e., in the vicinity of the ray connecting the source with the observation point. It is more convenient, therefore, to speark not of scattering but of wave *propagation* in random media with large inhomogeneities. According to the classification of Sect. III.2.1*[1.1], problems of this type, like those of wave scattering dealt with in Chap. III.4, fall into type (2).

Basically, there are three approximate approaches to fluctuations of shortwave ($\lambda \ll l_\epsilon$) fields in a random media: the geometrical optics method (GOM), the method of smooth perturbations (MSP) and the parabolic equation method (PEM).

This chapter is devoted to the simplest and most graphic of them − the geometrical optics method. Section 1.1 recalls the equations of geometrical optics and introduces related quantities (the eikonal and amplitude level). For any further progress to be possible, we again turn to the perturbation method, assuming now the fluctuations of the medium to be not only smooth but also *sufficiently weak*. Using this assumption we will, in Sects. 1.2–4, compute the statistical characteristics of fluctuations of the amplitude and phase, the arrival angle and group delay of a wave, and also the ray displacements. An important element in these calculations is that we take into account the possible presence in the medium of a *regular* (deterministic) refraction.

The general results are illustrated by examples, the most important of which is · wave propagation in a medium with locally homogeneous and isotropic turbulence (Sect. 1.5). In Sect. 1.6 the mean wave field and its coherence function are calculated.

1.1 Geometrical Optics Equations

To begin with, we will consider the simplest and most graphic of the above methods − the method of geometrical optics. Previous results obtained using this approach are given in [1.2, 3] and other publications, and a bibliography is given in [1.4–6]. The method is simple in that, unlike MSP and PEM, it *does not take into consider-*

---

* (Labels starting with a Roman number refer to the respective volume of the Principles, e.g. (I.2.1) stands for Eq. (2.1) of Vol. 1.)

ation diffraction effects, and is therefore not as versatile as the other two methods. However, within its scope the GOM offers definite advantages. First, it enables a number of effects to be analyzed (e.g., the influence of regular refraction and larger field fluctuations in the vicinity of caustics) that are not as easily described by the other techniques. Second, some of the results of the method hold even beyond the range of its validity; e.g., phase (but not amplitude) fluctuations and directions of wave propagation. It is this advantage that has largely contributed to the wide use of the GOM, although some strides have been made using the MSP and PEM. What is more, the geometrical optics approximation often was a heuristic foundation for both asymptotic methods, which take account of diffraction effects.

The following is a brief derivation of the geometrical optics equations for the simplest case of the scalar monochromatic wave propagating through a medium with stationary continuous inhomogeneities [1.7]. Suppose that the permittivity $\varepsilon(r)$ in the Helmholtz equation

$$\Delta u + k^2 \varepsilon(r)u = 0 \tag{1.1}$$

varies only slightly over the wavelength λ ($\lambda|\nabla\varepsilon| \ll \varepsilon$, i.e., the medium is smoothly inhomogeneous). It is natural to assume under these conditions that the field u at each point can be approximated by the plane wave

$$u = Ae^{iS} = Ae^{ik\varphi} \quad , \tag{1.2}$$

where the amplitude A and phase gradient ∇S are slow (in the scale of λ) functions of the coordinates.

Since A and ∇S vary slowly, we can easily derive the equations for A and S or for the quantity $\varphi = S/k$, which is the phase path of the wave and is known as the *eikonal*[1].

Debye suggested the following derivation of equations for A and φ. We expand A into a series in reciprocals of powers of the wave number[2]

$$u = \left(A_0 + \frac{A_1}{ik} + \frac{A_2}{(ik)^2} + \ldots \right) e^{ik\varphi} \quad . \tag{1.3}$$

The coefficients A_m in this expansion are generally complex and, therefore, also contribute to the phase of the resulting field.

Substituting (1.3) into the Helmholtz equation and equating to zero the coefficients at the same powers of k gives the set of equations for φ, A_0, A_1, A_2 ...

$$(k^2) \quad (\nabla\varphi)^2 = \varepsilon \quad , \tag{1.4}$$

$$(k) \quad 2(\nabla\varphi\nabla A_0) + A_0\Delta\varphi = 0 \quad , \tag{1.5}$$

---

[1] The eikonal φ is sometimes called the phase, but we will only use the term "phase" for the dimensionless quantity $S = k\varphi$.

[2] Strictly speaking, the expansion should be in the *dimensionless* small parameter $\mu \sim 1/kl_\varepsilon$, where $l_\varepsilon \sim \varepsilon/|\nabla\varepsilon| \gg \lambda$ is the characteristic scale of inhomogeneities (see. e.g., [1.7,8]). But to expand in $1/k$ means precisely expanding in $\mu = 1/kl_\varepsilon$.

(k^0) $2(\nabla\varphi\nabla A_1) + A_1\Delta\varphi = -\Delta A_0$,   (1.5a)

(k^{1-n}) $2(\nabla\varphi\nabla A_n) + A_n\Delta\varphi = -\Delta A_{n-1}$.   (1.5b)

Equation (1.4) is called the *eikonal equation,* and the subsequent equations for $A_n(n = 0, 1, 2, \ldots)$ are termed the *transport equations* for the amplitude of the nth approximation. The treatment is generally confined to the zeroth approximation of the GOM, where in (1.3) only A_0 is kept. The higher-order terms in (1.3) are discarded not only because of computational difficulties, but mainly because the series (1.3) is *asymptotic*[3]; and it is well known that, with asymptotic expansions, an approximation is not always improved by increasing the number of terms retained.

The eikonal equation (1.4) yields the characteristics (rays) along which the functional $\int \sqrt{\varepsilon}ds$ is extreme (Fermat's principle). Ray equations can be written in different forms. In this treatment it is convenient to represent them in the form [1.4, 5, 7]

$$\frac{dr}{ds} = t \quad , \quad \frac{dt}{ds} = \frac{1}{2\varepsilon}[\nabla\varepsilon - t(t\nabla\varepsilon)] \quad , \tag{1.6}$$

where ds is an element of the ray length, and t is the unit vector tangent to the ray, which doubles as the normal to the phase front $S = k\varphi = \text{const}$. Since $|\nabla\varphi| = \sqrt{\varepsilon}$, we have

$$t = \frac{\nabla S}{|\nabla S|} = \frac{\nabla\varphi}{|\nabla\varphi|} = \frac{\nabla\varphi}{\sqrt{\varepsilon}} \quad .$$

If in some way or other we have found a solution to the ray equations (1.6), the eikonal equation (1.4) and transport equation (1.5) can be integrated along the ray paths. The eikonal φ is found from the equation

$$\varphi = \int_0^s \sqrt{\varepsilon}\,ds = \int_0^s \sqrt{\varepsilon[r(s)]}\,ds \quad , \tag{1.7}$$

and A_0 from the conservation condition for the intensity $I = \sqrt{\varepsilon}A_0^2$ in the ray tube of infinitesimal thickness and with cross-sectional area $d\Sigma$ (Fig. 1.1)

$$Id\Sigma = \sqrt{\varepsilon}A_0^2 d\Sigma = \text{const} \quad . \tag{1.8}$$

This relationship follows directly from (1.5), if we write the latter in the form

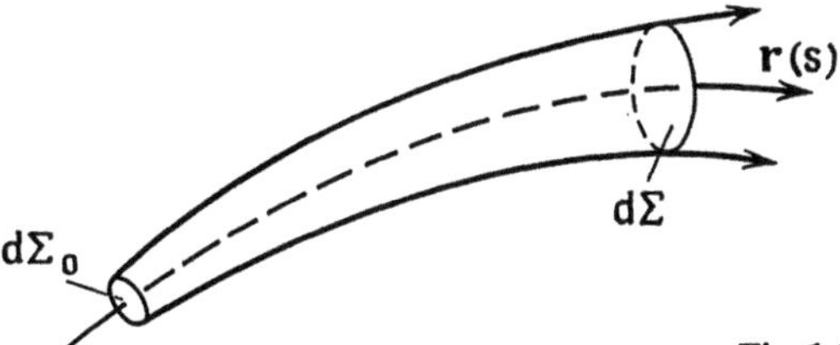

Fig. 1.1. A ray tube of infinitesimal thickness

---

[3] It reduces to an exact solution of the Helmholtz equation if the inclusion of more terms is accompanied by the limiting process $k \to \infty$, or rather $\mu \sim 1/kl_\varepsilon \to 0$.

$$\mathrm{div}(A_0^2 \nabla\varphi) = \mathrm{div}(\sqrt{\varepsilon}A_0^2 t) = \mathrm{div}(It) = 0$$

and integrate over the volume between the cross-sections of the infinitesimally thin ray tube.

One of the validity conditions of the GOM is the requirement that the medium parameters vary *smoothly*

$$l_\varepsilon \gg \lambda \quad . \tag{1.9}$$

In Sect. 2.1 we shall see that we can pass to the geometrical optics approximation [i.e., we can neglect all the terms in (1.3) save for the zeroth] only if the radius of the first Fresnel zone $\sqrt{\lambda L}$ is small (L is the distance covered by the wave) as compared to the characteristic scale of the inhomogeneities, l_ε,

$$\sqrt{\lambda L} \ll l_\varepsilon \quad . \tag{1.10}$$

If this condition is met, we can ignore the diffraction effects, which with the GOM are taken care of by the first order in $\mu = 1/kl_\varepsilon$ [1.7–11].

We will now turn to random media. It is impossible to obtain an analytic solution to the eikonal equation (1.4) or ray equation (1.6) for the permittivity ε, which depends on the coordinates in an arbitrary way. So here, too, we have to make use of approximate methods, notably of the *perturbation method*. Let $\varepsilon(r) = \overline{\varepsilon}(r) + \tilde{\varepsilon}(r)$, where the fluctuational component $\tilde{\varepsilon}$ is small as compared to the regular one ($\sigma_\varepsilon \ll \overline{\varepsilon}$). We represent the eikonal in the form

$$\varphi = \varphi_0 + \varphi_1 + \varphi_2 + \ldots \quad , \tag{1.11}$$

assuming that φ_0 satisfies the "unperturbed" eikonal equation

$$(\nabla\varphi_0)^2 = \overline{\varepsilon} \tag{1.12}$$

and $|\nabla\varphi_1| \sim \sigma_\varepsilon \ll |\nabla\varphi_0|$, $|\nabla\varphi_2| \sim \sigma_\varepsilon^2 \ll |\nabla\varphi_1|$, and so on. Substituting (1.11) into (1.4) and taking (1.12) into account, for the corrections φ_1, φ_2, $\ldots$ we get the following *linear* equations

$$2(\nabla\varphi_0 \nabla\varphi_1) = \tilde{\varepsilon} \quad ,$$
$$2(\nabla\varphi_0 \nabla\varphi_2) = -(\nabla\varphi_1)^2 \quad . \tag{1.13}$$

Given the unperturbed eikonal φ_0 and unperturbed rays $r_0 = r_0(s)$ and $t_0 = t_0(s)$ which obey the ray equations (1.6) at $\varepsilon = \overline{\varepsilon}$, the solutions to these equations can be expressed in quadratures. Note that in a zeroth approximation $\nabla\varphi_0 = \sqrt{\overline{\varepsilon}}t_0$, where $t_0 = dr_0/ds$ is the unit vector tangent to the unperturbed ray $r = r_0(s)$. The equation (1.13) for the first-order correction φ_1, which is generally the only one retained in calculations, will then be

$$2(\nabla\varphi_0 \nabla\varphi_1) = 2\sqrt{\overline{\varepsilon}}(t_0 \nabla\varphi_1) = 2\sqrt{\overline{\varepsilon}}\frac{d\varphi_1}{ds} = \tilde{\varepsilon} \quad , \quad \text{hence} \tag{1.14}$$

$$\varphi_1 = \frac{1}{2}\int\limits_0^s \frac{\tilde{\varepsilon}}{\sqrt{\overline{\varepsilon}}}ds' \quad . \tag{1.15}$$

The integration here is along the *unperturbed ray* $r = r_0(s)$, i.e., the integrand includes the functions $\tilde{\varepsilon} = \tilde{\varepsilon}[r_0(s')]$ and $\overline{\varepsilon} = \overline{\varepsilon}[r_0(s')]$.

4

In the simplest case of the plane wave propagating along the z-axis in a homogeneous (on average) medium with $\bar{\varepsilon} = \mathrm{const}$, the unperturbed value of the eikonal φ is $\varphi_0 = \sqrt{\bar{\varepsilon}}z$, the rays being straight lines parallel to the z-axis

$$x_0(s) = \mathrm{const} \quad , \quad y_0(s) = \mathrm{const} \quad , \quad z_0(s) = s \quad .$$

In that case

$$\varphi_1 = \varphi_1(x, y, z) = \frac{1}{2\sqrt{\bar{\varepsilon}}} \int_0^z \tilde{\varepsilon}(x, y, z')dz' \quad . \tag{1.16}$$

For the lower atmosphere, and also for the ionosphere in the case of ultrashort radio waves, we can put $\bar{\varepsilon} \approx 1$ and then

$$\varphi_1 = \frac{1}{2} \int_0^z \tilde{\varepsilon}(x, y, z')dz' \quad . \tag{1.17}$$

Later, we shall see that it is possible to develop analogously a perturbation theory for the amplitude, direction of wave propagation, deviations of the ray from its unperturbed path, etc.

Using the perturbation method to calculate the amplitude and phase fluctuations, we make use of the smallness of the permittivity fluctuations

$$\sigma_\varepsilon \ll \bar{\varepsilon} \tag{1.18}$$

and discard the terms of second order in σ_ε. The conditions which allow us to ignore the second-order terms reduce to the requirement that the variance of the *amplitude level* $\chi \equiv \ln(A/A^0)$ be small

$$\sigma_\chi^2 = \langle (\chi - \bar{\chi})^2 \rangle = \langle \tilde{\chi}^2 \rangle \ll 1 \quad , \tag{1.19}$$

which (for small σ_χ) is equivalent to the condition that $\sigma_A \ll \bar{A}$. It is obvious that noticeable fluctuations of χ occur where rays begin to intersect and form random foci and caustics [1.12]. The condition (1.19) is thus a constraint on the intensity of the fluctuations σ_ε^2 and distance L, the limiting values being those at which caustics are still unlikely.

The above validity conditions are essentially *sufficient* conditions. As for the necessary conditions, we can only provide here the qualitative consideration that diffraction effects exert a weaker influence on the behavior of the phase than on that of the amplitude, whereby the region of validity of the phase calculations may appear to be wider. In fact, a comparison with the method of smooth perturbations shows (Sect. 2.4) that some of the results of the geometrical optics approximation relating to the statistical characteristics of the phase and arrival angles are valid (up to a coefficient of about unity) beyond the confines defined by (1.10) and (1.19). But (1.9) and (1.18) must be obeyed in any event.

Later in the chapter we will calculate the various characteristics of the random wave (phase and level fluctuations, the statistics of arrival angles and the lateral displacements of rays, the mean field and the coherence function of a field) that are necessary when dealing with both direct and inverse problems of the statistical theory of wave propagation. The results of this chapter (and of the next two chapters) are applicable to treatments of a wide variety of physical phenomena (see, e.g., [1.4–6]).

1.2 Eikonal Fluctuations

In a first approximation to the perturbation theory the eikonal is $\varphi \approx \varphi_0 + \varphi_1$. The mean of the first-order correction φ_1 is zero ($\overline{\varphi}_1 = 0$), since in the approximation chosen the fluctuational component of the eikonal $\tilde{\varphi} = \varphi - \overline{\varphi}$ coincides with φ_1, and the covariance is

$$\psi_\varphi(r_1, r_2) = \langle \tilde{\varphi}(r_1)\tilde{\varphi}(r_2) \rangle = \langle \varphi_1(r_1)\varphi_1(r_2) \rangle \quad . \tag{1.20}$$

The covariance of the phase $S = k\varphi$ is clearly

$$\psi_S(r_1, r_2) = k^2 \psi_\varphi(r_1, r_2) \quad . \tag{1.21}$$

We will begin with the simplest case. Consider a plane wave $\exp(ikz)$ propagating through a statistically homogeneous medium with mean electric permittivity $\overline{\varepsilon} = 1$. From (1.17), the eikonal covariance is

$$\psi_\varphi(r_1, r_2) = \frac{1}{4} \int\limits_0^{z_1} dz' \int\limits_0^{z_2} dz'' \psi_\varepsilon(\varrho_1 - \varrho_2, z' - z'') \quad , \tag{1.22}$$

where $r_1 = (\varrho_1, z_1)$ and $r_2 = (\varrho_2, z_2)$ are the radius vectors of the observation points, and ψ_ε is the covariance of the fluctuations $\tilde{\varepsilon}$. We will now pass in (1.22) to new integration variables $\zeta = z' - z''$ and $\eta = (z' + z'')/2$. In these variables

$$\psi_\varphi(r_1, r_2) = \frac{1}{4} \iint\limits_\Sigma d\zeta \, d\eta \, \psi_\varepsilon(\varrho_1 - \varrho_2, \zeta) \quad , \tag{1.23}$$

where Σ is the region of integration in the plane (ζ, η). If $z_2 > z_1$, the region is a parallelogram, as shown in Fig. 1.2.

The integral (1.23) is dominated by the narrow layer $-l_\varepsilon \lesssim \zeta \lesssim l_\varepsilon$ (shaded in Fig. 1.2), where the covariance ψ_ε is markedly distinct from zero. Therefore, the integration limits with respect to ζ can be made infinite, and the integral with respect to η will be taken from 0 to z_1

$$\psi_\varphi(r_1, r_2) \approx \frac{1}{4} \int\limits_0^{z_1} d\eta \int\limits_{-\infty}^{\infty} d\zeta \psi_\varepsilon(\varrho_1 - \varrho_2, \zeta) = \frac{z_1}{2} \int\limits_0^{\infty} \psi_\varepsilon(\varrho_1 - \varrho_2, \zeta) d\zeta \quad .$$

Here we remember that ψ_ε is even in ζ. For $z_2 < z_1$ we would arrive at a similar expression, z_2 being substituted for z_1. Thus

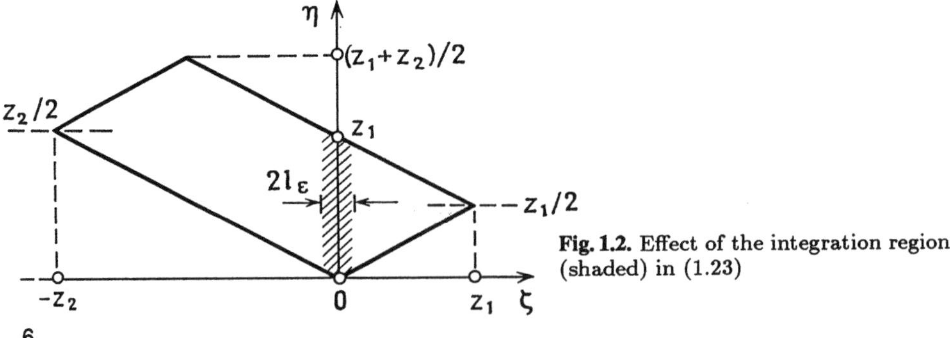

Fig. 1.2. Effect of the integration region (shaded) in (1.23)

$$\psi_\varphi(\mathbf{r}_1, \mathbf{r}_2) = \frac{z_<}{2} \int_0^\infty \psi_\varepsilon(\varrho, \zeta) d\zeta \quad , \tag{1.24}$$

where $z_< = \min\{z_1, z_2\}$ is the smaller of z_1 and z_2, and $\varrho = \varrho_1 - \varrho_2$. Specifically, for isotropic fluctuations

$$\psi_\varphi(\mathbf{r}_1, \mathbf{r}_2) = \frac{z_<}{2} \int_0^\infty \psi_\varepsilon\left(\sqrt{\varrho^2 + \zeta^2}\right) d\zeta \quad . \tag{1.25}$$

The eikonal covariance may also be expressed in terms of the spatial spectrum of $\tilde{\varepsilon}$. Substituting into (1.24) the spectral expansion

$$\psi_\varepsilon(\varrho, \zeta) = \int_{-\infty}^\infty \Phi_\varepsilon(\boldsymbol{\kappa}_\perp, \kappa_z) \exp\left(i\boldsymbol{\kappa}_\perp \cdot \varrho + i\kappa_z \zeta\right) d^2\kappa_\perp d\kappa_z$$

and integrating with respect to ζ gives

$$\psi_\varphi(\mathbf{r}_1, \mathbf{r}_2) = \frac{\pi z_<}{2} \int_{-\infty}^\infty \Phi_\varepsilon(\boldsymbol{\kappa}_\perp, 0) \exp\left(i\boldsymbol{\kappa}_\perp \cdot \varrho\right) d^2\kappa_\perp \quad . \tag{1.26}$$

If $\tilde{\varepsilon}$ is isotropic, then $\Phi_\varepsilon(\boldsymbol{\kappa}_\perp, 0) = \Phi_\varepsilon\sqrt{\kappa_\perp^2 + 0^2} = \Phi_\varepsilon(\kappa_\perp)$. If we now pass to the polar coordinates $\kappa = |\boldsymbol{\kappa}_\perp| = \sqrt{\kappa_x^2 + \kappa_y^2}$, $\alpha = \arctan(\kappa_y/\kappa_x)$ and integrate with respect to the angle α, we obtain

$$\psi_\varphi(\mathbf{r}_1, \mathbf{r}_2) = \pi^2 z_< \int_0^\infty \Phi_\varepsilon(\kappa) J_0(\kappa\varrho)\kappa \, d\kappa \quad , \tag{1.27}$$

where $J_0(\kappa\varrho)$ is the zeroth-order Bessel function.

The eikonal variance σ_φ^2 that follows from (1.24) or (1.26) at $\varrho = 0$ and $z_1 = z_2 = z$ grows linearly with the distance z

$$\sigma_\varphi^2(z) = \frac{z}{2} \int_0^\infty \psi_\varepsilon(0, \zeta) d\zeta = \frac{\pi z}{2} \int_{-\infty}^\infty \Phi_\varepsilon(\boldsymbol{\kappa}_\perp, 0) d^2\kappa_\perp \quad . \tag{1.28}$$

For isotropic fluctuations,

$$\sigma_\varphi^2(z) = \frac{z}{2} \int_0^\infty \psi_\varepsilon(\zeta) d\zeta = \pi^2 z \int_0^\infty \Phi_\varepsilon(\kappa)\kappa \, d\kappa \quad . \tag{1.29}$$

The integral of ψ_ε in these expressions equals the product of the fluctuation variance σ_ε^2 with the effective integral correlation radius $l_{\text{ef}} = \int_0^\infty K_\varepsilon(0, \zeta) d\zeta$. Therefore, (1.28) can be represented in the form

$$\sigma_\varphi^2(z) = \frac{z}{2} \int_0^\infty \sigma_\varepsilon^2 K_\varepsilon(0, \zeta) d\zeta = \frac{z}{2} \sigma_\varepsilon^2 l_{\text{ef}} \quad . \tag{1.30}$$

If, for instance, the covariance ψ_ε is Gaussian, i.e.,

$$\psi_\varepsilon(\varrho) = \sigma_\varepsilon^2 e^{-\varrho^2/2l_\varepsilon^2} \quad , \tag{1.31}$$

then the effective correlation radius l_{ef} is $\sqrt{\pi/2}\,l_\varepsilon$ and

$$\sigma_\varphi^2 = \frac{\sqrt{\pi}z}{2\sqrt{2}} l_\varepsilon \sigma_\varepsilon^2 \quad . \tag{1.32}$$

Consider the *longitudinal* covariance of the eikonal, assuming that the observation points r_1 and r_2 lie on the same ray ($\varrho = \varrho_1 - \varrho_2 = 0$)

$$\psi_\parallel(z_1, z_2) = \frac{z_<}{2} \int_0^\infty \psi_\varepsilon(0, \zeta)d\zeta = \frac{z_<}{2}\sigma_\varepsilon^2 l_{ef} \quad . \tag{1.33}$$

This suggests that the longitudinal covariance does not vary with $z_>$, i.e., the larger one of z_1 and z_2. However, it does not suggest that the longitudinal separation of the observation points also conserves the correlation coefficient $K_\parallel$. In fact, by (1.33),

$$K_\parallel(z_1, z_2) = \frac{\psi_\parallel(z_1, z_2)}{\sigma_\varphi(z_1)\sigma_\varphi(z_2)} = \frac{z_<}{\sqrt{z_1 z_2}} = \sqrt{\frac{z_<}{z_>}} = \begin{cases} \sqrt{z_2/z_1} & \text{for} \quad z_2 < z_1 \quad , \\ \sqrt{z_1/z_2} & \text{for} \quad z_2 > z_1 \quad . \end{cases}$$

Figure 1.3 represents the variation of $K_\parallel$ with z_2 at fixed z_1. It is seen that the longitudinal correlation of φ (and hence of the phase (S) occurs over the distance covered, i.e., $l_\parallel \sim z$.

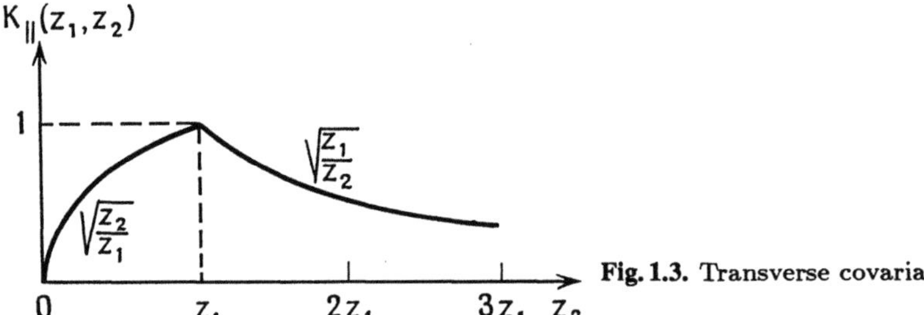

Fig. 1.3. Transverse covariance of the phase

The *transverse* correlation of the eikonal is characterized by other linear scales. Assuming that the observation points are only separated transversely ($z_1 = z_2 = L$); the transverse covariance will be

$$\psi_\perp(\varrho, L) = \frac{L}{2} \int_0^\infty \psi_\varepsilon(\varrho, \zeta)d\zeta \quad , \tag{1.34}$$

and for the isotropic fluctuations $\tilde\varepsilon$ it will be

$$\psi_\perp(\varrho, L) = \frac{L}{2} \int_0^\infty \psi_\varepsilon\left(\sqrt{\varrho^2 + \zeta^2}\right)d\zeta \quad . \tag{1.35}$$

It follows from (1.34, 35) that transverse correlation occurs within the electric permittivity correlation radius l_ε, i.e., $l_\perp \sim l_\varepsilon$. This is immediately seen on considering the Gaussian covariance (1.31), for which

8

$$\psi_\perp(\varrho, L) = \frac{\sqrt{\pi}}{2\sqrt{2}} L l_\varepsilon \sigma_\varepsilon^2 e^{-\varrho^2/2 l_\varepsilon^2} = \sigma_\varphi^2(L) e^{-\varrho^2/2 l_\varepsilon^2} \quad . \tag{1.36}$$

We will also need the *transverse structure function* of the eikonal

$$D_\perp(\varrho, L) = \langle [\varphi_1(\varrho_1, L) - \varphi_1(\varrho_2, L)]^2 \rangle \quad ,$$

which is expressed in terms of the transverse covariance $\psi_\perp$ (if it exists) in the following way:

$$D_\perp(\varrho, L) = 2[\psi_\perp(0, L) - \psi_\perp(\varrho, L)] = L \int_0^\infty [\psi_\varepsilon(0, \zeta) - \psi_\varepsilon(\varrho, \zeta)] d\zeta \quad .$$

But, by (III.1.69) [1.1],

$$\psi_\varepsilon(\varrho, \zeta) = \tfrac{1}{2}[D_\varepsilon(\infty) - D_\varepsilon(\varrho, \zeta)] \quad ,$$

and therefore

$$D_\perp(\varrho, L) = \frac{L}{2} \int_0^\infty [D_\varepsilon(\varrho, \zeta) - D_\varepsilon(0, \zeta)] d\zeta \quad . \tag{1.37}$$

Unlike (1.34), this equation is suitable not only for homogeneous, but also for *locally homogeneous* random fields $\tilde{\varepsilon}$, for which no covariance ψ_ε exists. For locally homogeneous, *isotropic* fields, instead of (1.37) we have

$$D_\perp(\varrho, L) = \frac{L}{2} \int_0^\infty [D_\varepsilon(\sqrt{\varrho^2 + \zeta^2}) - D_\varepsilon(\zeta)] d\zeta \quad . \tag{1.38}$$

We will also provide an expression for $D_\perp$ in terms of the spatial spectrum of the fluctuations

$$D_\perp(\varrho, L) = 2\pi^2 L \int_0^\infty \Phi_\varepsilon(\kappa)[1 - J_0(\kappa\varrho)]\kappa \, d\kappa \quad , \tag{1.39}$$

which is similar to (1.27). We will need these relations in Sect. 1.5, where we will look at eikonal fluctuations in the turbulent atmosphere.

We thus conclude our treatment of the plane wave in a statistically homogeneous medium. But, by confining ourselves to quasi-homogeneity, the geometrical optics approach enables us to derive the behavior of the fluctuations of nonplane (primarily spherical) waves, to take into account the regular refraction of the wave due to variations of the mean permittivity $\bar{\varepsilon}(r)$, and to dispense with the statistical homogeneity of fluctuations. Generally speaking, it is these possibilities that make the GOM advantageous over the other asymptotic methods, the MSP and PEM, in which the above generalizations either yield unwieldy expressions, or are not feasible at all.

The following is a qualitative consideration of some of these generalizations, using the results derived above for the plane wave. Clearly, quantitative arguments must be based on the initial expression (1.15).

Let us consider two unperturbed rays corresponding to a nonplane wave (Fig. 1.4). In the figure, L_1 and L_2 are the lengths of these rays which arrive at the observation points r_1 and r_2; t_{01} and t_{02} are the unit vectors along these rays, and δ is the

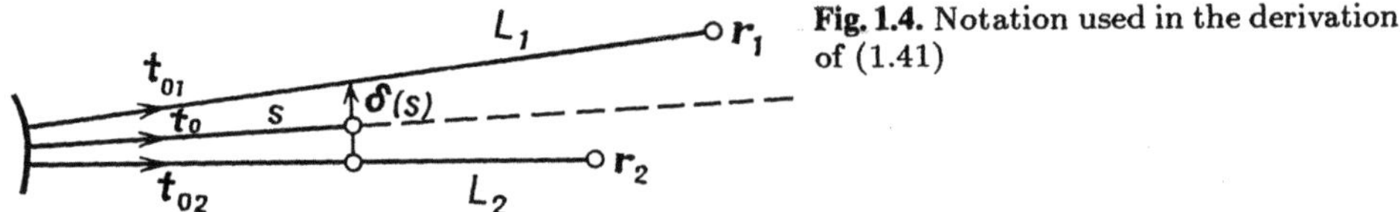

Fig. 1.4. Notation used in the derivation of (1.41)

vector connecting a point on one ray with the nearest point on the other ray. For simplicity, we will refer to δ as the "distance" between the rays. In the case of such a plane wave, L_1 and L_2 were the distances from the initial plane $z = 0$ to the observation points r_1 and r_2; t_{01} and t_{02} were equal and directed along the z-axis; and the distance between the observation points was $\delta = \varrho = \varrho_1 - \varrho_2$. Therefore, in the new notation, the expression (1.24) will be

$$\psi_\varphi(r_1, r_2) = \frac{L_<}{2} \int_0^\infty \psi_\varepsilon(\delta + t_0\zeta)d\zeta \quad , \tag{1.40}$$

where $L_< = \min\{L_1, L_2\}$.

The distinctive feature of the nonplane wave is that the distance between rays, δ, is a *variable* quantity depending on the distance s covered by the wave. If the transverse separation between the observation points is small as compared to the ray lengths L_1 and L_2, then the distance s can conveniently be measured along some middle ray coming, say, to the point $(r_1+r_2)/2$. In that case, the unit vectors t_{01} and t_{02} can be taken to be approximately equal to the unit vector t_0 along the middle ray (in Fig. 1.4 the middle ray is shown by a dashed line). To arrive at a correct result for a variable separation between the rays, in (1.40) we should substitute $\int_0^{L_<} ds$ for $L_<$ and treat t_0 as a unit vector along the middle ray. The covariance of the nonplane wave in a statistically homogeneous medium with $\bar\varepsilon = 1$ will then be

$$\psi_\varphi(r_1, r_2) = \frac{1}{2} \int_0^{L_<} ds \int_0^\infty \psi_\varepsilon[\delta(s) + \zeta t_0]d\zeta \quad . \tag{1.41}$$

If r_1 and r_2 coincide, i.e., $L_1 = L_2 = L$ and $\delta(s) = 0$, from (1.41) we find the eikonal variance

$$\sigma_\varphi^2(L) = \frac{L}{2} \int_0^\infty \psi_\varepsilon(t_0\zeta)d\zeta \quad . \tag{1.42}$$

We see that the eikonal variance σ_φ^2 is independent of the form of the primary wave and is, specifically, the same for the plane, spherical and cylindrical waves.

Using (1.41), we will consider the transverse covariance of the eikonal of the *divergent spherical wave,* assuming that the source and observation points lie within a randomly inhomogeneous medium (fluctuations within a layer of finite thickness are considered in Exercise 1.7.1, and phase fluctuations of the convergent spherical wave, in Exercise 1.7.2). Let r_1 and r_2 lie at the same distance from the source. If $\varrho = r_1 - r_2$ is the transverse separation between the *observation points,* the current distance between rays $\delta(s)$ will vary linearly from zero at $s = 0$ to ϱ at $s = L$

$$\delta(s) = \varrho \frac{s}{L} \quad .$$

Hence, for the spherical wave

$$\psi_\varphi^{\text{sph}}(\varrho, L) = \frac{1}{2} \int\limits_0^L ds \int\limits_0^\infty \psi_\varepsilon \left(\varrho \frac{s}{L} + t_0 \zeta \right) d\zeta \quad . \tag{1.43}$$

This relation can be written differently, if we consider that

$$\frac{1}{2} \int\limits_0^\infty \psi_\varepsilon \left(\frac{s}{L} \varrho + t_0 \zeta \right) d\zeta = \frac{1}{L} \psi_\varphi^{\text{P}} \left(\frac{s}{L} \varrho, L \right) \quad , \tag{1.44}$$

where $\psi_\varphi^{\text{P}}(\varrho, L)$ is the transverse covariance of the plane wave (1.34). This suggests the useful relation

$$\psi_\varphi^{\text{sph}}(\varrho, L) = \frac{1}{L} \int\limits_0^L ds \, \psi_\varphi^{\text{P}} \left(\varrho \frac{s}{L}, L \right) = \int\limits_0^1 d\gamma \, \psi_\varphi^{\text{P}}(\gamma \varrho, L) \quad , \tag{1.45}$$

which relates the covariances of the eikonal of the spherical and plane waves.

In the case of isotropic fluctuations of medium parameters, when ψ_φ^{P} depends on $\varrho = |\varrho|$, the covariance of the eikonal of the spherical wave will also depend only on ϱ, and (1.45) can be written as

$$\psi_\varphi^{\text{sph}}(\varrho, L) = \frac{1}{\varrho} \int\limits_0^\varrho \psi_\varphi^{\text{P}}(\varrho', L) d\varrho' \quad . \tag{1.46}$$

This relationship, like (1.43), suggests that the correlation radius of the eikonal of the spherical wave is larger than that of the plane wave. Clearly, for the convergent spherical wave, the inverse relation will hold (see Exercise 1.7.2).

Let us consider a specific example of the calculation of $\psi_\varphi^{\text{sph}}$. In the case of the Gaussian covariance of $\tilde{\varepsilon}$, when ψ_φ^{P} is given by (1.36), $\psi_\varphi^{\text{sph}}$ will be

$$\psi_\phi^{\text{sph}}(\varrho, L) = \frac{\pi}{4} L \sigma_\varepsilon^2 \frac{l_\varepsilon^2}{\varrho} \Phi \left(\frac{\varrho}{l_\varepsilon} \right) = \sigma_\varphi^2(L) \sqrt{\frac{\pi}{2}} \frac{l_\varepsilon}{\varrho} \Phi \left(\frac{\varrho}{l_\varepsilon} \right) \quad , \quad \text{where}$$

$$\Phi(t) = \sqrt{\frac{2}{\pi}} \int\limits_0^t e^{-t^2/2} dt$$

is the error function.

In Fig. 1.5, the correlation coefficients of the eikonal are given by curve 1 for the plane wave $[K_\varphi^{\text{P}}(\varrho) = \exp(-\varrho^2/2l_\varepsilon^2)]$, and curve 2 for the spherical wave $[K_\varphi^{\text{sph}}(\varrho) = \sqrt{\pi/2}(l_\varepsilon/\varrho)\Phi(\varrho/l_\varepsilon)]$. It is seen in the figure that the level $K_\varphi = 1/2$ for the plane wave is achieved at $\varrho = 1.18 l_\varepsilon$, and for the spherical wave at $\varrho = 2.40 l_\varepsilon$, i.e., the transverse correlation radius of the eikonal of the spherical wave is about twice as large as that of the plane wave.

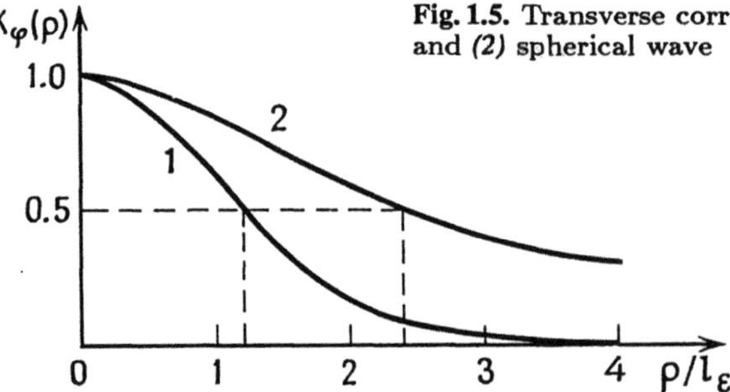

Fig. 1.5. Transverse correlation coefficients for *(1)* plane wave and *(2)* spherical wave

We can generalize (1.41) to a medium with a constant mean electric permittivity $\bar{\varepsilon}$ that is distinct from unity simply by changing the scales by a factor of $\sqrt{\bar{\varepsilon}}$, i.e., replacing $ds\,d\zeta$ by $ds\,d\zeta/\bar{\varepsilon}$

$$\psi_\varphi(\boldsymbol{r}_1,\boldsymbol{r}_2) = \frac{1}{2}\int\limits_0^{L_<}\frac{ds}{\bar{\varepsilon}}\int\limits_0^\infty \psi_\varepsilon[\delta(s)+t_0\zeta]d\zeta \quad . \tag{1.47}$$

However, this relationship also holds when $\bar{\varepsilon}$ varies slowly (in the scale of the correlation radius l_ε) along the middle ray: $\bar{\varepsilon} = \bar{\varepsilon}[\boldsymbol{r}_0(s)]$. As the wave progresses, regular refraction bends the middle ray so that the direction of the tangent unit vector t_0 is now dependent on s, i.e., $t_0 = t_0(s)$.

Lastly, without assuming the medium fluctuations to be statistically homogeneous, we can look at a more general model of statistically *quasi-homogeneous* fluctuations. The covariance of the quasi-homogeneous field $\tilde{\varepsilon}$ has the form (Sect. III.1.5)

$$\psi_\varepsilon(\boldsymbol{r}',\boldsymbol{r}'') = \psi_\varepsilon(\boldsymbol{r}'-\boldsymbol{r}'',\boldsymbol{R}) = \sigma_\varepsilon^2(\boldsymbol{R})K_\varepsilon(\boldsymbol{r}'-\boldsymbol{r}'',\boldsymbol{R}) \quad . \tag{1.48}$$

Here the variance $\sigma_\varepsilon^2(\boldsymbol{R})$ and the correlation coefficient $K_\varepsilon(\boldsymbol{r}'-\boldsymbol{r}'',\boldsymbol{R})$ vary slowly (in the scale l_ε) with the position of the center of mass $\boldsymbol{R} = (\boldsymbol{r}'+\boldsymbol{r}'')/2$. If we substitute (1.48) into (1.47), the values of $\boldsymbol{R}$ must be taken on the unperturbed middle ray $\boldsymbol{R} = \boldsymbol{r}_0(s)$, and it is advisable to represent $\boldsymbol{r}'-\boldsymbol{r}''$ as the sum $\delta(s)+t_0(s)\zeta$, where $\delta(s)$ is the transverse distance and $t_0(s)\zeta$ the longitudinal distance between the points $\boldsymbol{r}'$ and $\boldsymbol{r}''$ lying on adjacent rays (Fig. 1.6). As a result, the covariance of the eikonal will be

$$\psi_\varphi(\boldsymbol{r}_1,\boldsymbol{r}_2) = \frac{1}{2}\int\limits_0^{L_<}\frac{ds}{\bar{\varepsilon}[\boldsymbol{r}_0(s)]}\int\limits_0^\infty \psi_\varepsilon[\delta(s)+t_0(s)\zeta;\,\boldsymbol{r}_0(s)]d\zeta \quad , \tag{1.49}$$

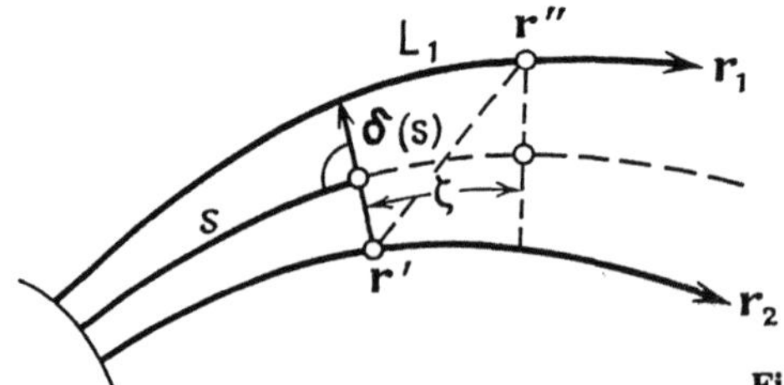

Fig. 1.6. Notation used in the derivation of (1.49)

where $L_<$ is the shorter of the lengths of the curvilinear rays L_1 and L_2. From (1.49) we obtain the variance at $L_1 = L_2 = L$ and $\delta = 0$

$$\sigma_\varphi^2(L) = \frac{1}{2} \int_0^L \frac{ds}{\overline{\varepsilon}[\boldsymbol{r}_0(s)]} \int_0^\infty \psi_\varepsilon[\zeta \boldsymbol{t}_0; \boldsymbol{r}_0(s)] d\zeta$$

$$= \frac{1}{2} \int_0^L \frac{ds\, \sigma_\varepsilon^2[\boldsymbol{r}_0(s)]}{\overline{\varepsilon}[\boldsymbol{r}_0(s)]} \int_0^\infty K_\varepsilon[\zeta \boldsymbol{t}_0; \boldsymbol{r}_0(s)] d\zeta \quad . \tag{1.50}$$

The internal integral in (1.50) is the effective correlation radius

$$l_{\mathrm{ef}}(s) = \int_0^\infty K_\varepsilon[\zeta \boldsymbol{t}_0(s); \boldsymbol{r}_0(s)] d\zeta \quad .$$

For anisotropic fluctuations, l_{ef} varies not only with the position of $\boldsymbol{r}_0(s)$ on the ray, but also with the direction of the ray $\boldsymbol{t}_0(s)$ at the point. If we write the eikonal variance as

$$\sigma_\varphi^2(L) = \frac{1}{2} \int_0^L \frac{\sigma_\varepsilon^2[\boldsymbol{r}_0(s)] l_{\mathrm{ef}}(s)}{\overline{\varepsilon}[\boldsymbol{r}_0(s)]} ds \quad , \tag{1.51}$$

we deduce that the largest contributions to σ_φ^2 come from those sections of the ray where $\sigma_\varepsilon^2 l_{\mathrm{ef}}/\overline{\varepsilon}$ is maximal. Specifically, this suggests that the eikonal fluctuations become more intense as the mean electric permittivity $\overline{\varepsilon}$ decreases. This intensification occurs, for instance, in the ionosphere where radio waves sent out by the Earth are reflected (see Exercise 1.7.4).

Let us consider the behavior of the transverse (relative to the unperturbed ray) covariance of the eikonal, assuming that $\boldsymbol{r}_1$ and $\boldsymbol{r}_2$ lie on the same unperturbed phase front $\varphi_0 = \mathrm{const}$

$$\psi_\varphi(\boldsymbol{r}_1, \boldsymbol{r}_2) = \frac{1}{2} \int_0^L \frac{ds}{\overline{\varepsilon}[\boldsymbol{r}_0(s)]} \int_0^\infty \psi_\varepsilon[\boldsymbol{\delta}(s) + \zeta \boldsymbol{t}_0; \boldsymbol{r}_0(s)] d\zeta \quad . \tag{1.52}$$

Clearly, the major contributors to (1.52) are those sections of the rays where the separation between them is minimal, i.e., the sections where the rays are closest together (we dealt with this already when we compared the fluctuations in the spherical and plane waves in a medium with $\overline{\varepsilon} = \mathrm{const}$). The influence of the separation between the rays on the eikonal covariance can be illustrated by the example of the oblique incidence of the plane wave on the reflecting layer with $\overline{\varepsilon} = \overline{\varepsilon}(z)$. Figure 1.7a, b shows two pairs of rays with the same separation ϱ between the observation points $\boldsymbol{r}_1$ and $\boldsymbol{r}_2$, normal to the plane of the drawing (Fig. 1.7a) and in the plane of the drawing (Fig. 1.7b). In the first case, both rays are parallel to each other and the separation δ is everywhere equal to the distance between the observation points, ϱ. But in the second case, where the rays lie in the same plane, the distance δ is always smaller than ϱ and even vanishes at the ray intersection. It is clear that the eikonal correlation radius is larger in the second case, where the rays are closer together. The difference in the correlation radii in cases a and b strongly suggests that the regular refraction makes the eikonal fluctuations anisotropic (anisomeric) in the plane normal to the ray.

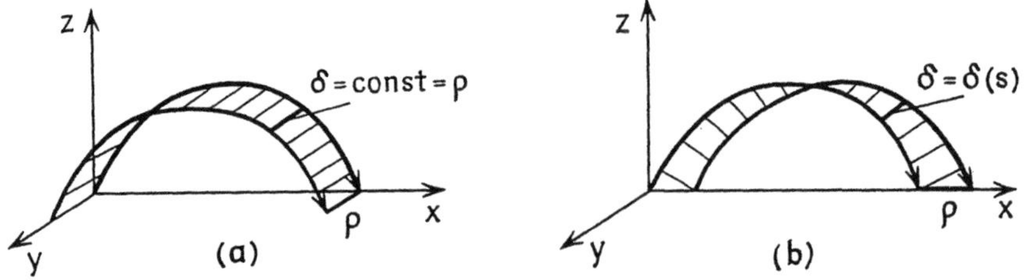

Fig. 1.7. Separation of rays (a) in parallel planes and (b) in an incidence plane

In the example considered, the rays reflected by the plane layer form a *caustic* tangent to the rays at turning points. Strictly speaking, the geometrical optics approach is illegitimate here, since it is known to give an infinite amplitude of the wave at caustic because the cross-section of the ray tube $d\Sigma$ in (1.8) vanishes. This disadvantage of the GOM can, however, be remedied by describing the field near caustic using improved asymptotic methods that provide a finite field at the caustic itself, and give the sum of the incident and reflected waves of the form $A \exp(ik\varphi)$ away from it [1.9, 13]. As a result, away from the caustic, the eikonal covariance for the reflected wave is determined by the same expression (1.52) as for the incident wave. But, in the immediate vicinity of the caustic, the GOM is clearly invalid.

1.3 Fluctuations of Arrival Angles, Ray Displacements and the Group Delay of Waves

1) Fluctuations of arrival angles. The arrival angles of waves are determined by the direction of the normal to the phase front, which in an isotropic medium coincides with the direction of the unit vector $t = \nabla\varphi/\sqrt{\varepsilon}$ tangent to the ray. Let us find the deviation of the vector from the unperturbed position $t_0 = \nabla\varphi_0/\sqrt{\varepsilon}$. In a first approximation, we have

$$t = \frac{\nabla\varphi}{\sqrt{\varepsilon}} = \frac{\nabla(\varphi_0 + \varphi_1 + \ldots)}{\sqrt{\varepsilon + \tilde{\varepsilon}}} = \frac{\nabla\varphi_0}{\sqrt{\varepsilon}} + \frac{1}{\sqrt{\varepsilon}}\left[\nabla\varphi_1 - \frac{\tilde{\varepsilon}\nabla\varphi_0}{2\varepsilon} + \ldots\right] \quad .$$

But, by (1.14), $\tilde{\varepsilon} = 2\sqrt{\varepsilon}(t_0\nabla\varphi_1)$, so that

$$t \approx t_0 + \frac{1}{\sqrt{\varepsilon}}[\nabla\varphi_1 - t_0(t_0\nabla\varphi_1)] \quad ,$$

and the first-order correction to the unperturbed direction t_0 will be

$$t_1 = t - t_0 = \frac{1}{\sqrt{\varepsilon}}[\nabla\varphi_1 - t_0(t_0\nabla\varphi_1] \equiv \frac{\nabla_\perp\varphi_1}{\sqrt{\varepsilon}} \quad , \tag{1.53}$$

where $\nabla_\perp$ is the operator of transverse (relative to the unperturbed ray) differentiation.

From (1.53), t_1 is normal to t_0 and lies in a plane tangent to the unperturbed phase front $\varphi_0 = \mathrm{const}$. Let α and β be two unit orthogonal vectors in this plane that, together with the vector t_0 tangent to the unperturbed ray, form an orthogonal

14

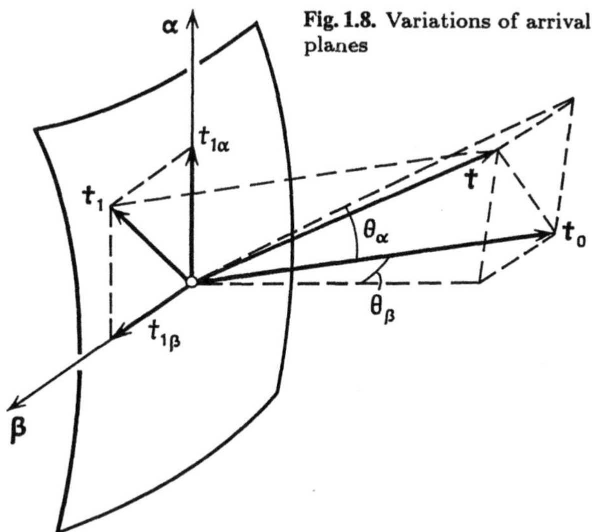

Fig. 1.8. Variations of arrival angles in two mutually orthogonal planes

frame of reference. We can decompose t_1 into two components along the directions α and β

$$t_1 = t_{1\alpha}\alpha + t_{1\beta}\beta \quad .$$

Up to second-order terms, the arrival angles of the ray θ_α and θ_β, measured from the direction of the unperturbed ray t_0 (Fig. 1.8), are $t_{1\alpha}$ and $t_{1\beta}$ respectively

$$\theta_\alpha \approx t_{1\alpha} = \frac{1}{\sqrt{\bar{\varepsilon}}}(\alpha\nabla_\perp\varphi_1) = \frac{1}{\sqrt{\bar{\varepsilon}}}\frac{\partial\varphi_1}{\partial\varrho_\alpha} \quad , \quad \theta_\beta \approx \frac{1}{\sqrt{\bar{\varepsilon}}}\frac{\partial\varphi_1}{\partial\varrho_\beta} \quad .$$

It follows that the mean arrival angles in both mutually orthogonal planes (t_0, α) and (t_0, β) are zero $(\bar{\theta}_\alpha = \bar{\theta}_\beta = 0)$, as $\overline{\varphi}_1 = 0$ and the elements of the correlation matrix for these angles are given by

$$\psi^\theta_{\alpha\beta}(\varrho_1, \varrho_2) \equiv \langle\theta_\alpha(\varrho_1)\theta_\beta(\varrho_2)\rangle = \frac{1}{\bar{\varepsilon}}\frac{\partial^2\psi_\perp(\varrho_1, \varrho_2)}{\partial\varrho_{1\alpha}\partial\varrho_{1\beta}} \quad . \tag{1.54}$$

We can simplify this equation somewhat by considering that the eikonal fluctuations in a plane tangent to the unperturbed phase front are quasi-homogeneous: the transverse covariance of the eikonal, $\psi_\perp(\varrho_1, \varrho_2) = \psi_\perp(\varrho, \varrho_+)$, varies "fast" (in the space scale $\sim l_\varepsilon$) with $\varrho = \varrho_1 - \varrho_2$, and "slowly" (in the space scale $L_\varepsilon \gg l_\varepsilon$) with the coordinate of the center of mass $\varrho_+ = (\varrho_1 + \varrho_2)/2$. Therefore, when we pass in (1.54) from ϱ_1 and ϱ_2 to ϱ and ϱ_+, we can only differentiate with respect to $\varrho = \varrho_1 - \varrho_2$. If, for simplicity, we discard ϱ_+ in $\psi^\theta_{\alpha\beta}$ and $\psi_\perp$, we will get

$$\psi^\theta_{\alpha\beta}(\varrho) = -\frac{1}{\bar{\varepsilon}}\frac{\partial^2\psi_\perp(\varrho)}{\partial\varrho_\alpha\partial\varrho_\beta} = \frac{1}{2\bar{\varepsilon}}\frac{\partial^2 D_\perp(\varrho)}{\partial\varrho_\alpha\partial\varrho_\beta} \quad . \tag{1.55}$$

The second form of this relationship holds not only for quasi-homogeneous, but also for locally homogeneous fluctuations $\tilde{\varepsilon}$, when there is no covariance ψ_φ.

We chose α and β such that, at small ϱ, the structure function $D_\perp$ has the form $D_\perp \approx c_\alpha \varrho_\alpha^2 + c_\beta \varrho_\beta^2$. The fluctuations of the arrival angles θ_α and θ_β at the same observation point, i.e., at $\varrho = 0$ will then be uncorrelated

$$\langle \theta_\alpha \theta_\beta \rangle = \psi_{\alpha\beta}^\theta(0) = \frac{1}{2\bar\varepsilon} \frac{\partial^2 D_\perp(0)}{\partial \varrho_\alpha \partial \varrho_\beta} = 0 \quad (\alpha \neq \beta) \quad . \tag{1.56}$$

But the variances of these angles are given by

$$\langle \theta_\alpha^2 \rangle = \psi_{\alpha\alpha}^\theta(0) = \frac{1}{2\bar\varepsilon} \frac{\partial^2 D_\perp(0)}{\partial \varrho_\alpha^2} \quad , \quad \langle \theta_\beta^2 \rangle = \frac{1}{2\bar\varepsilon} \frac{\partial^2 D_\perp(0)}{\partial \varrho_\beta^2} \quad , \tag{1.57}$$

and, in the general case, the anisotropic fluctuations of the eikonal are different. Thus, these variances are dissimilar when a plane wave is reflected from a plane layer where $\bar\varepsilon = \bar\varepsilon(z)$ (Fig. 1.7): even given isotropic fluctuations $\tilde\varepsilon$, the fluctuations of the arrival angle in the plane of the drawing are smaller than when normal to it.

For plane and spherical waves propagating through a statistically homogeneous, isotropic medium, the eikonal fluctuations in the plane normal to the ray are isotropic: $D_\perp(\varrho) = D_\perp(\varrho)$. In that case, according to the general properties of isotropic, vector random fields (see Exercise 4 to Chap. III.1), the covariance tensor $\psi_{\alpha\beta}^\theta$ is expressed through the unit tensor $\delta_{\alpha\beta}$ and the symmetric tensor $\varrho_\alpha \varrho_\beta / \varrho^2$ (we put $\bar\varepsilon = 1$)

$$\psi_{\alpha\beta}^\theta = \psi_\perp^\theta \left(\delta_{\alpha\beta} - \frac{\varrho_\alpha \varrho_\beta}{\varrho^2} \right) + \psi_\|^\theta \frac{\varrho_\alpha \varrho_\beta}{\varrho^2} \quad . \tag{1.58}$$

Here, $\psi_\perp^\theta$ is the transverse and $\psi_\|^\theta$ the longitudinal covariance of the isotropic, two-dimensional vector field $\theta = \{\theta_\alpha, \theta_\beta\} = \{t_{1\alpha}, t_{1\beta}\}$ (where "transverse" and "longitudinal" covariances refer to the components of the covariance tensor, and not to correlation along and across the ray). Comparing (1.58) with (1.55) gives

$$\psi_\|^\theta(\varrho) = \frac{1}{2\bar\varepsilon} D_\perp''(\varrho) \quad , \quad \psi_\perp^\theta = \frac{1}{2\varrho\bar\varepsilon} D_\perp'(\varrho) \quad . \tag{1.59}$$

These functions are related by the equation

$$\psi_\|^\theta(\varrho) = [\varrho \psi_\perp^\theta(\varrho)]' \quad ,$$

which is an obvious consequence of the potentiality of $\theta \approx t_1 = \nabla \varphi_1$. For the isotropic fluctuations $\tilde\varepsilon$, the variances of the angles θ_α and θ_β are equal

$$\langle \theta_\alpha^2 \rangle = \langle \theta_\beta^2 \rangle = \frac{1}{2\bar\varepsilon} D_\perp''(0) \quad . \tag{1.60}$$

For the plane wave, when $D_\perp$ is given by (1.38), the transverse and longitudinal covariances are

$$\psi_\|^\theta(\varrho) = \frac{L}{4\bar\varepsilon} \int_0^\infty \frac{\partial^2}{\partial\varrho^2} D_\varepsilon(\sqrt{\varrho^2 + \zeta^2})\,d\zeta = -\frac{L}{2\bar\varepsilon} \int_0^\infty \frac{\partial^2}{\partial\varrho^2} \psi_\varepsilon(\sqrt{\varrho^2 + \zeta^2})\,d\zeta \quad ,$$

$$\psi_\perp^\theta(\varrho) = \frac{L}{4\varrho\bar\varepsilon} \int_0^\infty \frac{\partial}{\partial\varrho} D_\varepsilon(\sqrt{\varrho^2 + \zeta^2})\,d\zeta = -\frac{L}{2\varrho\bar\varepsilon} \int_0^\infty \frac{\partial}{\partial\varrho} \psi_\varepsilon(\sqrt{\varrho^2 + \zeta^2})\,d\zeta \quad . \tag{1.61}$$

In particular, if the covariance has the Gaussian form (1.31), then

$$\psi_\perp^\theta(\varrho) = \frac{1}{\bar\varepsilon l_\varepsilon^2}\psi_\perp(\varrho) \quad , \quad \psi_\parallel^\theta(\varrho) = \frac{1}{\bar\varepsilon l_\varepsilon^2}\left(1 - \frac{\varrho^2}{l_\varepsilon^2}\right)\psi_\perp(\varrho) \quad ,$$

$$\langle\theta_\alpha^2\rangle = \langle\theta_\beta^2\rangle = \frac{\sigma_\varphi^2}{\bar\varepsilon l_\varepsilon^2} \quad ,$$

where $\psi_\perp(\varrho) = \sigma_\varphi^2 \exp(-\varrho^2/2l_\varepsilon^2)$, and σ_φ^2 is given by (1.32).

We can derive equations for the spherical wave in much the same way. We will only provide a useful relationship between the variances of the arrival angles for the plane and spherical waves[4]

$$\langle\theta_\alpha^2\rangle_{\text{sph}} = \tfrac{1}{3}\langle\theta_\alpha^2\rangle_{\text{p}} \quad , \tag{1.62}$$

which is readily obtained using (1.45)

$$[\psi_\perp''(0)]^{\text{sph}} = \frac{\partial^2}{\partial\varrho^2}\int_0^1 d\gamma\,\psi_\perp^{\text{P}}(\gamma\varrho)\bigg|_{\varrho=0} = [\psi_\perp''(0)]^{\text{P}}\int_0^1 \gamma^2 d\gamma = \frac{1}{3}[\psi_\perp''(0)]^{\text{P}} \quad .$$

The variances of the incident angles for the spherical wave are three times smaller than those for the plane wave that has travelled through the same distance in the medium. This could be qualitatively accounted for by the difference in the transverse correlation radius (recall that in the case of the spherical wave, $l_\perp$ is larger, since the rays are generally closer together than in that of the plane wave).

Experimentally, one generally measures not θ_α or θ_β, but some other quantity from which θ_α and θ_β are then derived (for the most part, approximately). Thus, an interferometer with base ϱ_α oriented along the α-axis measures the phase difference $\Delta S(\varrho_\alpha) = k\Delta\varphi(\varrho_\alpha)$. When related to the electric length of the base $k\varrho_\alpha$, this gives

$$\Theta_\alpha = \frac{\Delta\varphi(\varrho_\alpha)}{\varrho_\alpha} \quad ,$$

which at small ϱ_α, namely when $\varrho_\alpha \ll l_\varepsilon$, coincides with θ_α. Accordingly, the variance

$$\langle\Theta_\alpha^2\rangle = \langle[\Delta\varphi(\varrho_\alpha)]^2\rangle/\varrho_\alpha^2 = D_\perp(\varrho_\alpha)/\varrho_\alpha^2$$

at small ϱ may be a measure of $\langle\theta_\alpha^2\rangle$, since for $\varrho_\alpha \ll l_\varepsilon$

$$\langle\Theta_\alpha^2\rangle \approx \tfrac{1}{2}D_\perp''(0) \quad ,$$

which coincides with (1.60).

2) Statistics of Ray Displacements. In a medium containing random inhomogeneities, a ray is a wavy spatial curve. Let us calculate the root-mean-square displacement of the ray from its unperturbed position, confining ourselves for simplicity to the case of the *plane* wave propagating through a statistically homogeneous medium.

---

[4] We assume that, just as in deriving (1.45), the source of the spherical wave lies *within* a randomly inhomogeneous medium, and that both waves (plane and spherical) cover the same distance L in the medium.

For the ray $r(s)$, we will write the series of perturbation theory in $\tilde{\varepsilon}$

$$r(s) = r_0(s) + r_1(s) + r_2(s) + \ldots \quad .$$

From the ray equations (1.6), the first-order correction $q \equiv r_1$ will be (at $\tilde{\varepsilon} = 1$)

$$q = \int_0^L t_1 \, ds = \int_0^L \nabla_\perp \varphi_1 \, dz' \quad . \tag{1.63}$$

It follows that in the first order of perturbation theory the perturbed ray displaces only transversely with respect to the unperturbed ray. Thus, if the wave propagates along the z-axis, q will contain only the x- and y-components.

Substituting φ_1 from (1.17) into (1.63) and integrating by parts gives

$$q(\varrho, L) = \frac{1}{2} \int_0^L dz' \int_0^{z'} dz'' \nabla_\perp \tilde{\varepsilon}(\varrho, z'') = \frac{1}{2} \int_0^L (L - z') \nabla_\perp \tilde{\varepsilon}(\varrho, z') \, dz' \quad . \tag{1.64}$$

The covariance matrix for the displacements $(\varrho = \varrho_1 - \varrho_2)$ will be

$$\psi_{\alpha\beta}^q(\varrho, L) = \langle q_\alpha(\varrho_1, z) q_\beta(\varrho_2, z) \rangle$$

$$= \frac{1}{4} \iint_0^L (L - z')(L - z'') \frac{\partial^2}{\partial \varrho_{1\alpha} \partial \varrho_{2\beta}} \psi_\varepsilon(\varrho_1 - \varrho_2, z' - z'') \, dz' \, dz'' \quad , \tag{1.65}$$

where q_α and q_β are the components of q along two mutually perpendicular directions, α and β, in the plane $z = \text{const}$. Equation (1.65) can be simplified by passing over to the integration variables $\zeta = z' - z''$ and $\eta = (z' + z'')/2$. Arguing along the same lines as in the derivation of (1.25), we obtain

$$\psi_{\alpha\beta}^q(\varrho, L) = -\frac{1}{4} \iint_\Sigma d\zeta \, d\eta \left(L - \eta - \frac{\zeta}{2} \right) \left(L - \eta + \frac{\zeta}{2} \right) \frac{\partial^2 \psi_\varepsilon(\varrho, \zeta)}{\partial \varrho_\alpha \delta \varrho_\beta}$$

$$\approx -\frac{1}{2} \int_0^L d\eta (L - \eta)^2 \int_0^\infty \frac{\partial^2 \psi_\varepsilon(\varrho, \zeta)}{\partial \varrho_\alpha \partial \varrho_\beta} \, d\zeta$$

$$= -\frac{L^3}{6} \int_0^\infty \frac{\partial^2 \psi_\varepsilon(\varrho, \zeta)}{\partial \varrho_\alpha \partial \varrho_\beta} \, d\zeta \quad . \tag{1.66}$$

If the fluctuations $\tilde{\varepsilon}$ are isotropic, the two-dimensional vector field q will also be isotropic. The elements of the covariance matrix (1.66) will then take a form similar to (1.58)

$$\psi_{\alpha\beta}^q(\varrho, L) = \psi_\perp^q(\varrho, L) \left(\delta_{\alpha\beta} - \frac{\varrho_\alpha \varrho_\beta}{\varrho^2} \right) + \psi_\parallel^q(\varrho, L) \frac{\varrho_\alpha \varrho_\beta}{\varrho^2} \quad ,$$

where the longitudinal and transverse covariances of the ray displacements are given by

$$\psi_\parallel^q(\varrho, L) = -\frac{L^3}{6} \int_0^\infty \frac{\partial^2 \psi_\varepsilon(\sqrt{\varrho^2 + \zeta^2})}{\partial \varrho^2} \, d\zeta \quad ,$$

18

$$\psi_{\perp}^{q}(\varrho, L) = -\frac{L^3}{6\varrho} \int\limits_0^{\infty} \frac{\partial \psi_{\varepsilon}(\sqrt{\varrho^2 + \zeta^2})}{\partial \varrho} d\zeta \quad . \tag{1.67}$$

It is easily verified that, for the isotropic fluctuations $\tilde{\varepsilon}$, the mutually orthogonal components of the displacements $q_{\alpha}(\varrho_1)$ and $q_{\beta}(\varrho_2)$ at $\varrho_1 = \varrho_2$ are not correlated

$$\langle q_{\alpha} q_{\beta} \rangle = \psi_{\alpha\beta}^{q}(0) = 0 \quad ,$$

and that the mean squares q_{α} and q_{β} are equal

$$\langle q_{\alpha}^2 \rangle_{\mathrm{p}} = \langle q_{\beta}^2 \rangle_{\mathrm{p}} = \psi_{\alpha\alpha}^{q}(0) = -\frac{L^3}{6} \int\limits_0^{\infty} \frac{\psi_{\varepsilon}'(\zeta)}{\zeta} d\zeta \quad . \tag{1.68}$$

Specifically, for the isotropic Gaussian form of the permittivity fluctuation covariance (1.31), we have

$$\langle q_{\alpha}^2 \rangle_{\mathrm{p}} = \langle q_{\beta}^2 \rangle_{\mathrm{p}} = \sqrt{\frac{\pi}{2}} \frac{L^3 \sigma_{\varepsilon}^2}{12 l_{\varepsilon}} \quad . \tag{1.69}$$

If the wave is not plane, but the medium is still statistically homogeneous and $\bar{\varepsilon} = 1$, the covariance matrix (1.65) for the displacements is reduced to

$$\psi_{\alpha\beta}^{q}(\varrho, L) = -\frac{1}{2} \int\limits_0^{L} (L - s)^2 ds \int\limits_0^{\infty} \frac{\partial^2 \psi_{\varepsilon}[\delta(s) + \zeta t_0]}{\partial \varrho_{\alpha} \partial \varrho_{\beta}} d\zeta, \tag{1.70}$$

where, as before, $\delta(s)$ is the current distance between the rays, which depends on the transverse separation of the observation points, ϱ. In the special case of the *spherical wave*, $\delta = \varrho s/L$, so that $\partial/\partial\varrho_{\alpha} = s/L \partial/\partial\delta_{\alpha}$, and we can write

$$\psi_{\alpha\beta}^{q}(\varrho, L) = -\frac{1}{2} \int\limits_0^{L} (L - s)^2 \frac{s^2}{L^2} ds \int\limits_0^{L} \frac{\partial^2 \psi_{\varepsilon}(\delta + \zeta t_0)}{\partial \delta_{\alpha} \partial \delta_{\beta}} d\zeta \quad . \tag{1.71}$$

Hence, the mean square displacement q_{α} is

$$\langle q_{\alpha}^2 \rangle_{\mathrm{sph}} = \psi_{\alpha\alpha}^{q}(0) = -\frac{1}{2} \int\limits_0^{L} (L - s)^2 \frac{s^2}{L^2} ds \int\limits_0^{\infty} \frac{\partial^2 \psi_{\varepsilon}(\delta + \zeta t_0)}{\partial \delta_{\alpha}^2} \bigg|_{\delta=0} d\zeta \quad . \tag{1.72}$$

But, integrating with respect to s gives $L^3/30$. Therefore,

$$\langle q_{\alpha}^2 \rangle_{\mathrm{sph}} = -\frac{L^3}{60} \int\limits_0^{\infty} \frac{\partial^2 \psi_{\varepsilon}(\delta + \zeta t_0)}{\partial \delta_{\alpha}^2} \bigg|_{\delta=0} d\zeta = \frac{1}{10} \langle q_{\alpha}^2 \rangle_{\mathrm{p}} \quad , \tag{1.73}$$

i.e., the mean square displacement of the ray in the spherical wave is 0.1 times that in the plane wave.

Fluctuations of arrival angles and displacements could be calculated using another approach described in [1.5]. It relies on the fact that random deviations and deflections of the ray can, under certain conditions, be described as a Markov process occurring under the action of random "pushes" due to the gradients of electric permittivity (Sects. I.5.10 and I.6.4.). In this approach, the problem reduces to one of solving the Einstein-Fokker-Planck equation for the joint probability density of the displacements and directions of the ray. For the case of the plane wave propagating in a statistically homogeneous and isotropic medium, the solution of the Einstein-Fokker-Planck equation is the normal distribution law, the second moments, which completely characterize the normal distribution coinciding with those derived above. Therefore, we shall not dwell here on the treatment of ray statistics using the Einstein-Fokker-Planck equation. Suffice it to note that the regions of validity of this approach were established in [1.14, 15], and the generalization of the method to media with regular refraction was given in [1.16–18].

3) Group Path Fluctuations. The group path $\mathcal{L}$ in a homogeneous isotropic medium is given by

$$\mathcal{L} = c \int_0^L \frac{ds}{u_{\mathrm{gr}}} = \int_0^L \frac{\partial(\omega\sqrt{\varepsilon})}{\partial\omega} ds \quad , \quad \text{where} \tag{1.74}$$

$$u_{\mathrm{gr}} = c \left[\frac{\partial(\omega\sqrt{\varepsilon})}{\partial\omega} \right]^{-1}$$

is the group velocity of the wave. Clearly, the ratio $\mathcal{L}/c$ is the time of propagation of a signal. In a randomly refracting medium, the group path $\mathcal{L}$, and hence the propagation time, undergo fluctuations. The variance of the fluctuations of the group path can be calculated by the perturbation method. We will confine our discussion to the case of $\bar{\varepsilon} = $ const, i.e., where regular refraction is absent. Expanding $\mathcal{L}$ into a series in small fluctuations $\tilde{\varepsilon}$, we obtain the first-order correction $\mathcal{L}_1$ to the unperturbed group path $\mathcal{L}_0 = L\partial(\omega\sqrt{\bar{\varepsilon}})/\partial\omega$

$$\mathcal{L}_1 = \frac{1}{2} \int_0^L \frac{\partial}{\partial\omega} \left(\omega \frac{\tilde{\varepsilon}}{\sqrt{\bar{\varepsilon}}} \right) ds \quad . \tag{1.75}$$

Of course, the integration here is along the unperturbed ray.

In the nondispersive medium, both $\bar{\varepsilon}$ and $\tilde{\varepsilon}$ are independent of the frequency. In that case,

$$\mathcal{L}_1 = \frac{1}{2} \int_0^L \frac{\tilde{\varepsilon}}{\sqrt{\bar{\varepsilon}}} ds = \varphi_1 \quad , \tag{1.76}$$

i.e., the perturbation of the group path $\mathcal{L}_1$ coincides with the eikonal perturbation φ_1. This is because the group and phase velocities in the nondispersive medium are equal, so that the comments made in Sect. 1.2 regarding eikonal fluctuations can be fully applied to the fluctuations of the group path. In particular, according to (1.31), the variance of the group path is

$$\sigma_{\mathcal{L}}^2 = \sigma_{\varphi}^2 = \frac{1}{2} \int_0^L \frac{ds}{\bar{\varepsilon}} \int_0^\infty \psi_\varepsilon[t_0\zeta; \mathbf{r}_0(s)]d\zeta \quad . \tag{1.77}$$

An important special case of the dispersive medium is the isotropic cold plasma, for which $\bar{\varepsilon}$ and $\tilde{\varepsilon}$ are given by (III.4.191). Substituting (III.4.191) into (1.75) gives (at $\bar{\varepsilon} = \text{const}$)

$$\mathcal{L}_1 = -\frac{1}{2} \int_0^L \frac{\tilde{\varepsilon}}{\bar{\varepsilon}^{3/2}} ds \quad , \tag{1.78}$$

where the integrand includes $1/\bar{\varepsilon}^{3/2}$ instead of $1/\bar{\varepsilon}^{1/2}$ in the expression (1.15) for the eikonal φ_1. Bearing this in mind, we can write an expression similar to (1.50) for the variance of the group path in a plasma, but with $1/\bar{\varepsilon}^3$ substituted for $1/\bar{\varepsilon}$

$$\sigma_{\mathcal{L}}^2 = \frac{1}{2} \int_0^L \frac{ds}{(\bar{\varepsilon})^3} \int_0^\infty \psi_\varepsilon[\zeta t_0, \mathbf{r}_0(s)] d\zeta \quad . \tag{1.79}$$

It is obvious that the variances $\sigma_{\mathcal{L}}^2$ and σ_φ^2 can only coincide at $\bar{\varepsilon} \approx 1$, i.e., for sufficiently high-frequency waves, such that $\omega^2 \gg \omega_e^2 = 4\pi e^2 \overline{N}_e/m$.

1.4 Level Fluctuations

As has already been stated, the quantity

$$\chi = \ln \frac{A}{A_0} \quad ,$$

is called the *amplitude level* or simply the level. A^0 is a constant of the same dimensions as A, e.g. the amplitude of an unperturbed wave. Let us find the covariance of the level χ.

The level χ obeys the equation

$$2\nabla\chi\nabla\varphi + \Delta\varphi = 0 \quad , \tag{1.80}$$

which immediately follows from the transport equation (1.5). We will now represent χ as a series in $\tilde{\varepsilon}$

$$\chi = \chi_0 + \chi_1 + \chi_2 + \cdots \quad , \tag{1.81}$$

where χ_0 is the unperturbed value of the level, i.e., the solution to the equation

$$2\nabla\chi_0\nabla\varphi_0 + \Delta\varphi_0 = 0 \quad . \tag{1.82}$$

Substituting (1.81), and the similar expansion (1.11) for the eikonal φ, into (1.80) and taking account of (1.82), for the correction χ_1 we obtain

$$2\nabla\varphi_0\nabla\chi_1 + \Delta\varphi_1 + 2\nabla\chi_0\nabla\varphi_1 = 0 \quad . \tag{1.83}$$

In this equation, we can approximately substitute $\nabla\varphi_1$ and $\Delta\varphi_1$ for $\nabla_\perp\varphi_1$ and $\Delta_\perp\varphi_1$, where $\nabla_\perp = \nabla - t_0(t_0\nabla)$ and $\Delta_\perp = \nabla_\perp^2$ are the operators of transverse differentiation. In fact,

$$|\nabla_\perp\varphi_1| \sim \sqrt{|\Delta_\perp\psi_\varphi|} \sim \sqrt{\psi_\varphi/l_\varepsilon^2} \sim \sqrt{(\sigma_\varepsilon^2 l_\varepsilon)L/l_\varepsilon^2} \sim \sigma_\varepsilon\sqrt{L/l_\varepsilon} \quad ,$$

whereas $|\nabla_{\parallel}\varphi_1| = |d\varphi_1/ds| = |\tilde{\varepsilon}|/2 \sim \sigma_{\varepsilon}$. Since we are interested in the distances L which are large as compared to the correlation radius for the inhomogeneities, l_{ε}, we can put $|\nabla_{\parallel}\varphi_1| \ll |\nabla_{\perp}\varphi_1|$. Similarly, we can show that $|\Delta_{\parallel}\varphi_1| \ll |\Delta_{\perp}\varphi_1|$. As a result, (1.83) takes the form

$$2\nabla\varphi_0\nabla\chi_1 + \Delta_{\perp}\varphi_1 + 2\nabla_{\perp}\chi_0\nabla_{\perp}\varphi_1 = 0 \quad . \tag{1.84}$$

For the plane and the omnidirectional spherical waves $\nabla_{\perp}\chi_0 = 0$ since, at the phase front $\varphi_0 = \text{const}$, the amplitude of such waves is constant, and the last term in (1.84) vanishes. This term can also be ignored in the many other cases, e.g., in that of the wave beam or directed spherical wave where the ray width is larger than the correlation radius of the inhomogeneities, l_{ε}. In fact,

$$\nabla_{\perp}\chi \sim \frac{1}{a} \quad , \quad \nabla_{\perp}\varphi_1 \sim \frac{\sigma_{\varphi}}{l_{\varepsilon}} \quad , \quad \Delta_{\perp}\varphi \sim \frac{\sigma_{\varphi}}{l_{\varepsilon}^2} \quad .$$

This suggests that the third term in (1.84) is about l_{ε}/a times smaller than the second one. Discarding the term with $\nabla_{\perp}\chi_0$ and considering that

$$(\nabla\varphi_0\nabla\chi_1) = \sqrt{\bar{\varepsilon}}(t_0\nabla\chi_1) = \sqrt{\bar{\varepsilon}}\frac{d\chi_1}{ds} \quad ,$$

we write the equation for χ_1 in the following form:

$$\frac{d\chi_1}{ds} = -\frac{1}{2\sqrt{\bar{\varepsilon}}}\Delta_{\perp}\varphi_1 \quad . \tag{1.85}$$

The solution to (1.85) is

$$\chi_1 = -\frac{1}{2}\int_0^L \frac{\Delta_{\perp}\varphi_1}{\sqrt{\bar{\varepsilon}}}ds \quad , \tag{1.86}$$

so that the transverse covariance for the level, $\psi_{\chi}(\varrho, L) = \langle\chi_1(\varrho_1, L)\,\chi_1(\varrho_2, L)\rangle$, is expressed in terms of the eikonal covariance $\psi_{\varphi}(r', r'')$ in the following way:

$$\psi_{\chi}(\varrho, L) = \frac{1}{4}\iint_0^L ds'\,ds''\,\frac{\Delta'_{\perp}\Delta''_{\perp}\psi_{\varphi}(r', r'')}{\sqrt{\bar{\varepsilon}[r_0(s')]\bar{\varepsilon}[r_0(s'')]}} \quad , \tag{1.87}$$

where $r' = r'(s')$ and $r'' = r''(s'')$ lie on unperturbed rays arriving at the observation points r_1 and r_2 (Fig. 1.9), and $\Delta'_{\perp}$ and $\Delta''_{\perp}$ are the transverse Laplace operators in r' and r''. We will confine ourselves to an analysis of the level fluctuations for the plane and spherical waves propagating through a statistically homogeneous medium with $\bar{\varepsilon} = 1$.

According to (1.24), the eikonal covariance for the *plane* wave $\psi_{\varphi}(r', r'')$ is

$$\psi_{\varphi}(r', r'') = \frac{s_<}{2}\int_0^{\infty} \psi_{\varepsilon}(\varrho, \zeta)d\zeta \quad , \tag{1.88}$$

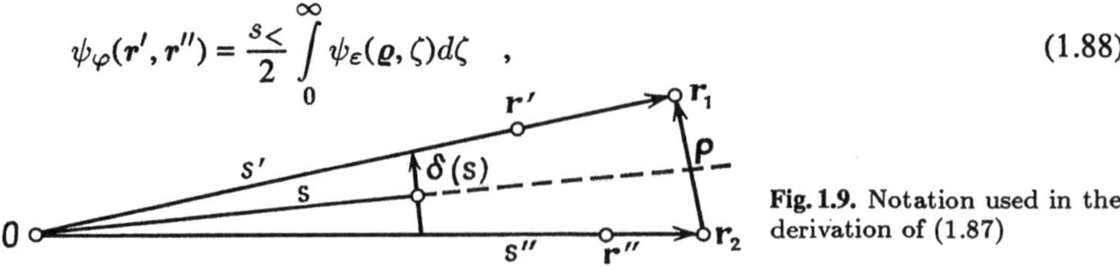

Fig. 1.9. Notation used in the derivation of (1.87)

where $s_<$ is the smaller of the distances s' and s'', and $\varrho = $ const is the distance between two parallel rays arriving at r_1 and r_2. Substituting (1.88) into (1.87), we get

$$\psi_\chi(\varrho, L) = \frac{1}{8} \int\limits_0^L\!\!\!\int s_< ds'\, ds'' \int\limits_0^\infty \Delta_\perp^2 \psi_\varepsilon(\varrho, \zeta) d\zeta \quad . \tag{1.89}$$

The two-fold integral in (1.89) is easily taken

$$\int\limits_0^L\!\!\!\int s_< ds'\, ds'' = \frac{L^3}{3} \quad .$$

Hence, for the plane wave,

$$\psi_\chi(\varrho, L) = \frac{L^3}{24} \int\limits_0^\infty \Delta_\perp^2 \psi_\varepsilon(\varrho, \zeta) d\zeta \quad . \tag{1.90}$$

Thus, the level fluctuations increase with the cube of the distance L covered by the wave, whereas the transverse covariance of the eikonal of the plane wave (1.34) varies with L

$$\psi_\perp(\varrho, L) = \frac{L}{2} \int\limits_0^\infty \psi_\varepsilon(\varrho, \zeta) d\zeta \quad . \tag{1.91}$$

Note that if the integral in (1.90) is expressed in terms of $\psi_\perp$, the covariance of the level of the plane wave will be

$$\psi_\chi(\varrho, L) = \frac{L^2}{12} \Delta_\perp^2 \psi_\perp(\varrho, L) \quad . \tag{1.92}$$

This formula simplifies the calculation of ψ_χ, if the eikonal covariance $\psi_\perp$ is known.

The level covariance can also be expressed through the spectral density of the electric permittivity fluctuations

$$\psi_\chi(\varrho, L) = \frac{\pi L^3}{24} \int\limits_{-\infty}^\infty \kappa_\perp^4 \Phi_\varepsilon(\kappa_\perp, 0) e^{i\kappa_\perp \varrho} d^2 \kappa_\perp \quad . \tag{1.93}$$

If in (1.90) or (1.93) we put $\varrho = 0$, we obtain two forms of the expression for the amplitude level variance

$$\sigma_\chi^2 = \frac{L^3}{24} \int\limits_0^\infty \Delta_\perp^2 \psi_\varepsilon(\zeta t_0) d\zeta = \frac{\pi L^3}{24} \int\limits_{-\infty}^\infty \kappa_\perp^4 \Phi_\varepsilon(\kappa_\perp, 0) d^2 \kappa_\perp \quad . \tag{1.94}$$

When the fluctuations $\tilde\varepsilon$ are isotropic, the level covariance only varies with the magnitude of the distance ϱ between the observation points

$$\psi_\chi(\varrho, L) = \frac{L^3}{24} \left(\frac{1}{\varrho} \frac{\partial}{\partial \varrho} \varrho \frac{\partial}{\partial \varrho} \right)^2 \int\limits_0^\infty \psi_\varepsilon(\sqrt{\varrho^2 + \zeta^2}) d\zeta \quad , \tag{1.95}$$

or, alternatively,

$$\psi_\chi(\varrho, L) = \frac{\pi^2 L^3}{12} \int_0^\infty \kappa^5 \Phi_\varepsilon(\kappa) J_0(\kappa\varrho) d\kappa \quad . \tag{1.96}$$

Specifically, either by (1.35) or (1.36), the Gaussian covariance (1.31) will be

$$\psi_\chi(\varrho) = \sigma_\chi^2 \left(1 - \frac{\varrho^2}{l_\varepsilon^2} + \frac{1}{8}\frac{\varrho^4}{l_\varepsilon^4}\right) e^{-\varrho^2/2l_\varepsilon^2} \quad , \quad \text{where} \tag{1.97}$$

$$\sigma_\chi^2 = \frac{2}{3}\frac{L^2}{l_\varepsilon^4}\sigma_\varphi^2 = \frac{\sqrt{\pi}}{3\sqrt{2}}\sigma_\varepsilon^2\frac{L^3}{l_\varepsilon^3} \quad . \tag{1.98}$$

It will be noticed that the ratio of the level variance σ_χ^2 to the phase variance $\sigma_S^2 = k^2\sigma_\varphi^2$ is about

$$\frac{\sigma_\chi^2}{\sigma_S^2} \sim \frac{L^2 \Delta_\perp^2 \psi_\varphi}{k^2 \psi_\varphi} \sim \frac{L^2}{k^2 l_\varepsilon^4} \sim \left(\frac{\sqrt{\lambda L}}{l_\varepsilon}\right)^4 \quad , \tag{1.99}$$

i.e., it is determined by the square of the wave parameter $D \sim L/k l_\varepsilon^2 \sim (\sqrt{\lambda L}/l_\varepsilon)^2$. But in the region of validity of geometrical optics, the condition $\sqrt{\lambda L} \ll l_\varepsilon$ holds [see (1.10)], whereby the wave parameter is small (the size of the first Fresnel zone is small as compared to the size of the inhomogeneities). Therefore, within the region of validity of the GOM, the level fluctuations must be small as compared to the phase fluctuations

$$\sigma_\chi^2 \ll \sigma_S^2 = k^2\sigma_\varphi^2 \quad . \tag{1.100}$$

This allows us to ignore amplitude as compared to phase fluctuations in a number of cases, e.g., in deriving the mean field and coherence function using the GOM (Sect. 1.6).

Let us now return to the expression (1.93), which can be viewed as the two-dimensional expansion of the covariance into the Fourier integral. Accordingly,

$$F_\chi(\boldsymbol{\kappa}_\perp, L) = \frac{\pi L^3}{24}\kappa_\perp^4 \Phi_\varepsilon(\boldsymbol{\kappa}_\perp, 0) \tag{1.101}$$

is the two-dimensional spectral density of the amplitude level fluctuations. For isotropic fluctuations,

$$F_\chi(\boldsymbol{\kappa}_\perp, L) = F_\chi(\kappa_\perp, L) = \frac{\pi L^3}{24}\kappa_\perp^4 \Phi_\varepsilon(\kappa_\perp) \quad . \tag{1.102}$$

The factor $\kappa_\perp^4$ in (1.102) reduces the contribution of the large-scale ($\kappa_\perp \longrightarrow 0$) components of the spectrum of $\tilde{\varepsilon}$. This implies that, in the geometrical optics approximation, the amplitude fluctuations are, for the most part, caused by the small-scale component of the spectrum of the permittivity fluctuations $\tilde{\varepsilon}$. The presence of $\kappa_\perp^4$ in (1.102) also allows this equation to be used for locally homogeneous isotropic fields, when the spatial spectrum $\Phi_\varepsilon(\kappa)$ may, as $\kappa \longrightarrow 0$, have a power singularity of the form $\kappa^{-\alpha}$ with $\alpha < 5$.

It should be stressed that the fact that the two-dimensional spectral density $F_\chi(\kappa_\perp, L)$ vanishes as $\kappa_\perp \longrightarrow 0$ is equivalent to the relation

$$(2\pi)^2 F_\chi(0) = \int\limits_{-\infty}^{\infty} \psi_\chi(\varrho) d^2\varrho = 0 \quad , \tag{1.103}$$

which, for isotropic fluctuations, takes the form

$$\int\limits_{0}^{\infty} \psi_\chi(\varrho)\varrho \, d\varrho = 0 \quad . \tag{1.104}$$

Equations (1.103, 104) are closely related to the energy conservation law. This question is considered in more detail in Sect. 3.4, where it is shown that, in the case of the plane wave, the integral of the fluctuational component $\tilde{I} = I - \overline{I}$ over the entire plane is zero

$$\int\limits_{-\infty}^{\infty} \tilde{I}(\varrho, z) d^2\varrho = 0 \quad . \tag{1.105}$$

However, since $I = |A|^2 = |A_0|^2 \exp(2\chi) \approx |A_0|^2(1 + 2\chi)$, for small level fluctuations (i.e., when the GOM is valid) we have

$$\tilde{I} = I - \overline{I} \approx 2|A_0|^2(\chi - \overline{\chi}) = 2|A_0|^2\tilde{\chi} \quad ,$$

so that from (1.105) we derive the conservation law for $\tilde{\chi}$

$$\int\limits_{-\infty}^{\infty} \tilde{\chi}(\varrho, z) d^2\varrho = 0 \quad . \tag{1.106}$$

Multiplying (1.106) by $\tilde{\chi}(\varrho', z)$ and averaging gives (1.103).

It follows from (1.103) that the level covariance assumes negative as well as positive values. This can also be illustrated by the special example of a medium that has a Gaussian covariance of $\tilde{\varepsilon}$: here (1.97) is negative in the interval $(4 - 2\sqrt{2})l_\varepsilon < \varrho < (4 + 2\sqrt{2})l_\varepsilon$.

The computation of the amplitude level covariance for the spherical wave is a somewhat more difficult exercise. Suppose that the source and observation point lie in a statistically homogeneous medium. According to (1.41),

$$\psi_\varphi^{\mathrm{sph}}(r', r'') = \frac{1}{2} \int\limits_{0}^{s_<} ds \int\limits_{0}^{\infty} \psi_\varepsilon(\delta + \zeta t_0) d\zeta \quad , \tag{1.107}$$

and the current separation between the rays arriving at r' and r'' (Fig. 1.9) will be

$$\delta(s) = \left(\frac{r'}{s'} - \frac{r''}{s''}\right) s \quad .$$

Differentiating (1.107), we get

$$\Delta'_\perp \Delta''_\perp \psi_\varphi^{\mathrm{sph}}(r', r'') = \frac{1}{2} \int\limits_{0}^{s_<} ds \left(\frac{s^2}{s's''}\right)^2 \int\limits_{0}^{\infty} \Delta^2_\perp \psi_\varepsilon(\delta + t_0\zeta) d\zeta \quad , \tag{1.108}$$

with the result that, by (1.87),

$$\psi_\chi^{\rm sph}(\varrho, L) = \frac{1}{8} \int_0^L \frac{ds'}{s'^2} \int_0^L \frac{ds''}{s''^2} \int_0^{2<} s^4 \, ds \int_0^\infty \Delta_\perp^2 \psi_\varepsilon(\delta + \zeta t_0) d\zeta \quad . \tag{1.109}$$

The current separation between the rays, δ, can be expressed here in terms of the distance $\varrho = r_1 - r_2$ between the observation points r_1 and r_2:

$$\delta = \varrho s/L \quad .$$

We will simplify (1.109) using the relationship

$$\int_0^L \frac{ds'}{s'^2} \int_0^L \frac{ds''}{s''^2} \int_0^{s<} s^4 F(s) ds = \frac{1}{L^2} \int_0^L s^2 (L-s)^2 F(s) ds \quad ,$$

which is easily proved by double integration by parts. In our case

$$F(s) = \frac{1}{8} \int_0^\infty \Delta_\perp^2 \psi_\varepsilon(\delta + \zeta t_0) d\zeta \quad ,$$

therefore

$$\psi_\chi^{\rm sph}(\varrho, L) = \frac{1}{8L^2} \int_0^L s^2 (L-s)^2 ds \int_0^\infty \Delta_\perp^2 \psi_\varepsilon(\delta + \zeta t_0) d\zeta \quad . \tag{1.110}$$

If we then take into account that the internal integral equals the covariance (1.90) for the plane wave multiplied by $24/L^3$, we get

$$\psi_\chi^{\rm sph}(\varrho, L) = \frac{3}{L^5} \int_0^L s^2 (L-s)^2 \psi_\chi^{\rm P}\left(\frac{\varrho s}{L}\right) ds \tag{1.111}$$

or, if we introduce the dimensionless integration variable $\gamma = s/L$,

$$\psi_\chi^{\rm sph}(\varrho, L) = 3 \int_0^1 \gamma^2 (1-\gamma)^2 \psi_\chi^{\rm P}(\gamma \varrho) d\gamma \quad . \tag{1.112}$$

At $\varrho = 0$, this suggests a universal relationship between the level variances for the plane and spherical waves. Since

$$3 \int_0^1 \gamma^2 (1-\gamma)^2 d\gamma = \frac{1}{10} \quad ,$$

the level variance for the spherical wave is 10 times smaller than that for the plane one and we have

$$[\sigma_\chi^2(L)]^{\rm sph} \equiv \psi_\chi^{\rm sph}(0, L) = 3 \int_0^1 \gamma^2 (1-\gamma)^2 \psi_\chi^{\rm P}(0) d\gamma = \frac{1}{10}[\sigma_\chi^2(L)]^{\rm P} \quad . \tag{1.113}$$

The factor $\gamma^2 (1-\gamma)^2 = s^2(L-s)^2/L^4$ in (1.112, 113) determines the relative contribution of the various ray sections to the total amplitude level fluctuations for the spherical wave. The presence of this factor, which vanishes at the beginning ($s = 0$) and at the end ($s = L$) of the line, is associated with the focusing and defocusing at inhomogeneities in the medium, i.e., the lens action of the inhomogeneities. It is well known that a lens placed near a point light source and near an observation point does not affect the intensity of the spherical wave. The lens effect is also responsible for the relatively small fluctuations of the spherical as compared to the plane wave: the inhomogeneities at the beginning of the path, for all practical purposes, exert no influence on the amplitude fluctuations of the spherical wave. However, it is precisely these inhomogeneities that have the strongest effect on the fluctuations of the plane wave.

1.5 Fluctuations of Wave Characteristics in a Turbulent Troposphere

We will now apply the general equations of geometrical optics to calculations of wave characteristics in a turbulent troposphere. For the troposphere we can, to a good accuracy, put $\bar{\varepsilon} = 1$ and regard the fluctuations $\tilde{\varepsilon}$ as a locally homogeneous isotropic field (Sect. III.1.4). Under these conditions, the structure function for the eikonal of the plane wave is given by (1.38) or (1.39).

We now insert the spectral density (III.1.79) into (1.39)

$$\Phi_\varepsilon(\kappa) = 0.033 C_\varepsilon^2 \kappa^{-11/3} \exp(-\kappa^2/\kappa_{\mathrm{m}}^2) \quad , \tag{1.114}$$

where C_ε^2 is the structure constant, $\kappa_{\mathrm{m}} = 5.92/l_0$, and l_0 the inner scale of turbulence. The integral (1.39) of this spatial spectrum is expressed in terms of the degenerated hypergeometrical function $_1F_1$ (see [Ref. 1.4, Sect. 42])

$$D_\perp(\varrho, L) = 2.2 C_\varepsilon^2 L \kappa_{\mathrm{m}}^{-5/3} \left[{}_1F_1\left(-\frac{5}{6}, 1, -\frac{\kappa_{\mathrm{m}}^2 \varrho^2}{4}\right) - 1 \right] \quad . \tag{1.115}$$

If this function is replaced by its asymptotic values at small and large values of its third argument, $D_\perp(\varrho, L)$ will then have the following limiting expressions:

$$D_\perp(\varrho, L) = \begin{cases} 0.73 C_\varepsilon^2 L \varrho^{5/3} & \text{for} \quad \varrho \gg l_0 \quad , \\ 0.82 C_\varepsilon^2 L l_0^{-1/3} \varrho^2 & \text{for} \quad \varrho \ll l_0 \quad . \end{cases} \tag{1.116}$$

We can also find analogous limiting expressions for $D_\perp(\varrho, L)$ directly from (1.38) if we apply the rough approximation for $D_\varepsilon(r)$, obtained by interpolating the limiting values of (III.1.74, 75) that hold for $r \gg l_0$ and $r \ll l_0$, respectively, to the region $r \sim l_0$

$$D_\varepsilon(r) = \begin{cases} C_\varepsilon^2 r^{2/3} & \text{for} \quad r \geq l_0 \quad , \\ C_\varepsilon^2 l_0^{-4/3} r^2 & \text{for} \quad r < l_0 \quad . \end{cases} \tag{1.117}$$

Having done so, we will see that the result obtained only differs in that, instead of the coefficient 0.82 in (1.116), we have unity.

For small separations of the observation points, equation (1.116) gives quadratic dependence of the structure function on ϱ. However, for large ϱ the behavior of $D_\perp$ is described by the "five-thirds" power law. The indefinite growth of $D_\perp$ with ϱ corresponds to the infinite value of the eikonal variance $\sigma_\varphi^2(L) = \psi_\perp(0, L) = D_\perp(\infty L)/2$. In reality σ_φ^2 is, of course, limited, for as $\varrho \longrightarrow \infty$ the structure function $D_\perp(\varrho, L)$ "saturates". To estimate σ_φ^2, we will approximate $D_\varepsilon(\varrho)$ by (III.1.81)

$$D_\varepsilon(r) = \begin{cases} C_\varepsilon r^{2/3} & \text{for} \quad r \leq L_0 \ , \\ C_\varepsilon^2 L_0^{2/3} = \text{const} & \text{for} \quad r > L_0 \ , \end{cases} \qquad (1.118)$$

where L_0 is the outer scale of turbulence [as opposed to (III.1.81), we ignore here the region $r \leq l_0$, which makes no significant contribution to σ_φ].

Since $\psi_\varepsilon(r) = [D_\varepsilon(\infty) - D_\varepsilon(r)]/2$, from (1.118) we find

$$\psi_\varepsilon(r) \approx \begin{cases} \sigma_\varepsilon^2[1 - (r/L_0)^{2/3}] & \text{for} \quad r \leq L_0 \ , \\ 0 & \text{for} \quad r > L_0 \ , \end{cases} \qquad (1.119)$$

where $\sigma_\varepsilon^2 = \psi_\varepsilon(0) = C_\varepsilon^2 L_0^{2/3}/2$ is the variance of the electric permittivity. It appears that the covariance (1.119), obtained by interpolating the limiting values for $r \ll L_0$ and $r \gg L_0$ to the intermediate region $r \sim L_0$, is not positive definite [the spectral density Φ_ε for (1.119) assumes negative values]. Nevertheless, by using (1.119) to compute the phase covariance, we can expect qualitatively correct results for $r \ll L_0$ and $r \gg L_0$. This is justified by the fact that equations (1.120, 121) below can also be derived in a more rigorous manner.

Substituting (1.119) into (1.36), we obtain the phase covariance

$$\psi_\perp(\varrho, L) = \begin{cases} \dfrac{L\sigma_\varepsilon^2}{2} \displaystyle\int\limits_0^{\sqrt{L_0 - \varrho^2}} \left[1 - \left(\dfrac{\varrho^2 + \zeta^2}{L_0^2}\right)^{1/3}\right] d\zeta & \text{for} \quad \varrho \leq L_0 \ , \\ 0 & \text{for} \quad \varrho > L_0 \ . \end{cases} \qquad (1.120)$$

In particular,

$$\sigma_\varphi^2 = \psi_\perp(0, L) = \frac{L\sigma_\varepsilon^2}{2} \int\limits_0^{L_0} \left[1 - \left(\frac{\zeta}{L_0}\right)^{2/3}\right] d\zeta = \frac{LL_0\sigma_\varepsilon^2}{5} = \frac{LC_\varepsilon^2 L_0^{5/3}}{10} \ . \qquad (1.121)$$

The mean square of the eikonal fluctuations is thus determined by the distance covered by the wave, the outer scale of turbulence and the fluctuation variance $\sigma_\varepsilon^2 = C_\varepsilon^2 L^{2/3}/2$, while the eikonal covariance vanishes when $\varrho > L_0$.

Putting into (1.21) $C_\varepsilon^2 = 10^{-15}\,\text{cm}^{-2/3}$, $L = 1\,\text{km}$, $L_0 = 1\,\text{m}$, we obtain a typical value of the eikonal variance $\sigma_\varphi^2 \sim 2.10^{-8}\,\text{cm}^2$. However, a typical value of the phase variance $\sigma_S^2 = k^2\sigma_\varphi^2$ is 3.10^2 at wavelength $\lambda = 5.10^{-5}\,\text{cm}$ (optical range), and 3.10^{-4} at $\lambda = 0.05\,\text{cm}$ (submillimeter range).

The statistical behavior of the arrival angles of the ray can be deduced from the equations of Sect. 1.3. Therefore, using (1.116), we find from (1.57) the variance of the arrival angle

$$\langle\theta_\alpha^2\rangle = \frac{1}{2}\frac{\partial^2 D_\perp(0, L)}{\partial \varrho_\alpha^2} = 0.82 C_\varepsilon^2 L l_0^{-1/3} \quad . \tag{1.122}$$

The level covariance, like the structure function of the phase, can be expressed, by (1.93), in terms of hypergeometric functions [1.4]. In particular, following [1.4], the level variance for the spectrum (1.114) is

$$\sigma_\chi^2 = 0.8 C_\varepsilon^2 L^3 l_0^{-7/3} \quad . \tag{1.123}$$

The correlation coefficient of the level, $K_\chi(\varrho) = \psi_\chi(\varrho, L)/\sigma_\chi^2(L)$, behaves as shown in Fig. 1.10 taken from [1.4], at $\kappa_m = 5.92/l_0$. As seen in the figure, the correlation coefficient $K_\chi(\varrho)$ takes on negative values, as is to be expected for the plane wave.

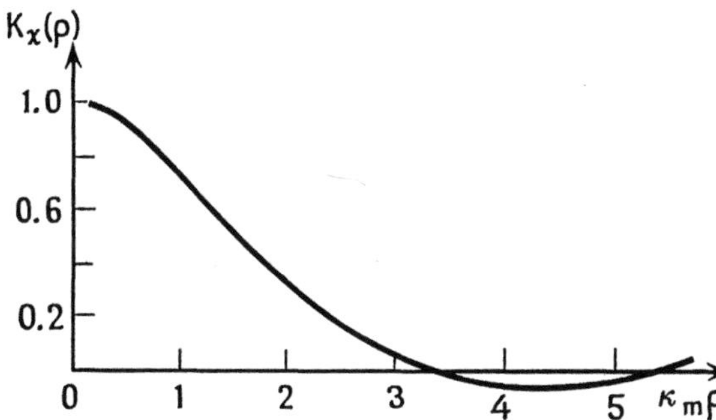

Fig. 1.10. Correlation coefficient of the amplitude level for an isotropic turbulent medium

The structure function of the eikonal of the spherical wave can be derived with the help of (1.45) or (1.46). In the limiting cases of $\varrho \ll l_0$ and $\varrho \gg l_0$, it is [1.4]:

$$D_\perp^{sph}(\varrho, L) = \begin{cases} \frac{1}{3}D_\perp^P(\varrho, L) = 0.27 C_\varepsilon^2 L l_0^{-1/3} \varrho^2 & \text{for} \quad \varrho \ll l_0 \quad , \\ \frac{3}{8}D_\perp^P(\varrho, L) = 0.27 C_\varepsilon^2 L \varrho^{5/3} & \text{for} \quad \varrho \gg l_0 \quad . \end{cases} \tag{1.124}$$

In the turbulent atmosphere, the general constraint on the GOM — the inequality (1.10) — reduces to the condition that the radius of the first Fresnel zone be small as compared to the *inner* scale of turbulence

$$\sqrt{\lambda L} \ll l_0 \quad . \tag{1.125}$$

For the lower atmosphere, where $l_0 \sim 1\,\text{cm}$, and the optical range ($\lambda = 5.10^{-5}\,\text{cm}$), according to (1.125), the distance L is limited by a value of order $200\,\text{m}$. But the diffraction treatment (see Sect. 2.4) shows that some of the results of the GOM are valid at much larger distances.

This is partly due to the fact that in the turbulence spectrum (1.114), the contribution of small-scale inhomogeneities with $\kappa \gtrsim \kappa_m$ is smaller than that of large-scale ones with $\kappa \lesssim \kappa_m$. Therefore, when at larger distances the size of the first Fresnel zone becomes comparable to the inner scale l_0 and (1.125) is violated, the inequality $\sqrt{\lambda L} \ll l$ will still hold for most of the inhomogeneities. This enables the GOM to be employed for larger inhomogeneities, with the scale $l > l_0$. These inhomogeneities chiefly influence the phase of the wave, not of its amplitude. As a result, the phase fluctuations in a turbulent medium are more pronounced (relative to level fluctuations) than in a medium with single-scale inhomogeneities. Indeed, by (1.121, 123) the ratio of level and phase variances is

$$\frac{\sigma_\chi^2}{\sigma_S^2} = \frac{\sigma_\chi^2}{k^2\sigma_\varphi^2} \sim \frac{C_\varepsilon^2 L^3 l_0^{-7/3}}{k^2 L L_0^{5/3} C_\varepsilon^2} \sim \frac{L^2}{k^2 (l_0^7 L_0^5)^{1/3}} \sim \left(\frac{\sqrt{\lambda L}}{l_0}\right)^4 \left(\frac{l_0}{L_0}\right)^{5/3} \quad . \tag{1.126}$$

As compared with (1.99), this equation contains the small factor $(l_0/L_0)^{5/3} \ll 1$, which reflects precisely the larger contribution of larger scales to phase fluctuations (large and small inhomogeneities in diffraction effects are discussed in Sect. 2.4).

1.6 Mean Field and Coherence Functions

In a first approximation to geometrical optics, the phase and level fluctuations of the wave are distributed by the normal (Gaussian) law, since in this approximation both quantities are expressed by integrals of $\tilde\varepsilon$ (or derivatives of $\tilde\varepsilon$), the ray meeting *many* inhomogeneities along the integration path L. Under these conditions, the normalization of the phase and level is a consequence of the Central Limit Theorem.

If the fluctuations of phase $\tilde S \approx S_1 = k\varphi_1$ and level $\tilde\chi \approx \chi_1$ obey the normal law, then the field

$$u = e^{iS+\chi} \approx e^{iS_0+\chi_0}\, e^{iS_1+\chi_1} = u_0 e^{iS_1+\chi_1} \tag{1.127}$$

[$u_0 = \exp(iS_0 + \chi_0)$ is the unperturbed field] is distributed according to the *log-normal* law. Considering that in the region of validity of the GOM the level variance σ_χ^2 is markedly smaller than the phase variance σ_S^2 [see (1.99) and (1.126)], we may, when calculating the moments of the field u, neglect the level fluctuations, i.e., we may put

$$u \approx u_0 e^{iS_1} . \tag{1.128}$$

Using the formula

$$\langle e^{i\xi}\rangle = e^{-\langle\xi^2\rangle/2} \quad , \tag{1.129}$$

which is valid for any normal quantity ξ with $\overline{\xi} = 0$, we find from (1.128)

$$\overline{u} = u_0\langle e^{iS_1}\rangle = u_0 e^{-\sigma_S^2/2} = u_0 e^{-k^2\sigma_\varphi^2/2} \quad ,$$

or, by (1.50),

$$\overline{u} = u_0\exp\left\{-\frac{k^2}{4}\int_0^L \frac{ds}{\overline{\varepsilon}[r_0(s)]}\int_0^\infty \psi_\varepsilon[\zeta t_0; r_0(s)]d\zeta\right\}$$

$$\equiv u_0\exp\left(-\frac{1}{2}\int_0^L \alpha(s)ds\right) \quad . \tag{1.130}$$

Here

$$\alpha(s) = \frac{k^2}{2\overline{\varepsilon}[r_0(s)]}\int_0^\infty \psi_\varepsilon[\zeta t_0; r_0(s)]d\zeta \tag{1.131}$$

acts as a sort of attenuation factor for the mean field along the ray, i.e., an extinction coefficient. According to (1.130), the mean field decays the faster, the larger the variance of the fluctuations σ_ε^2.

In a statistically homogeneous medium, the extinction coefficient is constant and the mean field decays exponentially

$$\overline{u} = u_0 e^{-\alpha L/2} . \tag{1.132}$$

At $\overline{\varepsilon} = 1$ and statistically homogeneous fluctuations, the extinction coefficient (1.131) can be written as

$$\alpha = \frac{k^2}{2} \int\limits_0^\infty \psi_\varepsilon(t_0\zeta) d\zeta . \tag{1.133}$$

This coincides with the total scattering cross-section of a unit volume, σ_0, which determines the field attenuation in the single-scattering approximation (see Exercise III.4.2.2).

The coherence function, i.e., the covariance $\Gamma_u(r_1, r_2) = \langle u(r_1)u^*(r_2)\rangle$, can be computed from (1.128, 129). We will provide the expression for the *transverse* coherence function, when the points r_1 and r_2 lie on the phase front ($\varrho = r_1 - r_2$)

$$\Gamma_u(\varrho, L) \approx |u_0|^2 \langle e^{i[S_1(r_1)-S_1(r_2)]}\rangle = |u_0|^2 \exp - \langle [S_1(r_1) - S_1(r_2)]^2\rangle/2$$
$$= I_0 \exp[-D_S(\varrho, L)/2] . \tag{1.134}$$

$I_0 \equiv |u_0|^2$ is the intensity of the unperturbed field and $D_S(\varrho, L) = k^2 D_\perp(\varrho, L)$ the transverse structure function of the phase. For a statistically homogeneous and isotropic medium ($\overline{\varepsilon} = 1$), $D_\perp$ is given by (1.38) or (1.39); expressions for $D_\perp$ in a turbulent atmosphere were given in the previous section. In the special case of the plane wave propagating through a turbulent medium [the structure function is given by (1.116)], the coherence function is

$$\Gamma_u(\varrho) = \begin{cases} I_0 \exp(-0.37 C_\varepsilon^2 k^2 L \varrho^{5/3}) & \text{for} \quad \varrho \gg l_0 , \\ I_0 \exp(-0.41 C_\varepsilon^2 k^2 L l_0^{-1/3} \varrho^2) & \text{for} \quad \varrho \ll l_0 , \end{cases} \tag{1.135}$$

and the covariance

$$\psi_u(\varrho, L) = \Gamma_u(\varrho, L) - |\overline{u}|^2 = e^{-D_S(\varrho,L)/2} - e^{-\sigma_S^2(L)}$$
$$= e^{-\sigma_S^2(L)}(e^{\psi_S(\varrho,L)} - 1) . \tag{1.136}$$

With small phase fluctuations ($\sigma_S^2 \ll 1$), the covariances (and hence the correlation radii) of the phase and the field itself coincide [$\psi_u(\varrho) \approx \psi_S(\varrho)$]. If $\sigma_S^2 \gg 1$, the second term in (1.136) can be ignored, and then

$$\psi_u(\varrho, L) \approx \Gamma_u(\varrho, L) = I_0 e^{-D_S(\varrho,L)/2} . \tag{1.137}$$

The correlation radius of the field (which can also be treated as the transverse coherence radius) can be estimated (when $\sigma_S^2 \gg 1$) from the condition $\psi_u(\varrho, L) \approx I_0/2$, or from the equivalent equation $D_S(\varrho, L) \approx 2\ln 2 \approx 1.4$. Since the phase structure function $D_S = k^2 D_\varphi$ is proportional to the distance covered by the wave, the correlation radius of the field decreases with L. For example, if a plane wave propagates in a turbulent troposphere and the separation between the observation points ϱ lies in the inertia interval $L^0 \gg \varrho \gg l_0$, then, from the first of the expressions (1.135), the correlation radius l_u of the field will be

$$l_u \approx (C_\varepsilon^2 k^2 L)^{-3/5} \quad .$$

In this case it decreases with distances as $L^{-3/5}$.

We will see below that the expressions (1.132) for the mean field, and (1.134) for the coherence function, hold beyond the region of validity of the GOM when level fluctuations are not small and diffraction effects cannot be ignored. As has already been said, this is because in the case of large-scale inhomogeneities, the spatial coherence of the field is violated for the most part due to phase fluctuations, whereas amplitude (level) fluctuations play a lesser role.

1.7 Exercises

1.7.1 In the general case, the transverse covariance of the eikonal is given by (1.52). In the simplest case of statistically homogeneous fluctuations and constant electric permittivity, unperturbed rays may be treated as straight lines diverging from a source.

Find the transverse covariance of the eikonal of the spherical wave that has passed through a randomly refracting layer of finite thickness (Fig. 1.11).

Solution. If L_0 is the thickness of the layer and R_0 the distance from the source to the layer, the distance between the rays varies according to the law $\delta(s) = \varrho s/(R_1 + L_1)$, and the integration with respect to s in (1.52) must be carried out from $R_1 = R_0/\cos\theta$ to $R_1 + L_1 = (R_0 + L_0)/\cos\theta$, where θ is the angle of incidence of the middle ray on the layer. Then

$$\psi_\varphi(\varrho) = \frac{1}{2\bar\varepsilon} \int\limits_{R_1}^{R_1+L_1} ds \int\limits_0^\infty \psi_\varepsilon\left(\frac{\varrho s}{R_1 + L_1} + t_0\zeta\right) d\zeta \quad ,$$

or, considering (1.44),

$$\psi_\varphi(\varrho) = \frac{1}{L_1} \int\limits_{R_1}^{R_1+L_1} ds\,\psi_\varphi^{\mathrm{P}}\left(\frac{\varrho s}{L_1 + R_1}, L_1 + R_1\right) \quad .$$

These expressions allow us to pass to the limiting cases of the plane wave ($R^0 \longrightarrow \infty$, i.e., the source at infinity) and of the source of the spherical wave at the layer boundary ($R_0 \longrightarrow 0$).

Expressions of this type make it possible to calculate, for example, the fluctuations of the phase of ultrashort radio waves that have passed through the statistically inhomogeneous ionosphere. In many cases, the amplitude (level) fluctuations can be ignored and the ionosphere replaced by an equivalent phase screen, so enabling the covariance of the field at the ionosphere output to be easily derived. The subsequent behavior of the field from the ionosphere to the ground is governed by the laws of wave diffraction in free space (Sect. III.2.3).

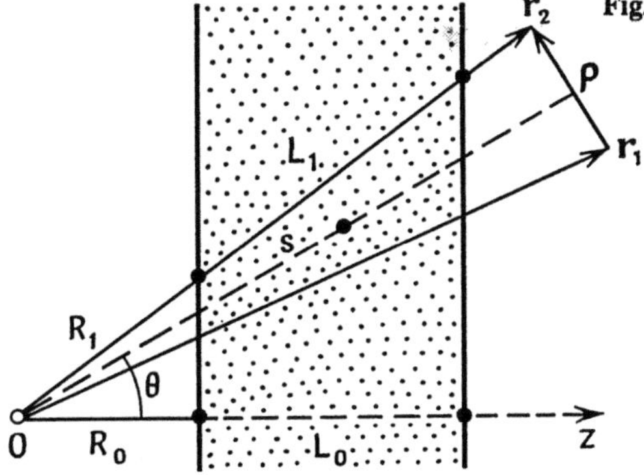

Fig. 1.11. Observation locations in exercise 1.7.1

1.7.2 Let a convergent spherical wave whose focus is at $z = F$ be incident on a randomly inhomogeneous half-space $z > 0$ (Fig. 1.12). Find the eikonal covariance in the plane $z = \text{const}$.

Solution.

$$\psi_\varphi^{\text{sph}}(\varrho) = \frac{1}{z} \int\limits_0^z ds\, \psi_\varphi^{\text{P}}\left(\varrho \left|\frac{F-s}{F-z}\right|\right)$$

$$= \begin{cases} \dfrac{F-z}{\varrho z} \displaystyle\int\limits_{\varrho}^{\varrho F/(F-z)} \psi_\varphi^{\text{P}}(\varrho')d\varrho' & \text{for } z < F \ , \\[4ex] \dfrac{z-F}{\varrho z}\left[\displaystyle\int\limits_0^{\varrho F/(z-F)} \psi_\varphi^{\text{P}}(\varrho')d\varrho' + \int\limits_0^{\varrho} \psi_\varphi^{\text{P}}(\varrho')d\varrho'\right] & \text{for } z > F \ . \end{cases}$$

(1.138)

These expressions suggest that the correlation radius for the convergent spherical wave is smaller than that for the plane one, and tends to zero as $z \longrightarrow F$ (of course, the above equation does not hold in the immediate vicinity of the focus). Beyond the focus, the correlation radius begins to increase and in the limit $z \gg F$ equation (1.138) describes the eikonal fluctuations of the divergent spherical wave.

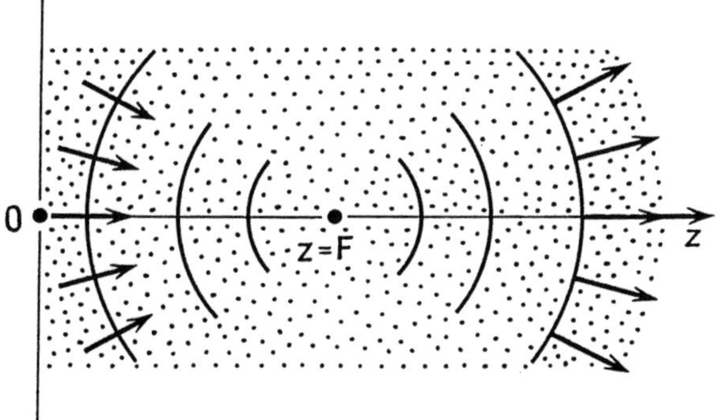

Fig. 1.12. Notation used in the derivation of the phase covariance in a convergent spherical wave (exercise 1.7.2)

1.7.3 Find the transverse covariance of the eikonal of the plane wave in a medium with anisotropic fluctuations described by the Gaussian covariance of the form (III.1.40), with scales a and $b = c$ $(a > b)$.

Solution. If the major axis of the inhomogeneities a lies in the plane (y, z) and is inclined at an angle γ to the direction of propagation of the wave (z-axis), then, by (1.34), we have

$$\psi_\varphi(x, y, L) = \sigma_\varphi^2(L) \exp\left(-\frac{x^2}{2b^2} - \frac{y^2}{2(a^2 \sin^2 \gamma + b^2 \cos^2 \gamma)}\right) \quad , \quad \text{where}$$

$$\sigma_\varphi^2 = \frac{\sqrt{\pi} L \sigma_\varepsilon^2 ab}{2\sqrt{2}\sqrt{a^2 \sin^2 \gamma + b^2 \cos^2 \gamma}} \quad .$$

As γ varies from 0 to $\pi/2$, the eikonal correlation scale along the y-axis $l_y = \sqrt{a^2 \sin^2 \gamma + b^2 \cos^2 \gamma}$ varies from b (at $\gamma = 0$) to a (at $\gamma = \pi/2$), whereas the correlation scale along the x-axis remains constant, $l_x = b$.

1.7.4 Compute the eikonal variance of the wave reflected from a layered medium where the mean electric permittivity varies linearly $\bar{\varepsilon} = \varepsilon_1 - \varepsilon_2 z$ (Fig. 1.13a), and the fluctuations are statistically homogeneous.

Solution. Let the random inhomogeneities lie above the level $z = h$. According to (1.51), the statistically homogeneous fluctuations are

$$\sigma_\varphi^2 = \frac{\sigma_\varepsilon^2 l_{ef}}{2} \int_0^L \frac{ds}{\bar{\varepsilon}[z(s)]} = \frac{\sigma_\varepsilon^2 l_{ef}}{2} \int_0^L \frac{ds}{\varepsilon_1 - \varepsilon_2 z(s)} \quad , \tag{1.139}$$

where L is the length of the ray path within the layer (Fig. 1.13b). In order to take the integral (1.139), a knowledge of the z *vs* s behavior is required. If θ is the incidence angle and $n_0 = \sqrt{\bar{\varepsilon}(h)} = \sqrt{\varepsilon_1 - \varepsilon_2 h}$ the refractive index at the beginning of the layer, then the ray equation can conveniently be written using the parameter $\tau = \int_0^s ds\sqrt{\bar{\varepsilon}}$ [1.7]

$$x(\tau) = x_0 + \tau n_0 \sin \theta \quad , \quad z(\tau) = h + n_0 \tau \cos \theta - \varepsilon_2 \tau^2/4 \quad ,$$

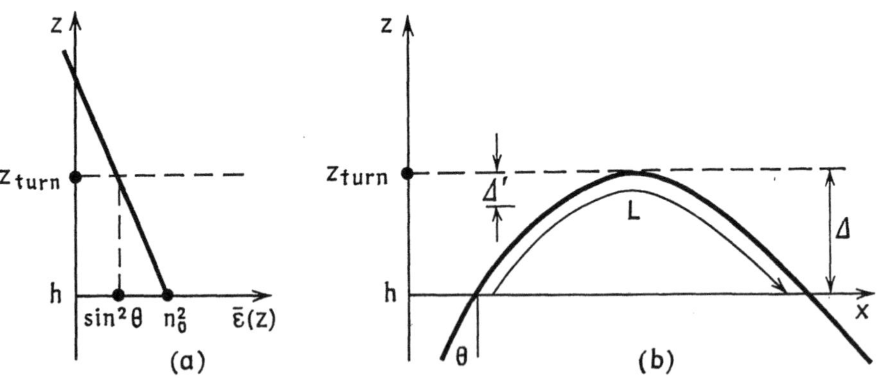

Fig. 1.13. (a) The linear profile of the permittivity of a layer. (b) The ray path in the layer

34

whereas $\bar{\varepsilon}[z(\tau)] = n_0^2 - \varepsilon_2 \tau n_0 \cos\theta + \varepsilon_2^2 \tau^2/4$. The rays turn at the points with the level $z_{\text{turn}} = h + (n_0^2 \cos^2\theta)/\varepsilon_2$ separated from the beginning of the ray $z = h$ by $\Delta = z_{\text{turn}} - h = (n_0^2 \cos^2\theta)/\varepsilon_2$.

If in (1.139) we go over to integration with respect to τ ($d\tau = ds/\sqrt{\bar{\varepsilon}}$) from 0 to $\tau = (4n_0 \cos\theta)/\varepsilon_2$, which corresponds to the ray returning to the initial level $z = h$, we get

$$\sigma_\varphi^2 = \frac{\sigma_\varepsilon^2 l_{\text{ef}}}{\varepsilon_2} \ln\frac{1+\cos\theta}{1-\cos\theta} \quad .$$

Using this expression, we can estimate the layer thickness $\Delta' = z_{\text{turn}} - h'$, which makes a 50 per cent contribution to σ_φ^2:

$$\Delta' = \frac{\Delta}{2}\frac{\sin\theta(1-\sin\theta)}{\cos^2\theta} \quad , \quad \Delta = z_{\text{turn}} - h = \frac{n_0^2 \cos^2\theta}{\varepsilon_2} \quad .$$

At the initial incidence angle $\theta = 45°$, we find $\Delta' = 0.21\Delta$, i.e., about a fifth of the layer accounts for half the eikonal variance σ_φ^2, and at $\theta = 30°$, this is a sixth ($\Delta' = \Delta/6$). Thus, the smaller part of the layer in the vicinity of the point of turn $z = z_{\text{turn}}$ [the mean permittivity $\bar{\varepsilon}(z)$ along the ray is minimal here] makes approximately the same contribution to the eikonal variance as the rest of the layer.

1.7.5 Derive the general expression for the transverse covariance of the level, taking regular refraction into account.

Solution. Let $\delta x_i(s_1)$ and $\delta x_k(s_2)$ be the increments of the coordinates $x_i(s_1)$ and $x_k(s_2)$ of two points, lying on the same unperturbed ray in transition to a nearest neighboring ray, so that the lengths s_1 and s_2 remain the same. If we introduce the notation

$$g_{ik}(s_1, s_2) = \left.\frac{\delta x_i(s_1)}{\delta x_k(s_2)}\right|_{s_{1,2}=\text{const}} \quad ,$$

$$s_< = \min\{s', s''\} \quad ,$$

$$F_{kl}(s, s') = [\delta_{ij} - t_i(s_<)t_j(s_<)]g_{ik}(s, s')g_{jl}(s, s') \quad ,$$

where t_i are the components of the unit vector tangent to the unperturbed ray, then, using (1.86, 41) we find the transverse covariance of the level

$$\psi_\chi(\varrho) = \frac{1}{8}\int\limits_0^L\!\!\int \frac{ds'\,ds''}{\sqrt{\bar{\varepsilon}'\bar{\varepsilon}''}}\int\limits_0^{s_<}\frac{ds}{\bar{\varepsilon}[r_0(s)]}F_{kl}(s, s')F_{pq}(s, s'')\int\limits_0^\infty\frac{\partial^4\psi_\varepsilon[\delta(s) + \zeta t_0]}{\partial x_k \partial x_l \partial x_p \partial x_q}d\zeta \quad ,$$

the summation being over the coincident subscripts.

1.7.6 If a wave passes twice through the same inhomogeneities (e.g., as a result of reflection from an obstacle), *double passage effects* occur [1.19, 20]. For example, for the plane wave that has covered a path L in both directions in a random medium, the phase variance is *twice as large* as that for the wave that has covered the distance $2L$ in the same medium, but only in one direction. Find the phase variance

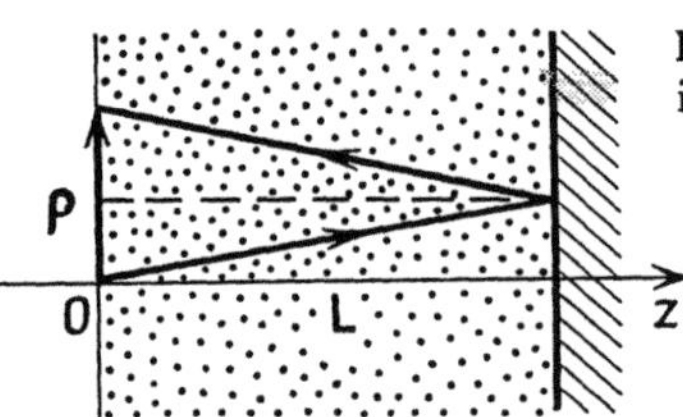

Fig. 1.14. Double passage of a ray through a randomly inhomogeneous layer

$\sigma^2_{\varphi d}$ of the spherical wave reflected from a plane separated from the source by L (Fig. 1.14).

Solution. Suppose that source lies at the origin of the coordinates. At $\bar{\varepsilon} = 1$, the phase fluctuations at a point $r = (\varrho, z)$, $\varrho \ll L$, in the plane $z = 0$, are given by

$$\varphi_1(\varrho, z) = \frac{1}{2} \int\limits_0^L \tilde{\varepsilon}\left(\frac{z'\varrho}{2L}, z'\right) dz' + \frac{1}{2} \int\limits_L^0 \tilde{\varepsilon}\left(\frac{\varrho}{2L}(2L - z'), z'\right) dz' \quad , \tag{1.140}$$

the first term corresponding to the outward, and the second to the return journey. Statistically averaging (1.140) gives (for $L \gg l_\varepsilon$)

$$\sigma^2_{\varphi d}(\varrho, L) = \int\limits_0^L d\eta \int\limits_0^\infty \left[\psi_\varepsilon(0, \zeta) + \psi_\varepsilon\left(\frac{\varrho\eta}{L}, \zeta\right)\right] d\zeta \quad . \tag{1.141}$$

At $\varrho = 0$, when the observation point coincides with the source, we have

$$\sigma^2_{\varphi d}(0, L) = 2L \int\limits_0^\infty \psi_\varepsilon(0, \zeta) d\zeta = 4\sigma^2_\varphi(L) = 2\sigma^2_\varphi(2L) \quad , \tag{1.142}$$

where $\sigma^2_\varphi(z)$ is, by (1.28), the phase variance for the single passage of the distance z. Doubling $\sigma^2_{\varphi d}(0, L)$ as compared to $\sigma^2_\varphi(2L)$ is due to the correlation of the phase fluctuations in both directions. If the forward and return rays pass for the most part through different inhomogeneities (i.e., $\varrho \gg l_\varepsilon$), the second term in (1.141) becomes negligible as compared to the first, and then

$$\sigma^2_{\varphi d}(\varrho, L) \approx 2\sigma^2_\varphi(L) = \sigma^2_\varphi(2L) \quad , \quad \text{for} \quad \varrho \gg l_\varepsilon \quad .$$

The correlation of the intensity fluctuations in both stages produces another interesting effect — backscattering enhancement [1.20].

2. Method of Smooth Perturbations

This chapter continues the discussion of the propagation of a scalar wave through a medium with large-scale inhomogeneities ($l_\varepsilon \gg \lambda$). However, here we use the approximate method of smooth perturbations (MSP). This is a more general technique than geometrical optics since, to a certain degree it takes into account *diffraction* effects. The MSP relies on the approximate, so-called *parabolic*, wave equation (Sect. 2.1) which is suitable for those cases where a wave that arrives at an inhomogeneity is largely scattered forward, i.e., with small deviations from its initial direction. By analyzing the errors introduced in going over to the parabolic equation, we discuss in Sect. 2.2 the energy conservation law in this approximation.

In Sect. 2.3, we pass from the parabolic equation for the wave field to the amplitude and phase equations on which MSP is based. The results derived using this method are treated in detail in Sect. 2.4, where some graphic interpretations of them are also provided. In Sect. 2.5, we establish theoretical boundaries for the applicability of MSP, and compare the results obtained with experiment.

2.1 The Parabolic Equation Approximation and Its Derivation

We have already mentioned that, when λ is small as compared to the size l_ε of the electric permittivity inhomogeneities, scattered waves concentrate within a narrow cone, with an angle θ at the vertex of the order of $\lambda/l_\varepsilon \ll 1$, i.e., they propagate essentially in the same direction as the primary wave, whereby multiple scattering comes to the fore.

One of the approximate methods used to deal with multiple scattering is that of geometrical optics (Chap. 1), but this completely ignores diffraction effects and has the constraint $\sqrt{\lambda L} \ll l_\varepsilon$. If this condition is violated, i.e., the distance L covered by the wave in an inhomogeneous medium is sufficiently large ($L \gg l_\varepsilon^2/\lambda \equiv L_d$), then diffraction effects become important. This can be elucidated by the following qualitative reasoning. If an inhomogeneity of size l_ε is exposed to a plane wave, the size of its geometrical "shadow", at any distance L, is also l_ε. Diffraction, however, "smears out" the sharp boundaries of the shadow by about $\sqrt{\lambda L}$ (this is the size of the light-shadow transitional region when diffraction occurs at the edge of an opaque screen). It is clear that the diffraction can only be ignored if $\sqrt{\lambda L} \ll l_\varepsilon$ [see (1.10)]. Not infrequently, this condition is a rigid constraint on the distance L.

As was already seen in Sect. 1.5, for the light in a turbulent atmosphere, we have $\sqrt{\lambda L} = l_\varepsilon$ even at $L \approx 200\,\mathrm{m}$, i.e., geometrical optics is no longer suitable at distances of several hundred meters.

In Chap. 1, we already mentioned more general approximate methods of handling diffraction problems using the parabolic equation. Some of these, like the GOM, rely on the smallness of the parameter λ/l_ε, while others also take account of diffraction effects. It will be recalled that these are the method of smooth perturbations (MSP), often referred to as the Rytov method, and the parabolic equation method (PEM). In this section, we give two derivations of the parabolic equation in order to bring out different aspects of the PEM. We will then obtain the equations of the MSP from the parabolic equation (Sect. 2.3).

We begin with the scalar Helmholtz equation

$$\Delta u + k^2[1 + \tilde{\varepsilon}(r)]u = 0 \quad , \tag{2.1}$$

where $k^2 = \omega^2\bar{\varepsilon}/c^2$ is the squared wave number (assuming that $\bar{\varepsilon} = $ const), and $\tilde{\varepsilon} = [\varepsilon(r) - \bar{\varepsilon}]/\bar{\varepsilon}$ the relative fluctuations of the electric permittivity, so that $\langle\tilde{\varepsilon}\rangle = 0$. First, we will provide the simplest derivation of the parabolic equation, based on physical arguments.

Suppose that a plane wave $A_0\exp(ikz)$ is incident on an inhomogeneous medium occupying the half-space $z > 0$. Since we assumed that $l_\varepsilon \gg \lambda$, scattered waves will for the most part propagate forwards, and the wave reflected from an inhomogeneous medium will be weak as compared to the incident wave. We will seek the field u in the medium in the form

$$u(\varrho, z) = v(\varrho, z)\exp(ikz), \quad \varrho = (x, y) \quad , \tag{2.2}$$

where $v(\varrho, x)$ is the amplitude of the wave (generally speaking, complex). If the medium were homogeneous ($\tilde{\varepsilon} = 0$), the amplitude v would be independent of the coordinates. In an inhomogeneous medium, we can expect that $v(\varrho, z)$ and its derivative with respect to z will only vary weakly along the wavelength, since v only varies due to the inhomogeneities, and their size is large as compared to λ. Therefore, the following conditions must hold:

$$\lambda\left|\frac{\partial v}{\partial z}\right| \ll |v| \quad , \quad \lambda\left|\frac{\partial^2 v}{\partial z^2}\right| \ll \left|\frac{\partial v}{\partial z}\right| \quad , \quad \text{or}$$

$$\left|\frac{\partial v}{\partial z}\right| \ll k|v| \quad , \quad \left|\frac{\partial^2 v}{\partial z^2}\right| \ll k\left|\frac{\partial v}{\partial z}\right| \tag{2.3}$$

[the factor $\exp(ikz)$ accounts for fast variation of u with z]. Conditions (2.3) are only met when the backscattered field is small. In fact, if u also includes the back wave $A\exp(-ikz)$, this implies that the amplitude v includes a term of the form $A\exp(-2ikz)$, which varies fast over distances of about λ. Consequently, conditions (2.3) already presuppose that the amplitude A of the back wave is sufficiently small as compared to that of the forward wave.

Substituting (2.2) into (2.1) gives

$$\frac{\partial^2 v}{\partial x^2} + \frac{\partial^2 v}{\partial y^2} + \frac{\partial^2 v}{\partial z^2} + 2ik\frac{\partial v}{\partial z} + k^2\tilde{\varepsilon}(\varrho, z)v = 0$$

But, by virtue of the second of equations (2.3) we may ignore the term $\partial^2 v/\partial z^2$ as compared to $2ik\partial v/\partial z$, which yields the parabolic equation for v

$$2ik\frac{\partial v}{\partial z} + \frac{\partial^2 v}{\partial x^2} + \frac{\partial^2 v}{\partial y^2} + k^2\tilde{\varepsilon}(\varrho, z)v = 0 \quad . \tag{2.4}$$

An equation of the parabolic type was first used by *Leontovich* [2.1] to solve the deterministic problem of radio wave diffraction around the Earth. The above derivation of (2.4) is clearly fairly crude. We will, therefore, provide another derivation which will bring out more distinctly the quantities to be discarded in order that we might arrive at (2.4) from (2.1) [2.2].

Equation (2.1), together with suitable boundary conditions, is equivalent to the following integral equation:

$$u(\boldsymbol{r}) = u_0(\boldsymbol{r}) - k^2 \int G(\boldsymbol{r} - \boldsymbol{r}')\tilde{\varepsilon}(\boldsymbol{r}')u(\boldsymbol{r}')d^3r' \quad , \tag{2.5}$$

where $u_0(\boldsymbol{r})$ is the primary wave (a wave in a medium without inhomogeneities), obeying the equation $\Delta u_0 + k^2 u_0 = 0$. $G(\boldsymbol{r} - \boldsymbol{r}')$ is Green's function

$$G(\boldsymbol{r} - \boldsymbol{r}') = -\frac{1}{4\pi|\boldsymbol{r} - \boldsymbol{r}'|}\exp\left(ik|\boldsymbol{r} - \boldsymbol{r}'|\right) \quad , \tag{2.6}$$

meeting the equation $\Delta G + k^2 G = \delta(\boldsymbol{r} - \boldsymbol{r}')$ and radiation conditions at infinity.

Consider the incident wave of the form $u_0(\varrho, z) = A\exp(ikz)$. We will break down the integration region with respect to z' in (2.5) into two sections: $0 < z' < z$ and $z' > z$

$$u(\boldsymbol{r}) = u_0(\boldsymbol{r}) - k^2 \int_0^z dz' \int d^2\varrho' G(\boldsymbol{r} - \boldsymbol{r}')\tilde{\varepsilon}(\boldsymbol{r}')u(\boldsymbol{r}')$$

$$- k^2 \int_z^\infty dz' \int d^2\varrho' G(\boldsymbol{r} - \boldsymbol{r}')\tilde{\varepsilon}(\boldsymbol{r}')u(\boldsymbol{r}') \quad . \tag{2.7}$$

Consider the first of the integrals. It is a sum of terms of the form

$$du_+(\boldsymbol{r}) = -k^2 G(\boldsymbol{r} - \boldsymbol{r}')\tilde{\varepsilon}(\boldsymbol{r}')u(\boldsymbol{r}')dz'\, d^2\varrho' \quad ,$$

and for each of these terms $z > z'$. The quantity du_+ is a spherical wave [the factor $G(\boldsymbol{r} - \boldsymbol{r}')$], with the center at the point $\boldsymbol{r}'$ and the amplitude $-k^2\tilde{\varepsilon}(\boldsymbol{r}')u(\boldsymbol{r}')dz'\, d^2\varrho'$ determined by the scattering of the wave $u(\boldsymbol{r}')$ from the inhomogeneity $\tilde{\varepsilon}(\boldsymbol{r}')$. Here, the longitudinal coordinate of the scattering point z' is always smaller than the longitudinal coordinate of the observation point z. This means that all the sources of the spherical waves in the first integral in (2.7) lie in the layer $0 < z' < z$, i.e., the scattered waves taken into account by this term come to the observation point from the same direction as the incident wave.

Similarly, we can easily see that the second integral in (2.7) is the total of all the waves arriving at $\boldsymbol{r}$ from the region $z' > z$. It is clear that for a wave to get to $\boldsymbol{r}$ from the region $z' > z$, it should be backscattered at least once (Fig. 2.1a).

39

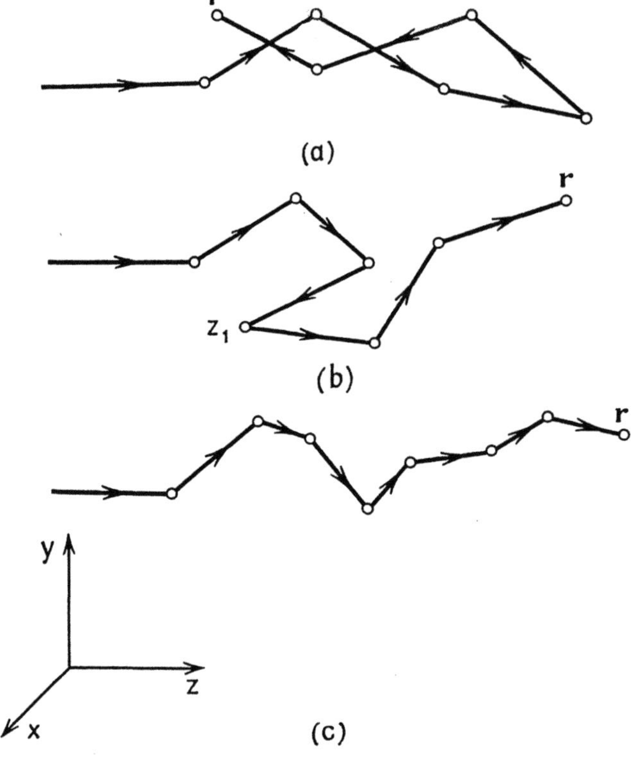

Fig. 2.1a–c. Propagation of a wave coming to an inhomogeneous medium along the positive z-axis: **(a)** The wave arrives at r from $z' > z$ and is backscattered at least once; **(b)** The wave arrives at r from $z' < z$ and is backscattered twice; **(c)** The wave arrives at r from $z' < z$ without undergoing backscattering. The solution of (2.7) takes into account all such waves

However, if the condition $\lambda \ll l_\varepsilon$ is met, then in each stage of scattering the bulk of the scattered energy is concentrated within a narrow, solid angle, near the initial direction of wave propagation. We can, therefore, expect that the intensity of the wave that has undergone at least one *back* scattering will be small as compared to the intensity of one that has undergone the same total number of *forward* scatterings. We may thus neglect the second term in (2.7) and write it in the form[1]

$$u^{(0)}(r) = u_0(r) - k^2 \int_0^z dz' \int d^2\varrho' G(r - r')\tilde{\varepsilon}(r')u^{(0)}(r') \quad . \tag{2.8}$$

In the approximation, described by (2.8), which must be met at *any* $z = \tilde{z}$, only waves from the region $z' < \tilde{z}$ arrive at *each* point. This means that, in going over from (2.7) to (2.8), we discard not only the waves shown in Fig. 2.1a, but aso the waves shown in Fig. 2.1b, where a wave comes from the right to one of the intermediate points z_1. The only type of wave that enters (2.8) corresponds to Fig. 2.1c. Here,

---

[1] The significance of the superscript (0) in (2.8) is clarified below.

a wave comes from the left to each of the intermediate points, and not only to r. This is easily seen if we write the formal solution to (2.8) in the form of an iterative series. Thus, only those waves that have not suffered *a single act of backscattering* along their propagation path within the layer $(0, z)$ satisfy (2.8). This is indicated by the superscript 0 in $u^{(0)}$. It can be shown [2.2] that the total wave field u is subdivided into the multitude of fields $u^{(0)}$, $u^{(1)}$, $u^{(2)}$, ..., where $u^{(l)}$ is the wave after the action of l backscattering.

Equation (2.8) can also be further simplified by more consistent application of the condition $\lambda \ll l_\varepsilon$. It was stated earlier that the characteristic angle of scattering at an inhomogeneity of scale l_ε has an order of $\theta \sim \lambda/l_\varepsilon$. If a primary wave propagates along the z-axis, then after a first scattering it will propagate farther at an angle θ to the z-axis. This means that, in (2.8), the ratio of the transverse component of $r - r'$ to its longitudinal component is about θ, i.e.,

$$\frac{|\varrho - \varrho'|}{|z - z'|} \sim \frac{\lambda}{l_\varepsilon} \ll 1 \quad . \tag{2.9}$$

Since this ratio is small, we can use the expansion of $|r - r'|$ into a Taylor series

$$\begin{aligned}
|r - r'| &= (z - z')\sqrt{1 + \frac{|\varrho - \varrho'|^2}{(z - z')^2}} \\
&= (z - z') + \frac{1}{2}\frac{|\varrho - \varrho'|^2}{(z - z')} - \frac{1}{8}\frac{|\varrho - \varrho'|^4}{(z - z')^3} + \cdots \quad . \end{aligned} \tag{2.10}$$

The exponent of Green's function G contains the phase shift $k|r - r'|$. We may substitute the above expansion for $|r - r'|$ and keep only the first two terms, if the following condition is met:

$$\frac{k}{8}\frac{|\varrho - \varrho'|^4}{(z - z')^3} \ll 1 \quad .$$

If, according to (2.9), we substitute $|\varrho - \varrho'| \sim \lambda(z - z')/l_\varepsilon$ here, we then arrive at the constraints[2]

$$\frac{\lambda z}{l_\varepsilon^2} \ll \left(\frac{l_\varepsilon}{\lambda}\right)^2 \quad , \quad \lambda \ll l_\varepsilon \quad . \tag{2.11}$$

If these constraints are obeyed, we can replace Green's function (2.6) by its Fresnel approximation

$$G_\mathrm{F}(r - r') = -\frac{1}{4\pi(z - z')}\exp\left[\mathrm{i}k(z - z') + \frac{\mathrm{i}k(\varrho - \varrho')^2}{2(z - z')}\right] \quad . \tag{2.12}$$

We may keep only the first term of (2.10) in the denominator of Green's function since the relative error, which only affects the amplitude, is about λ/l_ε.

Substituting into (2.8) the approximate expression (2.12) for Green's function, we arrive at

---

[2] Note that by virtue of the condition $\lambda \ll l_\varepsilon$, the quantity $\lambda z/l_\varepsilon^2$ must be small not as compared to unity, as in the GOM, but to the *large* parameter $(l_\varepsilon/\lambda)^2$. The first of (2.11) can thus also hold in the case where the applicability conditions for the GOM are violated.

$$u^{(0)}(\varrho, z) = u_0(\varrho, z)$$
$$+ \frac{k^2}{4\pi} \int_0^z \frac{\exp\left[ik(z - z')\right]}{z - z'} dz' \int d^2\varrho' \exp\left[\frac{ik(\varrho - \varrho')^2}{2(z - z')}\right] \tilde{\varepsilon}(r') u^{(0)}(\varrho', z') \quad ,$$

which is equivalent to the parabolic equation (2.4). In order to show that this is the case, we make use of the substitution $u^{(0)}(\varrho, z) = v(\varrho, z) \exp(ikz)$, which is similar to (2.2). Also, we will represent the incident wave in a similar form $u_0(\varrho, z) = A_0(\varrho) \exp(ikz)$. It will be noted that by the equation $\Delta u_0 + k^2 u_0 = 0$, the amplitude $A_0(\varrho)$ of the incident wave meets the condition $\Delta_\perp A_0(\varrho) = 0$, where $\Delta_\perp = \partial^2/\partial x^2 + \partial^2/\partial y^2$. Then, having cancelled out the common factor $\exp(ikz)$, we obtain

$$v(\varrho, z) = A_0(\varrho) + \frac{ik}{2} \int_0^z dz' \int d^2\varrho'$$
$$\times \left\{\frac{k}{2\pi i(z - z')} \exp\left[-\frac{k(\varrho - \varrho')^2}{2i(z - z')}\right]\right\} \tilde{\varepsilon}(r') v(\varrho', z') \quad . \tag{2.13}$$

In order to derive the differential equation for v, we must differentiate (2.13) with respect to z. But, differentiating an integral with respect to its upper limit results in an uncertainty — the value of the expression in brackets at $z' = z$. In order to establish the limit to which this expression tends, we will temporarily introduce the notation $a^2 = i(z - z')/k$. It will then take the form

$$\frac{1}{2\pi a^2} \exp\left[-\frac{(\varrho - \varrho')^2}{2a^2}\right] \quad .$$

This is recognizable as the two-dimensional Gaussian probability density, which corresponds to the variances a along both axes. But as $a \longrightarrow 0$, the Gaussian probability density tends to a delta-function, so that

$$\lim_{z' \longrightarrow z} \frac{k}{2\pi i(z - z')} \exp\left[\frac{ik(\varrho - \varrho')^2}{2(z - z')}\right] = \delta(\varrho - \varrho') \quad .$$

Differentiating (2.13) with respect to z and making use of the last equation, we get

$$\frac{\partial v}{\partial z} = \frac{ik}{2} \left\{ \tilde{\varepsilon}(r) v(r) + \int_0^z dz' \int d^2\varrho' \tilde{\varepsilon}(r') v(r') \right.$$
$$\left. \times \frac{\partial}{\partial z} \left[\frac{k}{2\pi i(z - z')} \exp\left(\frac{ik(\varrho - \varrho')^2}{2(z - z')}\right)\right] \right\} \quad .$$

We will now take into account the fact that the relationship

$$\frac{\partial}{\partial z} \left\{\frac{k}{2\pi i(z - z')} \exp\left[\frac{ik(\varrho - \varrho')^2}{2(z - z')}\right]\right\}$$
$$= \frac{i}{2k} \Delta_\perp \left\{\frac{k}{2\pi i(z - z')} \exp\left[\frac{ik(\varrho - \varrho')^2}{2(z - z')}\right]\right\}$$

holds, which can easily be checked by direct differentiation. By using this and bringing the operator $\Delta_\perp$ outside the integration sign, by (2.13), we obtain,

$$\frac{\partial v}{\partial z} = \frac{ik}{2}\tilde{\varepsilon}(r)v(r) + \frac{i}{2k}\Delta_\perp(v - A_0) \quad . \tag{2.14}$$

As we have already noted, $\Delta_\perp A_0 = 0$, whereby the last equation coincides with the parabolic equation (2.4).

In deriving (2.14) we established that in (2.8) we may go over to the Fresnel approximation for Green's function only if the conditions (2.11) are met. However, we have not yet discovered when the backscattered waves may be neglected. This is most easily done using the expression (III.4.109) for the effective cross-section of scattering from a unit volume (we are discussing here the simplest case of statistically isotropic fluctuations)

$$\sigma = \frac{\pi k^4}{2}\Phi_\varepsilon\left(2k\,\sin\frac{\theta}{2}\right) \quad . \tag{2.15}$$

If we integrate (2.15) over the back hemisphere ($\pi/2 < \theta < \pi$, $0 < \varphi < 2\pi$), we arrive at the effective cross-section of backscattering from a unit volume

$$\sigma_{\text{back}} = \frac{\pi k^4}{2}\int\limits_0^{2\pi} d\varphi \int\limits_{\pi/2}^{\pi} \sin\theta\,d\theta\,\Phi_\varepsilon\left(2k\,\sin\frac{\theta}{2}\right) \quad .$$

Integrating with respect to φ and introducing the new integration variable $\kappa = 2k\,\sin(\theta/2)$ instead of θ, we find

$$\sigma_{\text{back}} = \pi^2 k^2 \int\limits_{k\sqrt{2}}^{2k} \Phi_\varepsilon(\kappa)\kappa\,d\kappa \quad . \tag{2.16}$$

The effective cross-section σ_{back} is equal to that share of the incident energy which goes into the back wave due to scattering as the forward wave covers a unit distance. If we ignore the reduction in the incident energy due to this scattering, then, over the path z, the energy of the back wave will account for the portion $\sigma_{\text{back}}z$ (note that by ignoring the energy losses of the forward wave we thereby overestimate the energy of the back wave). Therefore, the condition for neglecting the backscattering and, thereby, discarding the second integral in (2.7) can be written as $\sigma_{\text{back}}z \ll 1$, or

$$\pi^2 k^2 z \int\limits_{k\sqrt{2}}^{2k} \Phi_\varepsilon(\kappa)\kappa\,d\kappa \ll 1 \quad . \tag{2.17}$$

Note now that if we integrated (2.15) over *all* angles ($0 \le \theta \le \pi$), we would obtain the *total* effective scattering cross-section σ_0 (in all directions) and, instead of (2.17), we would arrive at

$$\sigma_0 z = \pi^2 k^2 z \int\limits_0^{2k} \Phi_\varepsilon(\kappa)\kappa\,d\kappa \ll 1 \quad , \tag{2.18}$$

i.e., the applicability condition for the Born approximation. We see that the condition (2.17) is always less stringent than (2.18). If for $\kappa > k\sqrt{2}$ the function $\Phi_\varepsilon(\kappa)$ is small as compared to its value for $\kappa < k\sqrt{2}$, the condition (2.18) may be violated, even if (2.17) is met.

Suppose, for example, that $\psi_\varepsilon(r) = \sigma_\varepsilon^2 \exp(-r^2/2a^2)$. Then (Sect. III.1.3) $\Phi_\varepsilon(\kappa) = (2\pi)^{-3/2}\sigma_\varepsilon^2 a^3 \exp(-\kappa^2 a^2/2)$. Integrating in (2.17), we get

$$\sqrt{\pi}\sigma_\varepsilon^2 k^2 zae^{-k^2 a^2}(1 - e^{-k^2 a^2}) \ll 1 \quad . \tag{2.17a}$$

But in that case the applicability condition for the Born approximation has the form [see (III.4.85)]:

$$\sqrt{\pi/2}\sigma_\varepsilon^2 k^2 za(1 - e^{-2k^2 a^2}) \ll 1 \quad . \tag{2.18a}$$

If $ka \gg 1$ (large-scale inhomogeneities), (2.18a) will be $\sigma_\varepsilon^2 k^2 za \ll 1$, whereas from (2.17a) we get $\sigma_\varepsilon^2 k^2 za \ll \exp(k^2 a^2/2)$, i.e., the parameter $\sigma_\varepsilon^2 k^2 za$ is limited from above not by a unity, but by the *larger* number[3] $\exp(k^2 a^2/2)$.

On the other hand, if the Born approximation condition (2.18) is met, then (2.17) will be met *a fortiori*. In that case, the role of multiple scattering, including backscattering, is negligible.

As a second example, consider the spectral density

$$\Phi_\varepsilon(\kappa) = AC_\varepsilon^2 \kappa^{-11/3} \exp(-\kappa^2/\kappa_m^2)$$

corresponding to electric permittivity fluctuations due to turbulence [see (III.1.79)].

This form of the three-dimensional spectral density is applicable in the region $\kappa > 2\pi/L_0$, where L_0 is the outer scale of turbulence. We cannot estimate the region of validity of the Born approximation using this spectral density, since this form is not correct when $\kappa < 2\pi/L_0$. If, however, $k \gg 2\pi/L_0$, backscattering is only conditioned by the section of the spatial spectrum of inhomogeneities described by this formula. By estimating (2.17), we can then obtain the following applicability condition of the parabolic equation:

$$C_\varepsilon^2 k^{1/3} z \ll 1 \quad .$$

In the Earth's atmosphere, for example, $C_\varepsilon^2 \sim 10^{-15}\,\text{cm}^{-2/3}$, $\kappa_m \sim 10^2\,\text{cm}^{-1}$. If $k \sim 10^5\,\text{cm}^{-1}$ (visible light), the last condition leads to the constraint $z \ll 10^8$ km, which is always fulfilled since the path of a light ray in the Earth's atmosphere cannot be larger than 10^3 km. Applications of the PEM in optics are considered in some detail in [2.3].

To sum up, if the conditions

$$\lambda \ll l_\varepsilon \quad , \qquad \frac{\lambda z}{l_\varepsilon^2} \ll \left(\frac{l_\varepsilon}{\lambda}\right)^2 \quad , \qquad \pi^2 k^2 z \int\limits_{k\sqrt{2}}^{2k} \Phi_\varepsilon(\kappa)\kappa\,d\kappa \ll 1$$

---

[3] If, apart from the condition $ka \gg 1$, the condition $ka^2 \gg z$ is met, in which the results of the diffraction treatment coincide with those of the GOM, then the multiple forward scattering is nevertheless taken into account. This indicates that the GOM also takes care of the multiple forward scattering.

are met, then wave propagation through a random medium can be described by the parabolic equation

$$2\mathrm{i}k\frac{\partial v}{\partial z} + \Delta_\perp v + k^2\tilde{\varepsilon}(\varrho, z)v = 0 \quad , \quad \Delta_\perp = \frac{\partial^2}{\partial x^2} + \frac{\partial^2}{\partial y^2}$$

for the slowly varying complex amplitude related to the wave field u by $u(\varrho, z) = v(\varrho, z)\exp(\mathrm{i}kz)$. The parabolic equation approximation includes *multiple forward scattering* and ignores backscattering. The PEM takes into account diffraction in the Fresnel approximation. An important feature of the parabolic equation is the fact that it is of *first* order in z. It is, thus, sufficient to have only *one* boundary condition in the plane $z = $ const.

2.2 Energy Conservation Law in the Parabolic Equation Approximation

We begin with the wave equation

$$\Delta u + k^2[1 + \tilde{\varepsilon}(\boldsymbol{r})]u = 0 \quad , \tag{2.19}$$

from which the expression (2.4) for $v(\varrho, z)$ was derived. It has the energy integral (III.4.68, 69)

$$\mathrm{div}\vec{S} = 0 \quad , \tag{2.20}$$

$$\vec{S} = \frac{1}{2\mathrm{i}}(u^*\nabla u - u\nabla u^*) = \mathrm{Im}\,\{u^*\nabla u\} \quad . \tag{2.21}$$

[To simplify the notation, we have taken the factor a in (III.4.68) to be equal to k in subsequent equations.] The vector $\vec{S}$ is the energy flux density for the wave.

Suppose that $u = A\exp(\mathrm{i}S)$, where A is the amplitude and S the phase of the wave. Then, $\vec{S}$ will be

$$\vec{S} = A^2\nabla S \quad . \tag{2.22}$$

Let us now express $\vec{S}$ in terms of v. If the complex amplitude v is represented in the form $v = A\exp(\mathrm{i}S')$, then, from the expression $u = v\exp(\mathrm{i}kz)$, the total phase S will be related to the phase S' of the slowly varying function v by

$$S = kz + S' \quad . \tag{2.23}$$

Then

$$\vec{S}_\perp = A^2\nabla_\perp S' \quad , \quad S_z = A^2\left(k + \frac{\partial S'}{\partial z}\right) \quad ,$$

and the energy conservation law (2.20) takes the form

$$\frac{\partial}{\partial z}\left[A^2\left(k + \frac{\partial S'}{\partial z}\right)\right] + \nabla_\perp[A^2\nabla_\perp S'] = 0 \quad . \tag{2.24}$$

In the parabolic equation approximation we have

$$2ik\frac{\partial v}{\partial z} + \Delta_\perp v + k^2 \tilde{\varepsilon} v = 0 \quad .$$

Multiplying this by v^*

$$2ikv^*\frac{\partial v}{\partial z} + v^*\Delta_\perp v + k^2 \tilde{\varepsilon} vv^* = 0 \quad ,$$

and subtracting the complex conjugate equation

$$-2ikv\frac{\partial v^*}{\partial z} + v\Delta_\perp v^* + k^2 \tilde{\varepsilon} vv^* = 0 \quad ,$$

we arrive at

$$2ik\frac{\partial vv^*}{\partial z} + \nabla_\perp(v^*\nabla_\perp v - v\nabla_\perp v^*) = 0 \quad , \tag{2.25}$$

where we made use of the relation

$$v^*\Delta_\perp v - v\Delta_\perp v^* = \nabla_\perp(v^*\nabla_\perp v - v\nabla_\perp v^*) \quad .$$

Equation (2.25), derived in the parabolic equation approximation, is an analog of the conservation law (2.20). Substituting $v = A\exp(iS')$ into (2.25) gives

$$\frac{\partial}{\partial z}(kA^2) + \nabla_\perp(A^2\nabla_\perp S') = 0 \quad . \tag{2.26}$$

A comparison of (2.26) and (2.24) shows that, in going over to the parabolic equation, the component $\vec{S}_z$ of the energy flux density loses the term $A^2\partial S'/\partial z$. Consequently, we must have

$$\left|\frac{\partial S'}{\partial z}\right| \ll k \quad , \quad \text{or} \quad \left|\lambda\frac{\partial S'}{\partial z}\right| \ll 2\pi \quad , \tag{2.27}$$

which means that S' must vary slowly over the wavelength along the z-axis. This inequality is not an additional restriction; it follows naturally from the condition $\lambda \ll l_\varepsilon$, which has already been used in deriving (2.4). Indeed, purely geometrical phase shift is already accounted for by the term kz in (2.23). The term S' is, therefore, only associated with the inhomogeneities. However, the characteristic sizes of these latter are large as compared to the wavelength, whereby (2.27) is obeyed.

We can also interpret (2.8) in another way. The quantity $A^2 = vv^* = uu^* \equiv I$ is proportional to the energy density of the wave. Dividing (2.26) by k gives

$$\frac{\partial I}{\partial z} + \nabla_\perp\vec{S}_\perp = 0 \quad , \quad \vec{S}_\perp = A^2\frac{\nabla_\perp S'}{k} \quad . \tag{2.28}$$

This equation is analogous to the energy conservation law in a nonstationary field. The z-coordinate in (2.28) represents time, and the energy propagates in a plane (x, y). If we integrate (2.28) over a certain area Σ lying in the plane $z = \text{const}$ and bounded by a closed contour L, we get

46

$$\frac{d}{dz} \int_{\Sigma} I dx\, dy + \int_{\Sigma} \operatorname{div} \vec{S}'_{\perp} dx\, dy = 0$$

We now apply the two-dimensional Gauss theorem to the second term

$$\frac{d}{dz} \int_{\Sigma} I dx\, dy + \oint_{L} \vec{S}'_{n} dl = 0 \quad .$$

Thus, the energy concentrated in the area Σ varies according to the energy flux through the contour L that bounds Σ.

Let us consider a special case where the electric permittivity inhomogeneities are statistically homogeneous in the planes $z = \text{const}$, and the incident wave is plane. Symmetry suggests that all the averaged quantities may then depend on z, and not on the transverse radius vector ϱ. We now average (2.26)

$$\frac{\partial}{\partial z}(k\overline{A^2}) + \nabla_{\perp}\langle A^2 \nabla_{\perp} S'\rangle = 0 \quad .$$

Since as $\langle A^2 \nabla_{\perp} S'\rangle$ is independent of ϱ, we have

$$\nabla_{\perp}\langle A^2 \nabla_{\perp} S'\rangle = 0 \quad ,$$

and hence

$$\frac{d}{dz}\langle A^2\rangle = 0 \quad .$$

We see that for the plane wave propagating through a medium that is statistically homogeneous in the planes $z = \text{const}$, we have

$$\overline{uu^*} = \overline{vv^*} = \overline{A^2} = A_0^2 = \text{const} \quad . \tag{2.29}$$

This result follows directly from (2.26), but it appears to be approximate if we proceed from the exact equation (2.24). In fact, averaging (2.24) would give

$$\overline{A^2} + \frac{1}{k}\left\langle A^2 \frac{\partial S'}{\partial z}\right\rangle = \text{const} \quad .$$

It follows that (2.29) is only valid within a relative error of about λ/l_ε.

2.3 Method of Smooth Perturbations

We rewrite the parabolic equation (2.4)

$$2ik\frac{\partial v}{\partial z} + \Delta_{\perp} v + k^2 \tilde{\varepsilon}(\varrho, z)v = 0 \quad ,$$

$$\varrho = \{x, y\} \quad , \quad \Delta_{\perp} = \frac{\partial^2}{\partial x^2} + \frac{\partial^2}{\partial y^2} \quad , \tag{2.30}$$

which defines the complex amplitude v related to the wave field u by $u(\varrho, z) = v(\varrho, z)\exp(ikz)$. Like the Helmholtz equation (2.1), it permits of no exact solution

in the general case. We will, therefore, also have to utilize approximate methods
for (2.30). In this section, we shall be looking at the so-called method of smooth
perturbations (MSP) suggested for the deterministic problem of light diffraction at
an ultrasound wave by *Rytov* [2.4], and applied to statistic problems by *Obukhov*
[2.5]. The MSP is first of all suitable for examining such characteristics of the wave
as phase and level.

Let us write the complex amplitude $v(\varrho, z)$ in the form

$$v(\varrho, z) = A_0 \exp\left[iS'(\varrho, z) + \ln\frac{A(\varrho, z)}{A_0}\right] = A_0\exp\left[\Phi(\varrho, z)\right] \quad , \quad \text{whence} \quad (2.31)$$

$$\Phi = \ln(v/A_0)$$

Here, $S' = S - kz$ [see (2.23)] is the deviation of the phase from its regular shift kz
without inhomogeneities, and $\ln(A/A_0) \equiv \chi$ is the level. Thus,

$$\Phi = \chi + iS' \quad , \quad \chi = \text{Re}\{\Phi\} \quad , \quad S' = \text{Im}\{\Phi\} \quad .$$

Substituting (2.31) into (2.30) gives

$$2ik\frac{\partial\Phi}{\partial z} + \Delta_\perp\Phi + (\nabla_\perp\Phi)^2 + k^2\tilde{\varepsilon}(\varrho, z) = 0 \quad . \tag{2.32}$$

Unlike (2.30), this equation is *nonlinear,* but the random function $\tilde{\varepsilon}$ enters into it in
an additive manner, not as a coefficient.

We will seek the solution of (2.32) in the form of the expansion

$$\Phi = \Phi_1 + \Phi_2 + \dots \quad , \tag{2.33}$$

assuming that Φ_1 has the same order of smallness as $\sigma_\varepsilon = \sqrt{\langle\tilde{\varepsilon}^2\rangle}$; Φ_2 as σ_ε^2, and
so forth. If we substitute (2.33) into (2.32) and equate the groups of terms of the
same order of smallness to zero, we will obtain the following set of equations for
the successive approximations:

$$2ik\frac{\partial\Phi_1}{\partial z} + \Delta_\perp\Phi_1 = -k^2\tilde{\varepsilon}(\varrho, z) \quad , \tag{2.34a}$$

$$2ik\frac{\partial\Phi_2}{\partial z} + \Delta_\perp\Phi_2 = -(\nabla_\perp\Phi_1)^2 \quad , \tag{2.34b}$$

$$2ik\frac{\partial\Phi_3}{\partial z} + \Delta_\perp\Phi_3 = -2\nabla_\perp\Phi_1 \cdot \nabla_\perp\Phi_2 \quad . \tag{2.34c}$$

Using these equations, we will consider the level and phase fluctuations of the
wave in a statistically homogeneous random medium in the half-space $z > 0$, if a
plane wave $u(\varrho, z) = A_0 \exp(ikz)$ comes to it from the region $z < 0$. Since we have
ignored backscattering, only the field u, and not $\partial u/\partial z$, must be continuous at the
boundary $z = 0$

$$u(\varrho, 0) = A_0 \quad , \quad v(\varrho, 0) = A_0 \quad , \quad \Phi(\varrho, 0) = \ln\frac{v(\varrho, 0)}{A_0} = 0 \quad .$$

The boundary conditions for all the equations (2.34) have the same form

$$\Phi_l(\varrho, 0) = 0 \quad , \quad l = 1, 2, \ldots \quad ,$$

and the equations themselves differ only on the right-hand sides. The solution of any of these equations can, therefore, be represented in the form

$$\Phi_l(\varrho, z) = \int\limits_0^z dz' \int d^2\varrho' K(\varrho - \varrho', z - z') f_l(\varrho', z') \quad , \tag{2.35}$$

where K is Green's function of the operator $2ik\partial/\partial z + \Delta_\perp$, and f_l is the right-hand side of an appropriate equation for Φ_l.

It is generally possible to compute only the first several terms of the expansion (2.33) (usually, the first approximation is used, the next term Φ_2 being used only to estimate the error). In order for Φ to differ only slightly from Φ_1, it is necessary that the right-hand sides of (2.34) decrease sufficiently fast with number l, i.e., it is necessary that

$$\overline{|\nabla_\perp \Phi_1|^2} \ll k^2 \sigma_\varepsilon \quad , \quad \overline{|\lambda \nabla_\perp \Phi_1|^2} \ll \sigma_\varepsilon \quad . \tag{2.36}$$

Condition (2.36) implies that *transverse variations* of Φ_1 over distances of about the wavelength λ should be small as compared to σ_ε (the quantity Φ_1 itself has the same order of smallness as $\tilde{\varepsilon}$). This condition thus requires that Φ_1 vary sufficiently *smoothly*. It is for this reason that the method based on the application of perturbation theory to the complex phase was termed the method of *smooth* perturbations (MSP).

Green's function in (2.35) has the form (see Exercise 2)

$$K(\varrho - \varrho', z - z') = -\frac{1}{4\pi(z - z')} \exp\left[\frac{ik(\varrho - \varrho')^2}{2(z - z')}\right] \quad . \tag{2.37}$$

It differs from Green's function in the Helmholtz equation taken in the Fresnel approximation [see (1.12)] only in that it has no factor $\exp\left[ik(z - z')\right]$.

In the following, instead of (2.35), we will use the relationship relating the Fourier transforms to the appropriate transverse coordinates

$$\Phi_l(\varrho, z) = \int \varphi_l(\boldsymbol{\kappa}, z) \exp(i\boldsymbol{\kappa} \cdot \varrho) d^2\kappa \quad ,$$

$$f_l(\varrho, z) = \int \mu_l(\boldsymbol{\kappa}, z) \exp(i\boldsymbol{\kappa} \cdot \varrho) d^2\kappa \quad .$$

The equation equivalent to (2.35) has the form (see Exercise 2)

$$\varphi_l(\boldsymbol{\kappa}, z) = \frac{1}{2ik} \int\limits_0^z \exp(-i\kappa^2(z - z')/2k)\mu_l(\boldsymbol{\kappa}, z')dz' \quad . \tag{2.38}$$

Consider the first approximation Φ_1. According to (2.34a), $f_1(\varrho, z) = -k^2\tilde{\varepsilon}(\varrho, z)$, and from (2.38) we get (discarding the subscript 1 in φ_1)

$$\varphi(\boldsymbol{\kappa}, z) = \frac{ik}{2} \int\limits_0^z \exp(-i\kappa^2(z - z')/2k)\varepsilon(\boldsymbol{\kappa}, z')dz' \quad , \tag{2.39}$$

where $\varepsilon(\kappa, z)$ is the random two-dimensional Fourier amplitude of the electric permittivity $\tilde{\varepsilon}(\varrho, z)$. We will now go over to the level $\chi = \mathrm{Re}\{\Phi\}$ and phase $S' = \mathrm{Im}\{\Phi\}$.

Since

$$\Phi(\varrho, z) = \int \varphi(\kappa, z) \exp(i\kappa \cdot \varrho) d^2\kappa \quad ,$$

$$\Phi^*(\varrho, z) = \int \varphi^*(\kappa, z) \exp(-i\kappa \cdot \varrho) d^2\kappa = \int \varphi^*(-\kappa, z) \exp(i\kappa \cdot \varrho) d^2\kappa \quad ,$$

we can write

$$\chi(\varrho, z) = \int \chi(\kappa, z) \exp(i\kappa \cdot \varrho) d^2\kappa = \frac{\Phi + \Phi^*}{2}$$
$$= \int \frac{\varphi(\kappa, z) + \varphi^*(-\kappa, z)}{2} \exp(i\kappa \cdot \varrho) d^2\kappa \quad , \tag{2.40}$$

$$S'(\varrho, z) = \int S'(\kappa, z) \exp(i\kappa \cdot \varrho) d^2\kappa = \frac{\Phi - \Phi^*}{2i}$$
$$= \int \frac{\varphi(\kappa, z) - \varphi^*(-\kappa, z)}{2i} \exp(i\kappa \cdot \varrho) d^2\kappa \quad . \tag{2.41}$$

The spectral amplitude of the level and phase are expressed in terms of $\varphi(\kappa, z)$ as follows:

$$\chi(\kappa, z) = \frac{\varphi(\kappa, z) + \varphi^*(-\kappa, z)}{2} \quad , \quad S'(\kappa, z) = \frac{\varphi(\kappa, z) - \varphi^*(-\kappa, z)}{2i} \quad . \tag{2.42}$$

We will now proceed to find $\varphi^*(-\kappa, z)$. Since the field $\tilde{\varepsilon}(\varrho, z)$ is real, we have

$$\varepsilon^*(-\kappa, z') = \varepsilon(\kappa, z') \quad .$$

Therefore, by (2.39), we obtain

$$\varphi^*(-\kappa, z) = -\frac{ik}{2} \int_0^z \exp(i\kappa^2(z - z')/2k)\varepsilon(\kappa, z')dz' \quad . \tag{2.43}$$

Substituting (2.39) and (2.43) into (2.42) gives

$$\chi(\kappa, z) = \frac{k}{2} \int_0^z \sin\left[\frac{\kappa^2(z - z')}{2k}\right] \varepsilon(\kappa, z')dz' \quad , \tag{2.44}$$

$$S'(\kappa, z) = \frac{k}{2} \int_0^z \cos\left[\frac{\kappa^2(z - z')}{2k}\right] \varepsilon(\kappa, z')dz' \quad . \tag{2.45}$$

If we average these relationships, we will obtain $\overline{\chi}_1 = \overline{S}'_1 = 0$ (we have restored the subscript 1 here, since the mean level and phase are only zero in a first approximation). In a second approximation, the relations analogous to (2.44, 45) would include the spectral density of $(\nabla_\perp \Phi_1)^2$ [see (2.34b)], whose mean is nonzero, instead of $\varepsilon(\kappa, z)$. Thereby $\overline{\chi}_2$ and $\overline{S}'_2$ would be nonzero as well. Later in the book, we shall consider the calculation of these means in more detail. If we are interested

50

in the mean squares, e.g., $\overline{\chi^2}$ and $\overline{S'^2}$, then (2.44, 45) are sufficient for the derivation [if we retained the second-order terms in $\chi(\kappa, z)$, we would obtain third-order infinitesimals in $\langle \chi^2 \rangle$, which are negligible as compared to the leading term].

Consider the covariance of the level χ in the plane $z = $ const. By (2.44),

$$\psi_\chi(\varrho_1, \varrho_2; z) = \langle \chi(\varrho_1, z) \chi(\varrho_2, z) \rangle$$

$$= \int d^2\kappa_1 \int d^2\kappa_2 \exp\left[i(\kappa_1 \cdot \varrho_1 + \kappa_2 \cdot \varrho_2)\right]$$
$$\times \langle \chi(\kappa_1, z) \chi(\kappa_2, z) \rangle \quad . \tag{2.46}$$

In order to calculate ψ_χ, it is thus necessary to know the covariance of the spectral components that enters (2.46). Using (2.44) gives

$$\langle \chi(\kappa_1, z) \chi(\kappa_2, z) \rangle = \frac{k^2}{4} \int_0^z dz_1 \int_0^z dz_2 \, \sin\left[\frac{\kappa_1^2(z - z_1)}{2k}\right]$$
$$\times \sin\left[\frac{\kappa_2^2(z - z_2)}{2k}\right] \langle \varepsilon(\kappa_1, z_1) \varepsilon(\kappa_2, z_2) \rangle \quad . \tag{2.47}$$

However, for the statistically homogeneous medium we have

$$\langle \varepsilon(\kappa_1, z_1) \varepsilon(\kappa_2, z_2) \rangle = \delta(\kappa_1 + \kappa_2) F_\varepsilon(\kappa_1, z_1 - z_2) \quad , \tag{2.48}$$

where the two-dimensional spectral density F_ε is concentrated within the region

$$|\kappa||z_1 - z_2| < 2\pi \tag{2.49}$$

(see exercises to Chap. III.1). From (2.47, 48),

$$\langle \chi(\kappa_1, z) \chi(\kappa_2, z) \rangle = \delta(\kappa_1 + \kappa_2) F_\chi(\kappa_1, z) \quad , \quad \text{where} \tag{2.50}$$

$$F_\chi(\kappa, z) = \frac{k^2}{4} \int_0^z dz_1 \int_0^z dz_2 \, \sin\left[\frac{\kappa^2(z - z_1)}{2k}\right] \sin\left[\frac{\kappa^2(z - z_2)}{2k}\right] F_\varepsilon(\kappa, z_1 - z_2) \quad . \tag{2.51}$$

Substituting (2.51) into (2.46) gives

$$\psi_\chi(\varrho_1, \varrho_2; z) = \int d^2\kappa \exp\left[i\kappa \cdot (\varrho_1 - \varrho_2)\right] F_\chi(\kappa, z) \quad . \tag{2.52}$$

We see that $F_\chi(\kappa, z)$ is the two-dimensional (in the plane $z = $ const) spatial spectrum of the level fluctuations.

Equation (2.51), which relates the two-dimensional densities of the level fluctuations and the fluctuations ε, can be simplified substantially by using the property (2.49) of F_ε. To this end, we first introduce into (2.51) the new integration variables $z' = (z_1 + z_2)/2$ and $\zeta = z_1 - z_2$. In these variables, the boundaries of the integration regions will be $z' = \pm \zeta/2$ and $z' = z \pm \zeta/2$, and (2.51) will take the form

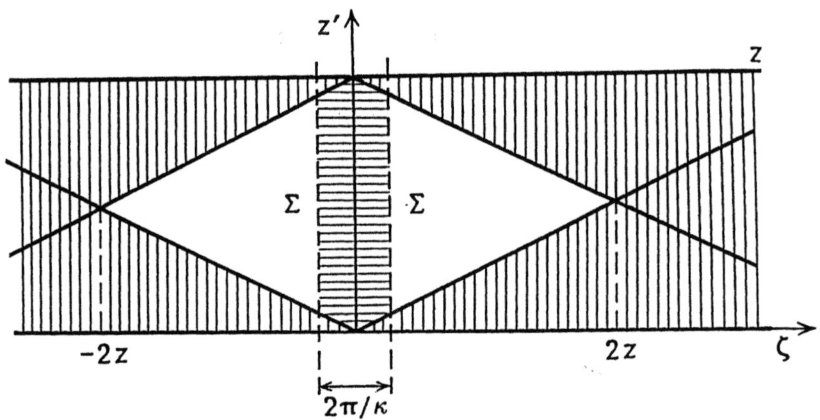

Fig. 2.2. Calculation of the level fluctuations. Integration region in z' and ζ

$$F_\chi(\kappa, z) = \frac{k^2}{4} \int_\Sigma dz' \, d\zeta \, \sin\left[\frac{\kappa^2(z - z')}{2k} - \frac{\kappa^2\zeta}{4k}\right]$$

$$\times \sin\left[\frac{\kappa^2(z - z')}{2k} + \frac{\kappa^2\zeta}{4k}\right] F_\varepsilon(\kappa, \zeta) \quad . \tag{2.53}$$

The integration region Σ is shown in Fig. 2.2.

By virtue of (2.49), in the region of importance for integration with respect to ζ, we have

$$\kappa\zeta < 2\pi \quad .$$

Therefore, the terms $\kappa^2\zeta/4k$ in the sines can be estimated in this region in the following way:

$$\frac{\kappa^2\zeta}{4k} < \frac{\kappa 2\pi}{4k} < \frac{\pi}{2}\frac{\kappa_{\mathrm{m}}}{k} \ll \frac{\pi}{2} \quad ,$$

since the maximal wave number κ_{m}, limiting the region where $F_\varepsilon(\kappa, \zeta)$ is concentrated, is assumed to be small as compared to the wave number k. In consequence, up to first-order corrections κ_{m}/k, we can discard the terms $\pm \kappa^2\zeta/4k$ in the sines, and (2.53) then becomes

$$F_\chi(\kappa, z) = \frac{k^2}{4} \int_\Sigma dz' \, d\zeta \, \sin^2\left[\frac{\kappa^2(z - z')}{2k}\right] F_\varepsilon(\kappa, \zeta) \quad . \tag{2.54}$$

We would now like to consider the spectral density $F_\chi(\kappa, z)$ for κ subject to

$$\kappa z \gg 1 \quad . \tag{2.55}$$

This relation means that the characteristic transverse scales of the level inhomogeneities, which we wish to consider, are small in comparison to the path length z ($2\pi/\kappa \ll z$). For such κ, the region that is of importance for integration with respect to ζ is, by virtue of (2.49), subject to the constraint $|\zeta| < 2\pi/\kappa \ll z$. Accordingly, in the entire integration region given in Fig. 2.2, the major contributor is the narrow band of width $2\pi/\kappa \ll z$ near the axis $\zeta = 0$ (in Fig. 2.2 it is shown by horizontal shading). Outside this band, $F_\varepsilon(\kappa, \zeta)$ is negligible, which enables additional regions shown in the figure by vertical shading to be added to the integration region Σ, i.e., the integration region Σ can be expanded to cover the infinite (in ζ) band $0 < z' < z$ (similar arguments were used in Sect. 1.2, Fig. 1.2). Since F is negligible in the additional integration region (e.g., it decays exponentially with ζ, as in the case of the power spectra Φ_ε), the permissible uncertainty here is insignificant. From (2.54), we obtain

$$F_\chi(\kappa, z) = \frac{k^2}{4} \int\limits_0^z \sin^2\left[\frac{\kappa^2(z - z')}{2k}\right] dz' \int\limits_{-\infty}^\infty F_\varepsilon(\kappa, \zeta)d\zeta \quad . \tag{2.56}$$

But, $F_\varepsilon(\kappa, \zeta)$ satisfies the relation [see (III.1.151)]

$$\int\limits_{-\infty}^\infty F_\varepsilon(\kappa, \zeta)d\zeta = 2\pi\Phi_\varepsilon(\kappa, 0) \quad , \tag{2.57}$$

where $\Phi_\varepsilon(\kappa, 0)$ is the three-dimensional spectral density of $\tilde{\varepsilon}$, the argument being the two-dimensional vector $(\kappa, 0)$. Using this equation and computing the first of the integrals in (2.56), we arrive at the final expression for F_χ:

$$F_\chi(\kappa, z) = \frac{\pi k^2 z}{4}\left[1 - \frac{k}{\kappa^2 z}\sin\frac{\kappa^2 z}{k}\right]\Phi_\varepsilon(\kappa, 0) \quad . \tag{2.58}$$

This equation has been derived for the region $\kappa z \gg 1$, i.e., $\kappa \gg z^{-1}$. However, in the region $\kappa \ll \sqrt{k/z}$, the function $1 - k/\kappa^2 z \sin(\kappa^2 z/k)$ tends to zero as κ^4. Therefore, if $\sqrt{k/z} \gg z^{-1}$ (i.e., $\sqrt{kz} \gg 1$), then essentially in the *entire* region where (2.58) is concentrated, the condition $\kappa z \gg 1$ is met. As a result, the constraint $\kappa z \gg 1$, used in deriving (2.58), is insignificant if the condition

$$\sqrt{kz} \gg 1$$

is satisfied, which means that the path length accommodates many wavelengths.

Let us turn to another purely formal consideration, which we will use later in the book. In deriving (2.58), we assumed that $F_\varepsilon(\kappa, z)$ is a "sharp" function of ζ for all the interesting values of κ. As a matter of fact, we have implicitly substituted the delta-function $\delta(\zeta)$ for $F_\varepsilon(\kappa, \zeta)$. We will now see that by replacing

$$F_\varepsilon(\kappa, \zeta) \longrightarrow 2\pi\Phi_\varepsilon(\kappa, 0)\delta(\zeta) \equiv F_\varepsilon^{\text{ef}}(\kappa, z) \quad , \tag{2.59}$$

we will immediately obtain from (2.51) the expression (2.58). In fact, substituting (2.59) into (2.51) and integrating with respect to z_2, we arrive at the expression (2.56), which uses the integral (2.57).

Just as, by (2.44), we have computed the two-dimensional spectral density of the level fluctuations, so we can also work out the two-dimensional spectral density of the phase $F_S(\kappa, z)$ based on (2.45). We will omit the computation here (see Exercise 3), and only provide the final result

$$F_S(\kappa, z) = \frac{\pi k^2 z}{4}\left[1 + \frac{k}{\kappa^2 z}\sin\frac{\kappa^2 z}{k}\right]\Phi_\varepsilon(\kappa, 0) \quad . \tag{2.60}$$

The covariance of the phase fluctuations in the plane $z = $ const is, as usual, expressed in terms of $F_S(\kappa, z)$

$$\psi_S(\varrho, z) = \int F_S(\kappa, z)\exp(i\kappa \cdot \varrho)d^2\kappa \quad . \tag{2.61}$$

If the field $\tilde{\varepsilon}(\varrho, z)$ is statistically isotropic at $z = $ const or in a three-dimensional space, then $\psi_\varepsilon(\varrho, z) = \psi_\varepsilon(\varrho, z)$ and the three-dimensional spectral density $\Phi_\varepsilon(\kappa_x, \kappa_y, \kappa_z)$ has the form $\Phi_\varepsilon(\sqrt{\kappa_x^2 + \kappa_y^2}, \kappa_z)$. In this case, $\Phi_\varepsilon(\kappa, 0) = \Phi_\varepsilon(\kappa, 0)$, i.e., it is only dependent on the magnitude of the two-dimensional vector κ. It follows from (2.58, 60) that the spectral densities $F_\chi(\kappa, z)$ and $F_S(\kappa, z)$ are also dependent only on $\kappa = |\kappa|$, and hence the random fields χ and S' are statistically isotropic in the planes $z = $ const. The expressions relating the covariances ψ_χ and ψ_S to F_χ and F_S become

$$\psi_\chi(\varrho, z) = \frac{\pi^2 k^2 z}{2}\int\limits_0^\infty J_0(\kappa\varrho)\left(1 - \frac{k}{\kappa^2 z}\sin\frac{\kappa^2 z}{k}\right)\Phi_\varepsilon(\kappa, 0)\kappa\,d\kappa \quad , \tag{2.62}$$

$$\psi_S(\varrho, z) = \frac{\pi^2 k^2 z}{2}\int\limits_0^\infty J_0(\kappa\varrho)\left(1 + \frac{k}{\kappa^2 z}\sin\frac{\kappa^2 z}{k}\right)\Phi_\varepsilon(\kappa, 0)\kappa\,d\kappa \quad , \tag{2.63}$$

where $J_0(x)$ is the Bessel function.

The phase structure function $D_S(\varrho, z) = \langle[S'(\varrho_1, z) - S'(\varrho_1 + \varrho, z)]^2\rangle$ is given by

$$D_S(\varrho, z) = \pi^2 k^2 z\int\limits_0^\infty [1 - J_0(\kappa\varrho)]\left(1 + \frac{k}{\kappa^2 z}\sin\frac{\kappa^2 z}{k}\right)\Phi_\varepsilon(\kappa, 0)\kappa\,d\kappa \quad . \tag{2.64}$$

2.4 Analysis of MSP Results

Consider the function

$$W_\chi(\kappa, z) = 1 - \frac{k}{\kappa^2 z}\sin\frac{\kappa^2 z}{k} \quad , \tag{2.65}$$

i.e., the factor used in changing from $\Phi_\varepsilon(\kappa, 0)$ to $F_\chi(\kappa, z)$. Expanding the sine into a Taylor series gives

$$W_\chi(\kappa, z) = \frac{1}{6}\left(\frac{\kappa^2 z}{k}\right)^2 - \frac{1}{120}\left(\frac{\kappa^2 z}{k}\right)^4 + \ldots \quad . \tag{2.66}$$

Thus, in the region

$$\frac{\kappa^2 z}{k} \ll 1 \quad , \quad \text{or} \quad \kappa \ll \kappa_F \equiv \sqrt{k/z} \quad ,$$

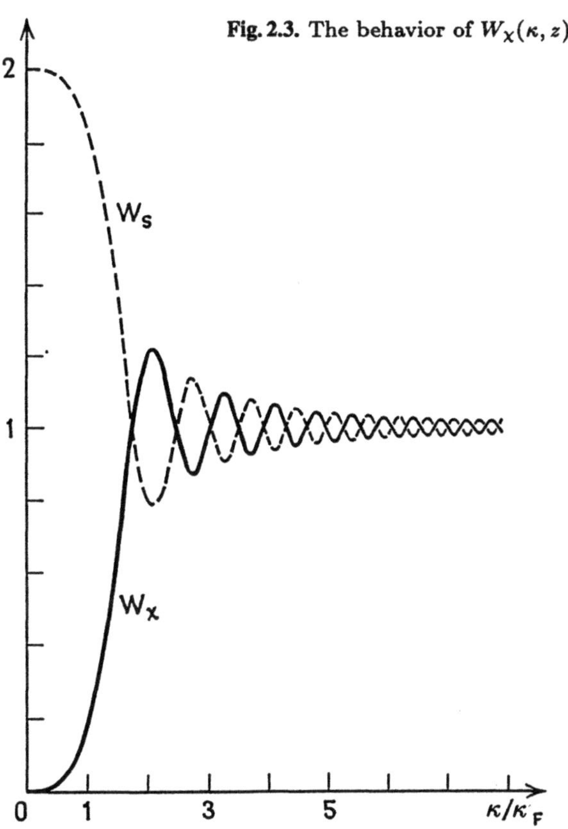

Fig. 2.3. The behavior of $W_\chi(\kappa, z)$ and $W_S(\kappa, z)$

$W_\chi(\kappa)$ varies with κ^4. The order of magnitude of the characteristic wave number $\kappa_F = \sqrt{k/z} = \sqrt{2\pi/\lambda z}$ equals that of the radius of the first Fresnel zone $\sqrt{\lambda z}$. For $\kappa \gg \kappa_F$, W_χ becomes independent of κ and tends to unity. The behavior of $W_\chi(\kappa, z)$ is given in Fig. 2.3. As to the function

$$W_S(\kappa, z) = 1 + \frac{k}{\kappa^2 z} \sin \frac{\kappa^2 z}{k} \tag{2.67}$$

which enters into (2.60), it equals 2 at $\kappa = 0$ and tends to unity when $\kappa \gg \kappa_F$, as does W_χ (Fig. 2.3).

The form of $\Phi_\varepsilon(\kappa, 0)$ may be different in different problems, but we have assumed that the wave number κ_m, corresponding to the size of the smallest inhomogeneities in the medium, is always small as compared to the wave number k of the radiation, with the result that, for $\kappa > \kappa_m$, $\Phi_\varepsilon(\kappa, 0)$ is negligible. Further, the quantity $\kappa_m/\kappa_F = \sqrt{\kappa_m^2 z/k} = \sqrt{2\pi\lambda z/l_0^2}$ varies with the ratio of the first Fresnel zone to the size of the smallest inhomogeneities of the electric permittivity. It will be recalled that the quantity $D = \kappa_m^2/\kappa_F^2 \sim \lambda z/l_0^2$ is said to be the *wave parameter* (Sect. III.2.3).

First, we will consider the case of $D \ll 1$, i.e., where $\kappa_m \ll \kappa_F$ or $l_0 \ll \sqrt{\lambda z}$ (Fig. 2.4). It has already been noted that the GOM is valid here. In the region to

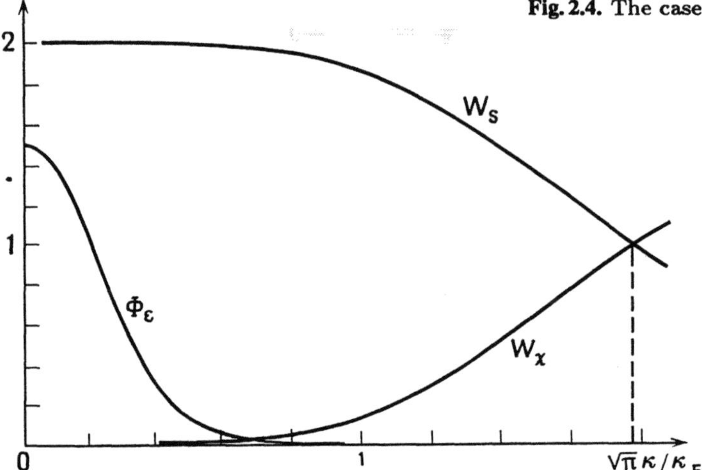

which most of the function $\Phi_\varepsilon(\kappa, 0)$ belongs, for $D \ll 1$ we can, to a sufficient accuracy, keep the leading terms of the expansions of $W_\chi(\kappa, z)$ and $W_S(\kappa, z)$

$$W_\chi(\kappa, z) \approx \frac{1}{6}\frac{\kappa^4 z^2}{k^2} \quad,$$

$$W_S(\kappa, z) \approx 2 \quad. \tag{2.68}$$

For the two-dimensional spatial spectra of the level and phase, from (2.58, 60), we have

$$F_\chi(\kappa, z) = \frac{\pi}{24} z^3 \kappa^4 \Phi_\varepsilon(\kappa, 0) \quad, \quad F_S(\kappa, z) = \frac{\pi k^2 z}{2} \Phi_\varepsilon(\kappa, 0) \quad. \tag{2.69}$$

Consequently, as has already been shown in Chap. 1, in the region of validity of the GOM, the mean square of the level fluctuations grows with distance as z^3, and the mean square of the phase fluctuations, as z. The wave number k does not enter the expression for $F_\chi(\kappa, z)$, i.e., in the geometrical-optics limit, the level fluctuations are independent of the wavelength.

The mean square of the level fluctuations, in terms of F_χ, is given by

$$\overline{\chi^2} = \int F_\chi(\kappa, z) d^2\kappa = \frac{\pi z^3}{24} \int \kappa^4 \Phi_\varepsilon(\kappa, 0) d^2\kappa \quad. \tag{2.70}$$

It follows that the contribution of the spectral components corresponding to the large-scale inhomogeneities, which are in turn described by small κ, is offset by the factor κ^4, which becomes zero at $\kappa = 0$. The major contribution to $\overline{\chi^2}$ comes from the spectral interval corresponding to small-scale inhomogeneities. In this region, the product $\kappa^4 \Phi_\varepsilon(\kappa, 0)$ shows a maximum, whereby the transverse correlation radius for the fluctuations is also of order l_0 (Fig. 2.5).

Equation (2.70) can also be written as

$$\overline{\chi^2} = \frac{z^3}{24} \int_0^\infty [\Delta_\perp^2 \psi_\varepsilon(\varrho, \zeta)]_{\varrho=0} d\zeta \tag{2.71}$$

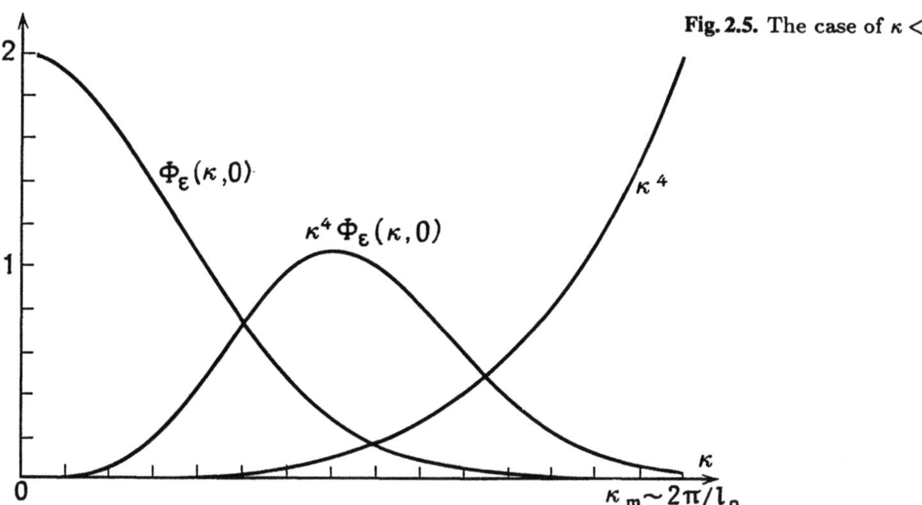

Fig. 2.5. The case of $\kappa < \kappa_m$

(see Exercise 2.6.5). This form of expression for $\overline{\chi^2}$ has already been derived in Chap. 1, using the GOM [see (1.90)].

From (2.60, 69), the mean square of the phase fluctuation will be

$$\overline{S'^2} = \frac{\pi k^2 z}{2} \int \Phi_\varepsilon(\kappa,0)d^2\kappa \quad . \tag{2.72}$$

Again, the main contributor here is the same interval of the inhomogeneities spectrum for which $\Phi_\varepsilon(\kappa,0)$ is maximal. The amplitude and phase fluctuations are thus controlled by *different* intervals of the spectrum of the electric permittivity inhomogeneities. We can also write (2.72) in the form (see Exercise 5)

$$\overline{S'^2} = \frac{k^2 z}{2} \int_0^\infty \psi_\varepsilon(0,\zeta)d\zeta \quad , \tag{2.73}$$

which corresponds to equation (1.28), derived using the GOM, for the eikonal $\varphi = S'/k$. We see, accordingly, that the relationships obtained using the MSP, in the limiting case of $D \ll 1$ ($\sqrt{\lambda z} \ll l_0$), go over into respective equations of geometrical optics.

Let us now consider another limiting case, viz. that of $D \gg 1$, corresponding to the Fraunhofer diffraction zone.

If the three-dimensional spectrum $\Phi_\varepsilon(\kappa)$ has several characteristic scales κ_n ($n = 1, 2, \ldots$), we will assume that the condition $D = \kappa_n^2/\kappa_F^2 \gg 1$ is met for the smallest of these wave numbers κ_l, i.e., the region of space in question corresponds to the Fraunhofer zone for all the scales of inhomogeneities available.

The mutual arrangement of the curves W_χ, W_S and Φ_ε for this case is shown in Fig. 2.6. We see that in most of the region of importance for the integration, $W_\chi \approx W_S \approx 1$. Therefore, if the three-dimensional spectral density $\Phi_\varepsilon(\kappa,0)$ displays no unintegrable singularity at zero [and hence the behavior of the weight function $W_\chi(\kappa,z)$ at zero is immaterial] we can assume that

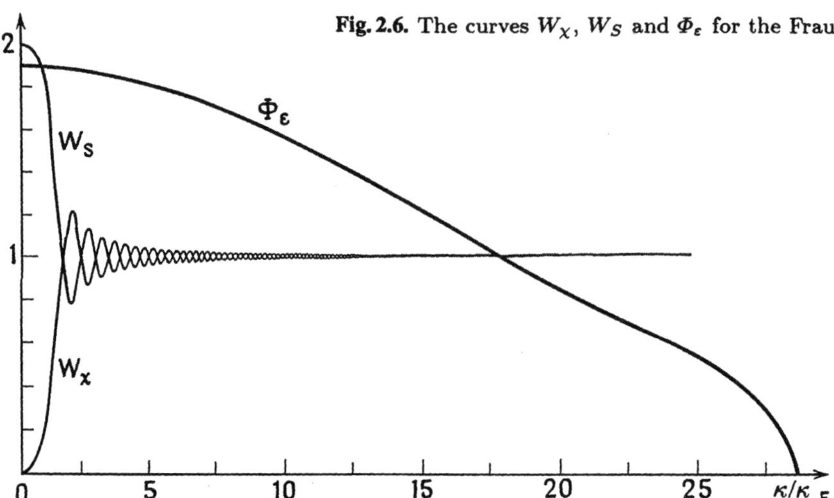

Fig. 2.6. The curves W_χ, W_S and Φ_ε for the Fraunhofer zone

$$F_\chi(\kappa, z) \approx F_S(\kappa, z) \approx \frac{\pi k^2 z}{4} \Phi_\varepsilon(\kappa, 0) \quad \text{for} \quad D \gg 1 \quad . \tag{2.74}$$

Consequently, in the limiting case of large D, the spectral densities of the amplitude and phase fluctuations appear to be about equal, with the result that the covariances are also about equal.

$$\psi_\chi(\varrho, z) \approx \psi_S(\varrho, z) \approx \frac{\pi k^2 z}{4} \int \exp(i\boldsymbol{\kappa} \cdot \boldsymbol{\varrho}) \Phi_\varepsilon(\kappa, 0) d^2\kappa \quad \text{for} \quad D \gg 1 \tag{2.75}$$

For the mean squares of the level and phase fluctuations where $D \gg 1$, we have

$$\overline{\chi^2} \approx \overline{S'^2} \approx \frac{\pi k^2 z}{4} \int \Phi_\varepsilon(\kappa, 0) d^2\kappa \quad \text{for} \quad D \gg 1 \quad . \tag{2.76}$$

Note that the mean square of the phase fluctuations for $D \gg 1$ only differs from equation (2.72), which holds for $D \ll 1$, by a coefficient that is smaller by a factor of two. As for the level fluctuations, the $\overline{\chi^2}$ vs z behavior changes as we go over to the case of $D \gg 1$: for $D \ll 1$, by (2.70), $\overline{\chi^2} \sim z^3$, whereas for $D \gg 1$ $\overline{\chi^2} \sim z$. In other words, at the beginning (in the geometrical optics region, where $D \sim \lambda z / l_\varepsilon^2 \ll 1$) $\overline{\chi^2}$ grows as z^3, and then (in the Fraunhofer zone, where $\lambda z / l_\varepsilon^2$ becomes large) the growth slows down to the linear dependence

$$\overline{\chi^2} = \begin{cases} \dfrac{\pi z^3}{24} \int \kappa^4 \Phi_\varepsilon(\kappa, 0) d^2\kappa & \text{for} \quad z \ll l_\varepsilon^2/\lambda \quad , \\[3mm] \dfrac{\pi k^2 z}{4} \int \Phi_\varepsilon(\kappa, 0) d^2\kappa & \text{for} \quad z \gg l_\varepsilon^2/\lambda \quad . \end{cases} \tag{2.77}$$

By way of example, we will now consider the medium in which the covariance of the electric permittivity fluctuations is Gaussian

$$\psi_\varepsilon(r) = \sigma_\varepsilon^2 \exp(-r^2/2a^2) \quad , \tag{2.78}$$

and the appropriate three-dimensional spectral density is

58

$$\Phi_\varepsilon(\kappa) = \frac{\sigma_\varepsilon^2 a^3}{(2\pi)^{3/2}} \exp\left(-\frac{\kappa^2 a^2}{2}\right) \ . \tag{2.79}$$

The inhomogeneities here are characterized by the only space scale a.

A straightforward calculation (see Exercise 2.6.6) for such a spectrum $\Phi_\varepsilon(\kappa)$ yields

$$\overline{\chi^2} = \frac{\sqrt{2\pi}}{8}\sigma_\varepsilon^2 k^2 az\left[1 - \frac{\arctan D}{D}\right] = \frac{\sqrt{2\pi}}{16}\sigma_\varepsilon^2(ka)^3(D - \arctan D) \ , \tag{2.80}$$

$$\overline{S'^2} = \frac{\sqrt{2\pi}}{8}\sigma_\varepsilon^2 k^2 az\left[1 + \frac{\arctan D}{D}\right] = \frac{\sqrt{2\pi}}{16}\sigma_\varepsilon^2(ka)^3(D + \arctan D) \ , \tag{2.81}$$

where $D = 2z/ka^2$. For $D \longrightarrow 0$, we can use the expansion $\arctan D = D - D^3/3 + \ldots$, which, by (2.70,71), gives a cubic dependence of $\overline{\chi^2}$ on z

$$\overline{\chi^2} = \frac{1}{3}\sqrt{\frac{\pi}{2}}\sigma_\varepsilon^2\left(\frac{z}{a}\right)^3 \ , \quad \text{for} \quad D \ll 1 \ , \tag{2.82}$$

and, by (2.72), a linear dependence for $\overline{S'^2}$

$$\overline{S'^2} = \tfrac{1}{4}\sqrt{2\pi}\sigma_\varepsilon^2 k^2 az \ , \quad \text{for} \quad D \ll 1 \ . \tag{2.83}$$

But, when $D \gg 1$, we can neglect $\pm(\arctan D)/D$ in (2.80, 81), and then, according to (2.76), we obtain similar expressions for $\overline{\chi^2}$ and $\overline{S'^2}$

$$\overline{\chi^2} \approx \overline{S'^2} \approx \tfrac{1}{8}\sqrt{2\pi}\sigma_\varepsilon^2 k^2 az \ , \quad \text{for} \quad D \gg 1 \ . \tag{2.84}$$

The variation with D of $\overline{\chi^2}$ and $\overline{S'^2}$ divided by $\sqrt{2\pi}\sigma_\varepsilon^2(ka)^3/16$ is shown in Fig. 2.7.

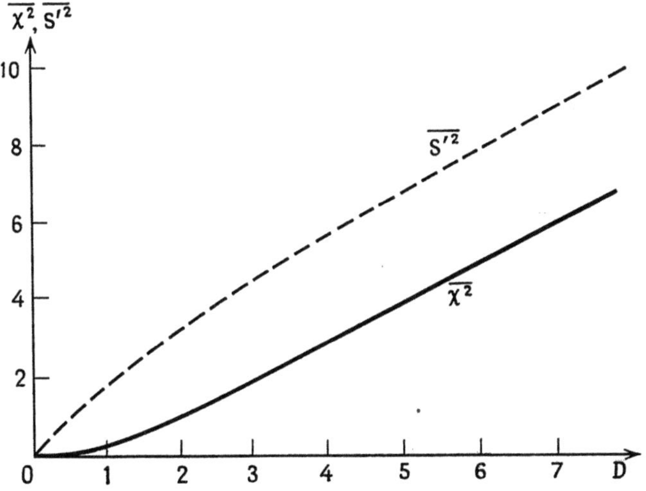

Fig. 2.7. Variation of χ^2 and S'^2 with the wave parameter (or distance) for the Fraunhofer zone

It should be noted that the relationships of the form (2.82–84) also hold in the more general case, where the three-dimensional spectral density of the inhomogeneities $\Phi_\varepsilon(\kappa)$ is limited throughout $[\Phi_\varepsilon(\kappa) < C]$ and shows a maximum at the point $\kappa = 0$. Starting from the general equations (2.70), (2.72) and (2.76), it can be shown that, when $D \ll 1$,

$$\overline{\chi^2} \sim \sigma_\varepsilon^2 (z/l_\varepsilon)^3 \quad , \quad \overline{S'^2} \sim \sigma_\varepsilon^2 k^2 l_\varepsilon z \quad ,$$

and, when $D \gg 1$,

$$\overline{\chi^2} \approx \overline{S'^2} \sim \sigma_\varepsilon^2 k^2 l_\varepsilon z \quad ,$$

where l_ε is the correlation radius for the fluctuations $\tilde{\varepsilon}$.

Consider another example, which is not covered by the general limiting relationships (2.74–76). This is the exponential spectrum corresponding to electric permittivity fluctuations caused by turbulence

$$\Phi_\varepsilon(\kappa) = \mathcal{A} C_\varepsilon^2 \kappa^{-11/3} \quad , \tag{2.85}$$

where $\mathcal{A} \approx 0.033$, and C_ε^2 is the structure characteristic that enters into the two-thirds law for electric permittivity fluctuations. In fact, the spectrum $\Phi_\varepsilon(\kappa)$ is only exponential in the limited range of wave numbers

$$\kappa_0 \ll \kappa \ll \kappa_m \quad , \quad \kappa_0 = 2\pi/L_0 \quad , \quad \kappa_m = 2\pi/l_0 \quad , \tag{2.86}$$

where l_0 and L_0 are the inner and outer scales of turbulence, respectively. However, in many cases of practical interest, the wave parameter $D_0 = \kappa_0^2/\kappa_F^2$ "composed" of the radius of the first Fresnel zone and the *outer* scale of turbulence, L_0, does not exceed unity, even for sufficiently large distances z^4.

At the same time, the wave parameter $D = \kappa_m^2/\kappa_F^2$ "composed" of the *inner* scale of turbulence and the radius of the first Fresnel zone can be both smaller and larger than unity. In the latter case, the GOM is already inapplicable owing to the presence of inhomogeneities that are smaller than the radius of the first Fresnel zone. On the other hand, since $D_0 \ll 1$, there are always inhomogeneities which are larger than $\sqrt{\lambda z}$ and, therefore, characterized not by the Fraunhofer diffraction, but by the Fresnel diffraction, or even geometrical optics.

Let us suppose for the moment that the constraints (2.86) are removed. The function $\kappa F_\chi(\kappa, z)$, whose integral controls $\overline{\chi^2}$, will then be given by the curve shown in Fig. 2.8. In the region $\kappa \ll \kappa_F$, we have $\kappa F_\chi(\kappa, z) \sim \kappa \kappa^4 \kappa^{-11/3} = \kappa^{4/3}$, i.e., $F_\chi(0, z) = 0$, and for $\kappa \gg \kappa_F$, we have $\kappa F_\chi(\kappa, z) \sim \kappa^{-8/3}$. Therefore, the area under the curve $\kappa F_\chi(\kappa, z)$ in Fig. 2.8 is finite, and the integral

$$\overline{\chi^2} = 2\pi \int_0^\infty F_\chi(\kappa, z) \kappa \, d\kappa \tag{2.87}$$

converges.

---

[4] For example, when light propagates in the atmosphere, typical values of the radius of the first Fresnel zone are several tens of centimeters, since the outer scale of turbulence, L_0, is of the order of meters. At several hundred kilometers, D_0 still remains small in comparison to unity.

60

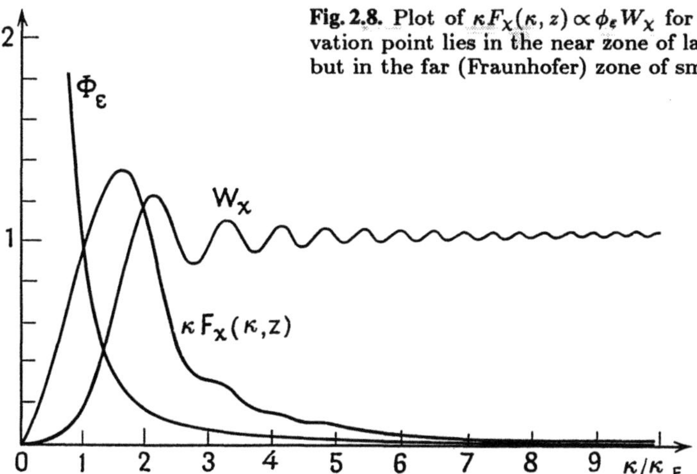

Fig. 2.8. Plot of $\kappa F_\chi(\kappa, z) \propto \phi_\varepsilon W_\chi$ for the case where the observation point lies in the near zone of large-scale inhomogeneities, but in the far (Fraunhofer) zone of small-scale inhomogeneities

We will now examine the consequences of taking into account the real behavior of $\Phi_\varepsilon(\kappa)$ beyond the interval (κ_0, κ_m), where the law (2.85) is valid. For $\kappa < \kappa_0$, the three-dimensional spectral density $\Phi_\varepsilon(\kappa)$ grows more slowly than $\kappa^{-11/3}$ with decreasing κ, or does not grow at all.

Therefore, if $\kappa_0 \ll \kappa_F$, then for $\kappa < \kappa_0$ the product $\kappa W_\chi \Phi_\varepsilon$ is already small and exerts no significant influence on the integral (2.87). Accordingly, if the "real" function $\Phi_\varepsilon(\kappa)$ in the region $\kappa < \kappa_0$ decreases as compared to the purely exponential function (2.85), this will exert no noticeable influence on (2.87). Thus, if $\kappa_0 \ll \kappa_F$, we may ignore deviations of the "real" spectrum $\Phi_\varepsilon(\kappa)$ from the purely exponential one.

Equally, if $\kappa_m \gg \kappa_F$, the point κ_m already lies in the region where the product $\kappa F_\chi \Phi_\varepsilon$ is small and falls off with increasing κ, so that the integral

$$\int\limits_{\kappa_m}^{\infty} \kappa^{-11/3} \kappa \, d\kappa = \frac{3}{5} \kappa_m^{-5/3} \tag{2.88}$$

for large κ_m is so small that its contribution to (2.87) can also be ignored. Since the integral of the product of the "real" spectrum by W_χ differs in the high-frequency part from the integral with the purely exponential spectrum by the small quantity (2.88), we can also overlook this difference when $\kappa_m \gg \kappa_F$.

Therefore, if the conditions $\kappa_0 \ll \kappa_F \ll \kappa_m$ are met, in computing $\overline{\chi^2}$ we can consider that $\Phi_\varepsilon(\kappa) \sim \kappa^{-11/3}$ for all κ, and in that case the substitution of (2.85) into (2.62) gives

$$\psi_\chi(\varrho, z) = \frac{\pi^2 A}{2} C_\varepsilon^2 k^2 z \int\limits_0^{\infty} J_0(\kappa\varrho)\left(1 - \frac{k}{\kappa^2 z} \sin\frac{\kappa^2 z}{k}\right)\kappa^{-8/3}\, d\kappa \quad . \tag{2.89}$$

Replacing the integration variable $\kappa^2 z/k = t^2$, we obtain

$$\psi_\chi(\varrho, z) = \frac{\pi^2 A}{2} C_\varepsilon^2 k^{7/6} z^{11/6} \int\limits_0^{\infty} J_0\left(\varrho\sqrt{\frac{k}{z}}\,t\right)\left(1 - \frac{\sin t^2}{t^2}\right) t^{-8/3}\, dt \quad . \tag{2.90}$$

61

Putting $\varrho = 0$ [recall that $J_0(0) = 1$], we arrive at

$$\overline{\chi^2} = N C_{\varepsilon}^2 k^{7/6} z^{11/6} \quad \text{for} \quad \kappa_0 \ll \kappa_F \ll \kappa_m \quad , \tag{2.91}$$

where N is a numerical constant

$$N = \frac{\pi^2 A}{2} \int_0^{\infty} \left(1 - \frac{\sin t^2}{t^2} \right) t^{-8/3} dt \approx 0.077 \quad . \tag{2.92}$$

The exponential variation of $\overline{\chi^2}$ with z (exponent 11/6) appears to be intermediate between the individual cases $\overline{\chi^2} \sim z^3$ and $\overline{\chi^2} \sim z$, which correspond to the geometrical optics and Fraunhofer zones, respectively. This is because, for $\kappa_0 \ll \kappa_F \ll \kappa_m$, there are always inhomogeneities for which the observation point lies within the Fresnel diffraction zone.

The separation ϱ between the observation points only enters into (2.90) via $\varrho\sqrt{k/z} = \sqrt{2\pi}\varrho/\sqrt{\lambda z}$. This implies that the correlation radius of the level fluctuations is of the order of $\sqrt{\lambda z}$. The behavior of the covariance (2.90) is depicted in Fig. 2.9 (see [2.6]).

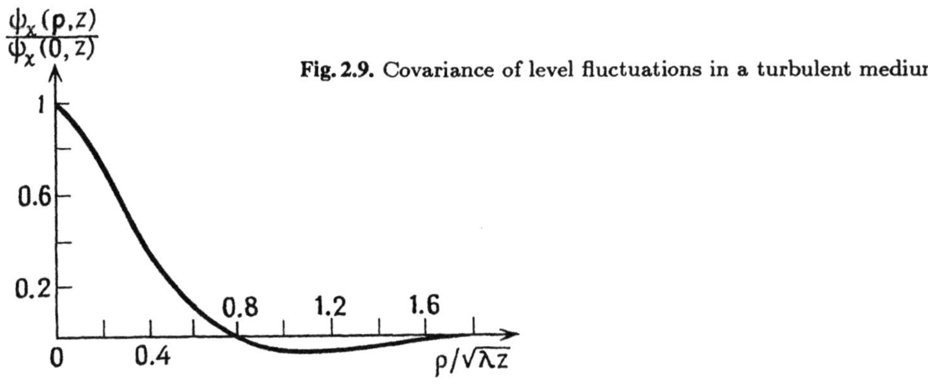

Fig. 2.9. Covariance of level fluctuations in a turbulent medium

Turning now to phase fluctuations, if we substitute the three-dimensional spectral density of the form (2.90) into (2.63) and neglect the constraints (2.86), the integral in (2.63) will diverge at $\kappa = 0$. The point is that, unlike $W_\chi(\kappa, z)$, the weight function $W_S(\kappa, z)$ does not vanish at $\kappa = 0$. As a result, whereas the spectral components associated with large-scale inhomogeneities make a negligible contribution to the amplitude fluctuations, their role for phase fluctuations appears to be comparable to that of small-scale inhomogeneities. This does not allow us to extrapolate the spectral density of the form (2.85) to the region $\kappa < \kappa_0$. To calculate $\overline{S'^2}$, we need to know the true three-dimensional spectral density in the region of wave numbers mentioned above, i.e., we have to take into consideration the influence of large inhomogeneities. The simplest way to take into account the "saturation" of fluctuations in the large-scale region is to replace (2.85) by the spectrum (2.79), which has no singularities as $\kappa \longrightarrow 0$. The eikonal variance for such a spectrum was estimated earlier using the GOM [see (1.121)].

We now consider the *structure function* of the phase $D_S(\varrho, z)$ in the plane $z = \text{const}$. Here we can expect that the large-scale inhomogeneities (much larger than ϱ) will be of no consequence, since they make equal contribution to the phase shifts along both rays and, therefore, no longer enter into the difference $S'(\varrho_1, z) - S'(\varrho_2, z)$. In (2.64), this is manifested by the fact that the factor $1 - J_0(\kappa\varrho)$ in the region $\kappa \ll \varrho^{-1}$ varies as κ^2. Therefore, if we substitute the exponential three-dimensional spectrum (2.85) into (2.64) and remove the constraints (2.86), we will arrive at the converging integral

$$D_S(\varrho, z) = \pi^2 A C_\varepsilon^2 k^2 z \int_0^\infty [1 - J_0(\kappa\varrho)]\left(1 + \frac{k}{\kappa^2 z}\sin\frac{\kappa^2 z}{k}\right)\kappa^{-8/3}d\kappa \quad . \quad (2.93)$$

Here, the singularity at zero has the form $\kappa^{2-8/3} = \kappa^{-2/3}$, i.e., it is integrable. Reasoning along the same lines as when deriving (2.88), we can see that, subject to the conditions

$$\kappa_0 \ll 1/\varrho \ll \kappa_m \quad \text{or} \quad l_0 \ll \varrho \ll L_0 \quad , \tag{2.94}$$

(2.93) holds. In this treatment, the wave number $2\pi/\varrho$ will play the same role as κ_F in the earlier case, since the weight function $[1 - J_0(\kappa\varrho)]\,W_S(\kappa, z)$ for $\kappa \ll 2\pi/\varrho$ has the form κ^2, and is approximately constant for $\kappa \gg 2\pi/\varrho$.

If in (2.93) we change the variable $\kappa\varrho = t$, we obtain

$$D_S(\varrho, z) = \pi^2 A C_\varepsilon^2 k^2 \varrho^{5/3} z \int_0^\infty [1 - J_0(t)]\left(1 + \frac{k\varrho^2}{t^2 z}\sin\frac{t^2 z}{k\varrho^2}\right)t^{-8/3}dt \quad . \quad (2.95)$$

An examination of this integral (see Exercise 7) yields the following results. If $l_0 \ll \varrho \ll L_0$ and $k\varrho^2 \gg z$, then

$$D_S(\varrho, z) = 2M C_\varepsilon^2 k^2 z \varrho^{5/3} \quad ,$$

$$M = \pi^2 A \int_0^\infty [1 - J_0(t)]t^{-8/3}dt \approx 0.73 \quad , \tag{2.96}$$

and for $k\varrho^2 \ll z$

$$D_S(\varrho, z) = M C_\varepsilon^2 k^2 z \varrho^{5/3} \quad . \tag{2.97}$$

In order to derive $\psi_\chi(\varrho, z)$ and $D_S(\varrho, z)$ for $\varrho \ll l_0$, we should take into account the fact that the three-dimensional spectral density $\Phi_\varepsilon(\kappa)$ for $\kappa \gg \kappa_m$ falls off quickly. For the region $\varrho \ll l_0$, we can obtain (see Exercise 8)

$$\psi_\chi(\varrho, z) = \overline{\chi^2}(z)[1 - \alpha_1(\varrho/l_0)^2 + \ldots] \quad , \tag{2.98}$$

$$D_S(\varrho, z) = \alpha_2(\varrho/l_0)^2 + \ldots \quad . \tag{2.99}$$

To conclude this section, we will derive the above relationships, proceeding from simple qualitative arguments (however, we will not work out the numerical coefficients entering into these equations).

First, consider phase fluctuations. We saw earlier in this section that the geometrical optics approximation will suffice in dealing with them, since the inclusion of diffraction effects only leads to other numerical coefficients. We take a ray of length z. If the inhomogeneities have the scale l_ε, then the ray will be long enough to accommodate $N \sim z/l_\varepsilon$ independent inhomogeneities. After a single inhomogeneity, the phase shift of the wave will be

$$l_\varepsilon \frac{\omega}{c} \sqrt{\varepsilon} \approx l_\varepsilon k \left(1 + \frac{\tilde{\varepsilon}}{2} \right) \quad ,$$

i.e., the random component of this progression will be of order $l_\varepsilon k \tilde{\varepsilon}$. Since the random quantities corresponding to the different inhomogeneities are statistically independent, the total mean square of the phase shift is

$$\overline{S'^2} \sim N \overline{(l_\varepsilon k \tilde{\varepsilon})^2} = \frac{z}{l_\varepsilon} l_\varepsilon^2 k^2 \sigma_\varepsilon^2 = \sigma_\varepsilon^2 k^2 l_\varepsilon z \quad ,$$

which corresponds to (2.83).

In the same approximation, we will now estimate the amplitude fluctuations, which can be treated as a consequence of random focusing and defocusing. Suppose, as before, the inhomogeneity to have the size l_ε, and the deviation of the electric permittivity from the mean to be $\tilde{\varepsilon}$. The focal distance of the lens bounded by a spherical surface with radius R is $F = nR/(n - \bar{n})$, where n and $\bar{n}$ are the refractive indices of the lens and the medium, respectively. For our inhomogeneity, the focal distance will, therefore, be $F \sim l_\varepsilon/\tilde{\varepsilon}$, and the angle of ray deviation α (Fig. 2.10) will be $\alpha \sim (l_\varepsilon/F) \sim \tilde{\varepsilon}$. The diameter of the beam incident on the "lens" at a distance z behind it will change by $\Delta d = z\alpha \sim z\tilde{\varepsilon}$. But the wave amplitude A is related to the beam diameter d by the relationship $A^2 d^2 = A_0^2 l_\varepsilon^2$, whence $\Delta A/A = -\Delta d/d$. If the intensity fluctuations are *weak*, the focal distance F is very large and d, for $z \ll F$, will be about l_ε, i.e., $A \sim A_0$, whereby

$$\frac{\Delta A}{A_0} \sim \frac{z\tilde{\varepsilon}}{l_\varepsilon} \quad . \tag{2.100}$$

We have thus worked out the relative change in the amplitude due to one "lens". The total focal power $1/F$ of several *weak* "lenses", closely spaced as compared to their respective focal distances, equals the sum of the focal powers of the component lenses, so that the total change in the amplitude will be

$$\frac{\delta A}{A_0} = \sum_j \frac{z \tilde{\varepsilon}_j}{l_{\varepsilon j}} \quad .$$

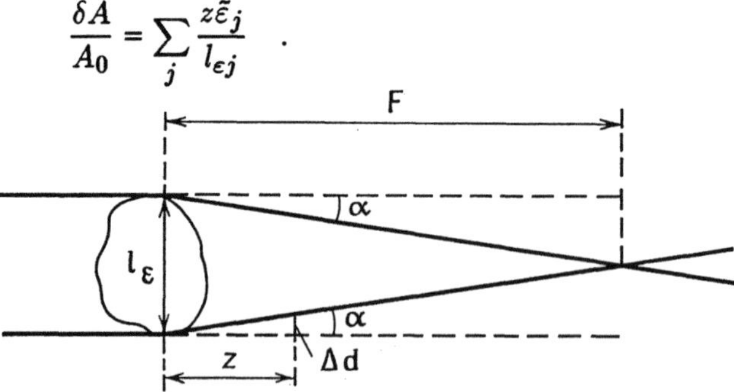

Fig. 2.10. Change in the cross-section of the ray tube after one inhomogeneity

This sum has a zero mean, and the mean square is

$$\overline{\left(\frac{\delta A}{A_0}\right)^2} \approx \frac{z^2}{l_\varepsilon^2} \sum \overline{\tilde{\varepsilon}_j^2} = \frac{z^2}{l_\varepsilon^2} \sigma_\varepsilon^2 N \quad ,$$

where $N \approx z/l_\varepsilon$ is the mean number of inhomogeneities over the distance z. Thus,

$$\overline{\left(\frac{\delta A}{A_0}\right)^2} \sim \sigma_\varepsilon^2 \left(\frac{z}{l_\varepsilon}\right)^3 \quad ,$$

which corresponds to (2.82).

Let us now consider the case where the medium contains inhomogeneities of different size obeying the three-dimensional spectral density (2.85) described by the structure function $D_\varepsilon(\varrho) = C_\varepsilon^2 \varrho^{2/3}$, an inhomogeneity of size l being characterized by the electric permittivity fluctuations $\tilde{\varepsilon}_l \sim C_\varepsilon l^{1/3}$.

First, consider the amplitude fluctuations. From (2.100),

$$\frac{\Delta A}{A} \sim \frac{z C_\varepsilon l^{1/3}}{l} \sim C_\varepsilon \frac{z}{l^{2/3}} \quad . \tag{2.101}$$

We thus see that the effect is larger, the smaller the inhomogeneity. If the size of the smallest inhomogeneities, l_0, meets the condition $l_0 \gg \sqrt{\lambda z}$ (whereby we can ignore the diffraction effects), then, in (2.101), we can take l_0 instead of l. The number of such inhomogeneities along the path of wave propagation is $N \sim z/l_0$, and the total mean square of the amplitude fluctuations will be

$$\overline{\left(\frac{\delta A}{A_0}\right)^2} \sim C_\varepsilon^2 \left(\frac{z}{l_0^{2/3}}\right)^2 \frac{z}{l_0} \sim C_\varepsilon^2 z^3 l_0^{-7/3} \quad \text{for} \quad \sqrt{\lambda z} \ll l_0 \quad . \tag{2.102}$$

Now let $l_0 \ll \sqrt{\lambda z}$, but at the same time $\sqrt{\lambda z} \ll L_0$, where L_0 is the outer scale of the inhomogeneities. This means that there are inhomogeneities of size $l > \sqrt{\lambda z}$ to which geometrical optics, and hence (2.101), is applicable. However, the inhomogeneities whose size l lies within the interval $(l_0, \sqrt{\lambda z})$ are not covered by geometrical optics and their treatment must take into account diffraction effects. The role of diffraction here consists only in that the lenses of size smaller than $\sqrt{\lambda z}$ no longer focus or defocus, and in both cases they produce divergent beams with an angle of about λ/l. Consequently, the inhomogeneities of size smaller than $\sqrt{\lambda z}$ can be discarded. But, the action of the larger inhomogeneities is still described by (2.101). We can, therefore, derive the final result directly from (2.102), if we replace the size l_0 by the size of the largest inhomogeneities that are still capable of focusing (or defocusing) the wave, i.e., by $\sqrt{\lambda z}$. We will thus arrive at

$$\overline{\left(\frac{\delta A}{A_0}\right)^2} \sim C_\varepsilon^2 \lambda^{-7/6} z^{11/6} \sim C_\varepsilon^2 k^{7/6} z^{11/6} \quad , \quad \text{for} \quad L_0 \gg \sqrt{\lambda z} \gg l_0 \quad ,$$

$$\tag{2.103}$$

which corresponds to (2.91).

We will now turn to fluctuations of the phase difference at two points, the coordinates (ϱ_1, z) and (ϱ_2, z) being separated by $\varrho = |\varrho_1 - \varrho_2|$.

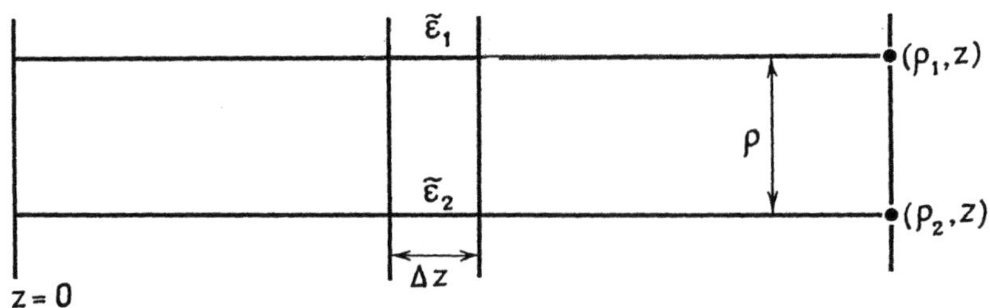

Fig. 2.11. Variation of the phase shift of two rays with distance

Suppose two rays come to the points (ϱ_1, z) and (ϱ_2, z) in Fig. 2.11. We next choose a layer of such thickness Δz that the quantities $\tilde{\varepsilon}(\varrho_i, z)$ change little over Δz. This section of the path introduces the phase shift $\Delta S \sim k\Delta\varepsilon(\varrho)\Delta z$, where $\Delta\varepsilon(\varrho)$ is the difference of the values $\tilde{\varepsilon}$ on the rays. There are two possibilities here.

If $\varrho \ll l_0$, both rays lie within the confines of one inhomogeneity of size l_0 and we can select $\Delta z \sim l_0$. For the section of length l_0, we then have

$$\overline{\Delta S_i^2} \sim k^2 \overline{[\Delta\varepsilon(\varrho)]^2} l_0^2 \quad .$$

But, $\overline{\Delta\varepsilon^2(\varrho)}$ is the structure function $\tilde{\varepsilon}$. As was shown in Sect. III.1.4, if $\varrho \ll l_0$, then

$$D_\varepsilon(\varrho) \sim C_\varepsilon^2 (\varrho/l_0)^2 l_0^{2/3} \sim C_\varepsilon^2 l_0^{-4/3} \varrho^2 \quad ,$$

and hence,

$$\overline{\Delta S_i^2} \sim C_\varepsilon^2 k^2 l_0^{-4/3} \varrho^2 l_0^2 \quad .$$

The number of independent inhomogeneities along the path z in this case is $N \sim z/l_0$, so that the total mean square of phase difference is

$$\overline{\delta S^2} \sim C_\varepsilon^2 k^2 \varrho^2 l_0^{2/3} \frac{z}{l_0} \sim C_\varepsilon^2 k^2 l_0^{-1/3} \varrho^2 z \quad \text{for} \quad \varrho \ll l_0 \quad . \tag{2.104}$$

Now let $L_0 \gg \varrho \gg l_0$. In this case, the main contribution to ΔS is made by inhomogeneities of size ϱ. In fact, smaller inhomogeneities have smaller values of $\tilde{\varepsilon}$, and larger ones make the same contribution to the phase shifts of both rays and are, therefore, of no consequence for the phase difference ΔS. Consequently, we can choose $\Delta z \sim \varrho$, and then

$$\Delta S_i \sim kC_\varepsilon \varrho^{1/3} \varrho \sim C_\varepsilon k\varrho^{4/3} \quad ,$$

[since $\Delta\varepsilon(\varrho) \sim C_\varepsilon \varrho^{1/3}$]. The total mean square of the phase difference due to all $N \sim z/\varrho$ inhomogeneities will be of the order of

$$\overline{\delta S^2} \sim C_\varepsilon^2 k^2 \varrho^{8/3} \frac{z}{\varrho} \sim C_\varepsilon^2 k^2 \varrho^{5/3} z \quad , \tag{2.105}$$

which corresponds to (2.96).

It follows that the basic relationships of the MSP for the turbulent medium can be obtained from the geometrical optics approach, if we take into account in addition that the inhomogeneities with scales smaller than the radius of the first Fresnel zone are of no importance for amplitude fluctuations.

66

As in the limiting case $D \gg 1$, described by (2.84), geometrical optics is completely ineffective here. Equations (2.84) can also be derived by qualitative arguments, although it is then necessary to use the purely diffraction concept of the effective scattering cross-section for the Fraunhofer diffraction zone [2.7].

2.5 Distribution of Amplitude and Phase Fluctuations. Energy Conservation and Limits of Applicability of MSP

We will now examine the distributions of the level and phase of the wave. If, in calculating the complex phase Φ, we can confine ourselves to the first approximation Φ_1, then, by (2.34a, 35)

$$\Phi_1(\varrho, z) = -k^2 \int d^2\varrho' \int_0^z dz' K(\varrho - \varrho', z - z')\tilde{\varepsilon}(\varrho', z') \quad , \tag{2.106}$$

where K is defined by (2.37). Hence the level χ and phase S' are

$$\chi_1(\varrho, z) = -k^2 \int d^2\varrho' \int_0^z dz' \operatorname{Re}\{K(\varrho - \varrho', z - z')\}\tilde{\varepsilon}(\varrho', z') \quad , \tag{2.107a}$$

$$S_1'(\varrho, z) = -k^2 \int d^2\varrho' \int_0^z dz' \operatorname{Im}\{K(\varrho - \varrho', z - z')\}\tilde{\varepsilon}(\varrho', z') \quad . \tag{2.107b}$$

Using these relationships, we could find the moments $\langle \chi_1^n \rangle$, $\langle S'^n \rangle$, $\langle \chi_1^l S_1'^m \rangle$ and from them obtain the probability densities for χ and S', and the joint probability density for (χ_1, S_1'). This procedure, however, is too complicated.

At the end of the previous section we obtained the basic equations for the mean squares of the level and phase fluctuations, proceeding from simple qualitative arguments whereby the inhomogeneous medium was broken down into individual volume elements which make *statistically independent* contributions to χ and S'. We can use a similar device in (2.107a,b) by representing the integrals as sums of integrals over the layers (z_l, z_{l+1}), whose longitudinal size Δz is much larger than the correlation radius of the electric permittivity fluctuations ($\Delta z \gg l_\varepsilon$). In that case, χ and S' will be

$$\chi = \sum_l \chi^{(l)}, \quad S' = \sum_l S'^{(l)} \quad , \tag{2.108}$$

where the terms can be treated as statistically independent. If Δz can be selected so that

$$l_\varepsilon \ll \Delta z \ll z \quad ,$$

then, over the wave path, the number of independent terms in (2.108) will be large: $N = z/\Delta z \gg 1$. Then, χ and S' appear to be represented as sums of a large number of statistically independent terms and, according to the Central Limit Theorem, can be regarded as normally distributed.

This reasoning is not very rigorous. In fact, the contributions of individual layers are not uncorrelated to so great an extent, since the inhomogeneities adjacent to the boundaries of these layers make a correlated contribution to the neighboring terms of (2.108). Further, the fact that individual terms in (2.108) are *uncorrelated* does not by any means suggest that they are statistically *independent*. The rigorous proof of the statement that the distributions of χ and S' tend to normal should therefore rely on the limit theorems for the linear functionals of random functions[5] (see, e.g., [2.8]). We will not, however, discuss this issue in more detail. We only note that experimental data support the assumption of normal distribution of χ and S' in those cases where a first approximation in the MSP is applicable [2.9].

The amplitude A is related to the level χ by

$$\chi = \ln\frac{A}{A_0} \quad , \quad \text{hence} \tag{2.109a}$$

$$A = A_0 \exp(\chi) \quad . \tag{2.109b}$$

It follows from (2.109a) that the logarithm of the amplitude is normally distributed, and hence the amplitude itself has a log-normal distribution. Finding the moments $\overline{A^n}$ is a straightforward exercise. We represent χ in the form $\chi = \overline{\chi} + \tilde{\chi}$. As was noted above, in a first approximation of the MSP, $\overline{\chi}$ is zero and so we need a second approximation. Equation (2.109b) takes the form $A = A_0 \exp\overline{\chi}\exp\tilde{\chi}$. Hence

$$\overline{A^n} = A_0^n e^{n\overline{\chi}}\langle e^{n\tilde{\chi}}\rangle \quad .$$

Since, for the normal random variable $n\tilde{\chi}$ with zero mean we have

$$\langle\exp(n\tilde{\chi})\rangle = \exp\left[\tfrac{1}{2}\overline{(n\tilde{\chi})^2}\right] = \exp(n^2\sigma_\chi^2/2) \quad , \quad \text{then}$$

$$\overline{A^n} = A_0^n \exp(n\overline{\chi} + n^2\sigma_\chi^2/2) \quad . \tag{2.110}$$

The plane wave propagation through a statistically homogeneous medium obeys the conservation law $\overline{A^2} = \text{const} = A_0^2$ [see (2.29)]. Setting into (2.110) $n = 2$, we obtain

$$\overline{A^2} = A_0^2 \exp\left(2\overline{\chi} + 2\sigma_\chi^2\right) \quad .$$

It follows from this and the relation $\overline{A^2} = A_0^2$ that

$$\overline{\chi} = -\sigma_\chi^2 \quad , \tag{2.111}$$

which is a corollary of energy conservation.

---

[5] For example, for the random processes $\xi(t)$, the conditions under which the distribution of $\eta = \int_{t_1}^{t_2} \xi(t)f(t)dt$ tends to normal as the interval (t_2, t_1) increases also include the so-called strong mixing condition

$$W_2\left(\int_{-\infty}^{t}\xi(\tau)f(\tau)d\tau \quad , \quad \int_{t+T}^{\infty}\xi(\tau)f(\tau)d\tau\right)$$

$$-W_1\left(\int_{-\infty}^{t}f(\tau)\xi(\tau)d\tau\right)W_1\left(\int_{t+T}^{\infty}f(\tau)\xi(\tau)d\tau\right) \xrightarrow[T\to\infty]{} 0 \quad .$$

Of course, $\overline{\chi}$ could also be found directly from a second approximation of the MSP equations. The treatment is given, e.g., in [2.10], and the results obtained there agree with (2.111).

If we substitute (2.111) into (2.110), the latter takes the form

$$\overline{A^n} = A_0^n \exp\left(n(n-2)\sigma_\chi^2/2\right) \quad . \tag{2.112}$$

Let us find the mean square of the intensity fluctuations $I = A^2$. Clearly, $\overline{I} = \overline{A^2} = A_0^2$. From (2.112) we obtain for $\overline{I^2} = \overline{A^4}$

$$\overline{I^2} = A_0^4 e^{4\sigma_\chi^2} \quad .$$

Therefore, the intensity variance $\sigma_I^2 = \overline{I^2} - \overline{I}^2$ will be

$$\sigma_I^2 = A_0^4(e^{4\sigma_\chi^2} - 1) \quad . \tag{2.113}$$

Note that, like the amplitude, the intensity $I = A^2 = A_0^2 \exp 2\chi$ has a log-normal distribution.

We will now turn to the region of validity of the MSP. It should be noted first of all that, since the relationships of the MSP have been derived from the approximate parabolic equation (2.4), the validity conditions for the latter also hold for the MSP. But, the equations of the MSP are only solved to a first approximation, therefore we should identify the conditions under which the corrections to the quantities σ_χ^2, $\overline{S'^2}$, $D_S(\varrho)$, etc., found from a second approximation, will be small. Approximate computations are given in a number of works (see, e.g., [2.10, 11]), and the results of these treatments may be summarized as follows.

Above all, the level fluctuations found to a first approximation of the MSP must be small, i.e.,

$$\overline{\chi_1^2} \ll 1 \quad . \tag{2.114}$$

If this condition is met, the second-order corrections to σ_χ^2 will be insignificant. For example, if the three-dimensional spectral density of the fluctuations ε has a purely exponential form, as is the case with turbulent inhomogeneities, then σ_χ^2 (including the higher-order approximations of the MSP) is subject to

$$\sigma_\chi^2 = \overline{\chi_1^2} + \alpha_2(\overline{\chi_1^2})^2 + \alpha_3(\overline{\chi_1^2})^3 + \ldots = f(\overline{\chi_1^2}) \quad , \tag{2.115}$$

where $f(\overline{\chi_1^2}) \approx \overline{\chi_1^2}$ for $\overline{\chi_1^2} \ll 1$.

If we seek $D_S(\varrho, z)$, we will obtain for exponential spectra and to higher-order approximations

$$D_S(\varrho, z) = D_{S1}(\varrho, z) + \alpha_2'[D_{S1}(\varrho, z)]^2 + \ldots \quad , \tag{2.116}$$

where $D_{S1}(\varrho, z)$ is the structure function of the phase derived to a first approximation of the MSP. Accordingly, $D_S(\varrho, z)$ may be obtained using the MSP if

$$D_{S1}(\varrho, z) \ll 1 \quad . \tag{2.117}$$

It should be noted that the conditions (2.114, 117) are independent: instances can arise where one of them is met, while the other is not. In that case, the first approximation of the MSP is suitable for computing one quantity, but not the other. The constraint (2.117) seems to be excessively rigid. The treatment in the parabolic equation approximation (Chap. 3) imposes a much weaker constraint than (2.117) on $D_S(\varrho, z)$.

A word of caution is required here. If the condition $\overline{\chi_1^2} \ll 1$ is violated, say $\overline{\chi_1^2} \approx 1$, the inclusion of a second approximation is no remedy, since, by (2.115), *all* the terms of the expansion become significant.

In the region of validity of the MSP we can, by virtue of the condition $\sigma_\chi^2 \ll 1$, expand the exponential factors in the expressions for $\overline{A^n}$ into a series and keep only the first terms. For example, instead of (2.113), we can write

$$\frac{\overline{\sigma_1^2}}{(\overline{I})^2} = e^{4\sigma_\chi^2} - 1 \approx 4\sigma_\chi^2 \quad . \tag{2.118}$$

At this point one may well ask: since the MSP is only valid for weak fluctuations of the level (and amplitude) of the wave, has it any advantages over the method of small perturbations?

If the field $u(\mathbf{r})$ is sought in the form $u = u_0 + u_1 + \ldots$, the fluctuations of the amplitude, $\tilde{A}$, and phase, $\tilde{S}$, can be represented in terms of u_0 and u_1: $\tilde{A}/A_0 = \mathrm{Re}\ \{u_1/u_0\}$, and $\tilde{S} = \mathrm{Im}\ \{u_1/u_0\}$. Consequently, given the second moments for $\mathrm{Re}\ \{u_1\}$ and $\mathrm{Im}\ \{u_1\}$, we can find the appropriate quantities for the amplitude and phase. The expressions thus derived for the amplitude and phase fluctuations coincide with those found from the first-approximation equations, with the exception that here $\ln A_0 + \tilde{A}/A_0$ must be substituted for $\ln A$, i.e., the first term in the expansion of $\ln A$ in $\tilde{A}/A_0$ must be taken into account.

Turning to the distributions of the amplitude, the results obtained from both techniques will differ drastically.

We apply perturbation theory to amplitude fluctuations and find the generalized Rayleigh law for the amplitude distribution [see (I.5.38)]. For this law, $\langle \tilde{A}^2 \rangle / A_0^2 = 4/\pi - 1 = 0.27$, i.e., large amplitude fluctuations are left unexplained. At the same time, the MSP leads to the log-normal distribution law for A, which is not subject to this constraint. Although formally (2.114) must hold, it appears in actual fact that the expressions for σ_χ^2 derived to a first approximation of the MSP agree well with experiment in the region $\sigma_\chi^2 \leq 1$ [2.9, 11]. Also, the distributions for A found experimentally for $\sigma_\chi^2 \leq 1$ fit the log-normal distribution, and cannot thus be approximated by the Rayleigh distribution.[6]

---

[6] In some later works there has been a revival of interest in the scope of the MSP. Among the problems considered are those for which the MSP is not valid *a fortiori* (e.g., ones having sharp jumps of electric permittivity). In some cases, what is meant by the MSP is not the approximation based on the PEM, which is discussed here, but rather an approximation in which the total field is defined as $u(\mathbf{r}) = \exp[\psi(\mathbf{r})]$, which obeys the Helmholtz equation. Of course, in this case a significant role must be played by backscattering. But, for both PEM and MSP to be applicable, it must be small. Therefore, consideration of the applicability of a linearized equation for ψ has no direct connection with the scope of the MSP.

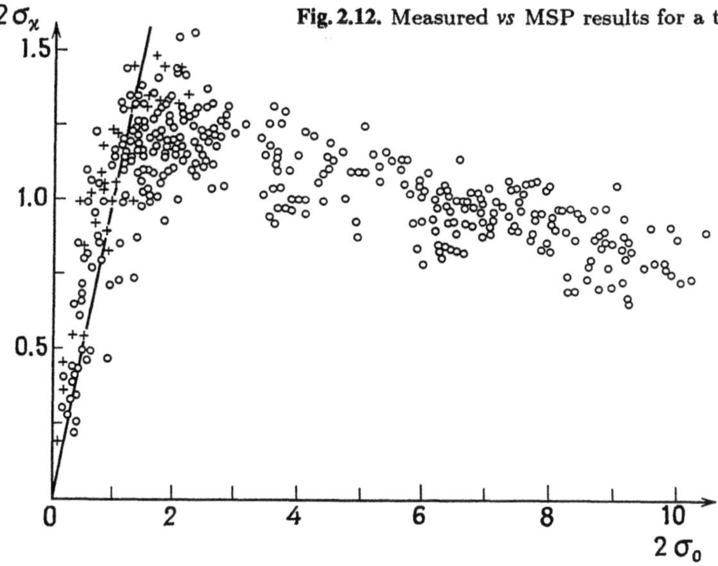

Fig. 2.12. Measured *vs* MSP results for a turbulent atmosphere

But in the region where σ_χ^2 (found by the MSP) is larger than unity, the experimental evidence disagrees markedly with predictions. Figure 2.12 is a comparison of the MSP results with the experimental data [2.9]. The ordinate is the measured σ_χ, the abscissa − the values of $\sigma_0 \equiv \sqrt{\overline{\chi_1^2}}$ computed for a first approximation of the MSP using the values of C_ε^2 found independently (from micrometeorological measurements).

Thus, it follows that in order to cater for the region $\overline{\chi_1^2} > 1$, methods are required which lie beyond the scope of perturbation theory and the MSP.

2.6 Exercises

2.6.1 Green's function $G(r)$ obeys the equation

$$\Delta G(r) + k^2 G(r) = \delta(r) \tag{2.119}$$

and the radiation condition at infinity. Derive the two-dimensional spectral expansion for $G(r)$.

Solution. Substituting into (2.119) the three-dimensinal Fourier transform

$$G(r) = G(\varrho, z) = \int\limits_{-\infty}^{\infty} d^2\kappa \int\limits_{-\infty}^{\infty} dp\,g(\kappa, p) \exp\left[i(\kappa \cdot \varrho + pz)\right] \quad , \tag{2.120}$$

we get

$$(k^2 - \kappa^2 - p^2)g(\kappa, p) = \frac{1}{8\pi^3} \quad ,$$

and hence

$$G(\boldsymbol{r}) = \frac{1}{8\pi^3} \int\limits_{-\infty}^{\infty} d^2\kappa \int\limits_{-\infty}^{\infty} dp \frac{\exp\left[\mathrm{i}(\boldsymbol{\kappa}\cdot\boldsymbol{\varrho} + pz)\right]}{k^2 + \mathrm{i}\varepsilon - \kappa^2 - p^2} \ , \tag{2.121}$$

In the integrand, the denominator contains the infinitesimal ($\varepsilon \longrightarrow +0$) imaginary term that describes the wave damping (Im $\{k\} > 0$) and yields, in computing (2.121), Green's function for *divergent* waves.

If in (2.121) we integrate with respect to p, we will obtain the desired representation of $G(\boldsymbol{r})$ in the form of the two-dimensional Fourier integral. Consider the integral with respect to p:

$$J = \int\limits_{-\infty}^{\infty} \frac{e^{\mathrm{i}pz}\, dp}{(k^2 - \kappa^2 + \mathrm{i}\varepsilon) - p^2} \ . \tag{2.122}$$

The poles $p_{1,2} = \pm\sqrt{k^2 - \kappa^2 + \mathrm{i}\varepsilon}$ lie both in the upper half-plane (Im $\{p_1\} > 0$) and in the lower one (Im $\{p_2\} < 0$). If $z > 0$, the integration path can be closed by an infinite half-circle in the upper half-plane of the complex variable p, and the residue at the upper pole gives

$$J = -2\pi\mathrm{i}\frac{e^{\mathrm{i}p_1 z}}{p_1 - p_2} = -\pi\mathrm{i}\frac{\exp\left(\mathrm{i}\sqrt{k^2 - \kappa^2}\,z\right)}{\sqrt{k^2 - \kappa^2}} \ , \quad \text{where} \tag{2.123}$$

$$\sqrt{k^2 - \kappa^2} = \begin{cases} \sqrt{k^2 - \kappa^2} & \text{for} \quad k^2 > \kappa^2 \ , \\ \mathrm{i}\sqrt{\kappa^2 - \kappa^2} & \text{for} \quad k^2 < \kappa^2 \ , \end{cases}$$

(the sign in the last expression is selected such that the condition Im $\{p_1\} > 0$ is satisfied both when $k^2 > \kappa^2$ and when $k^2 < \kappa^2$). For $z > 0$, substituting (2.123) into (2.121) gives

$$G(\varrho, z) = \frac{1}{8\mathrm{i}\pi^2} \int\limits_{-\infty}^{\infty} d^2\kappa \frac{\exp\left[\mathrm{i}(\boldsymbol{\kappa}\cdot\boldsymbol{\varrho} + \sqrt{k^2 - \kappa^2}\,z)\right]}{\sqrt{k^2 - \kappa^2}} \quad \text{for} \quad z > 0 \ . \tag{2.124}$$

For $z < 0$, when the residue must be taken at the pole p_2, the integration is carried out in much the same way. The final result, which applies in both cases, differs from (2.124) in that here we use $|z|$ instead of z.

2.6.2 Find Green's function for the equations (2.34).
Solution. We will seek the solution of the equation

$$2\mathrm{i}k\frac{\partial\Phi}{\partial z} + \Delta_\perp \Phi = f(\varrho, z) \tag{2.125}$$

with the initial condition $\Phi(\varrho, 0) = 0$ in the form of the two-dimensional Fourier integral

$$\Phi(\varrho, z) = \int \exp\left(\mathrm{i}\boldsymbol{\kappa}\cdot\boldsymbol{\varrho}\right)\varphi(\boldsymbol{\kappa}, z)d^2\kappa \ . \tag{2.126}$$

If we substitute into (2.125) an analogous expansion for $f(\varrho, z)$

$$f(\varrho, z) = \int \exp\left(\mathrm{i}\boldsymbol{\kappa}\cdot\boldsymbol{\varrho}\right)\mu(\boldsymbol{\kappa}, z)d^2\kappa \ ,$$

$$\mu(\kappa, z) = \frac{1}{4\pi^2} \int \exp(-i\kappa \cdot \varrho')f(\varrho', z)d^2\varrho' \quad , \tag{2.127}$$

we will have

$$2ik\frac{\partial\varphi(\kappa, z)}{\partial z} - \kappa^2\varphi(\kappa, z) = \mu(\kappa, z) \quad , \quad \text{or}$$

$$2ik\exp\left(\frac{\kappa^2 z}{2ik}\right)\frac{\partial}{\partial z}\left[\exp\left(-\frac{\kappa^2 z}{2ik}\right)\varphi(\kappa, z)\right] = \mu(\kappa, z) \quad .$$

Multiplying this equation by $\exp(-\kappa^2 z/2ik)/2ik$ and integrating with respect to z from 0 to z, subject to the boundary condition $\varphi(\kappa, 0) = 0$, gives

$$\varphi(\kappa, z) = \frac{1}{2ik} \int_0^z \exp\left[-\frac{i\kappa^2(z - z')}{2k}\right]\mu(\kappa, z')dz' \quad . \tag{2.128}$$

Substituting (2.128) into (2.126) gives

$$\Phi(\varrho, z) = \frac{1}{2ik} \int_0^z dz' \int d^2\kappa \exp\left[i\kappa \cdot \varrho - \frac{i\kappa^2(z - z')}{2k}\right]\mu(\kappa, z') \quad .$$

We arrive at the final expression for Φ by using the inverse Fourier transformation for $\mu(\kappa, z')$ [the second of equations (2.127]

$$\Phi(\varrho, z) = \frac{1}{8\pi^2 ik} \int_0^z dz' \int d^2\varrho' f(\varrho', z) \int d^2\kappa \exp\left[i\kappa \cdot (\varrho - \varrho') - \frac{i\kappa^2(z - z')}{2k}\right]$$

Taking the integral with respect to κ, we obtain a formula of the form (2.35), where $K(\varrho, z)$ is given by (2.37).

2.6.3 Find the two-dimensional spatial spectral density $F_S(\kappa, z)$ of the phase fluctuations (2.45).

Solution. Just as for the level fluctuations [see (2.50)], we have

$$\langle S(\kappa, z)S(\kappa', z)\rangle = \delta(\kappa + \kappa')F_S(\kappa, z) \quad . \tag{2.129}$$

It follows from (2.45) that

$$\langle S(\kappa, z)S(\kappa', z)\rangle = \frac{k^2}{4} \int_0^z dz' \int_0^z dz'' \cos\left[\kappa^2(z - z')/2k\right]$$
$$\times \cos\left[\kappa'^2(z - z'')/2k\right]\langle\varepsilon(\kappa, z')\varepsilon(\kappa', z'')\rangle \quad , \tag{2.130}$$

and since, by (2.48),

$$\langle\varepsilon(\kappa, z')\varepsilon(\kappa', z'')\rangle = \delta(\kappa + \kappa')F_\varepsilon(\kappa, z' - z'') \quad ,$$

the spectral densities F_S and F_ε will be related by

$$F_S(\kappa, z) = \frac{k^2}{4} \int_0^z dz' \int_0^z dz'' \cos\left[\kappa^2(z - z')/2k\right]$$

$$\times \cos\left[\kappa^2(z - z'')/2k\right] F_\varepsilon(\kappa, z' - z'') \quad . \tag{2.131}$$

If we now make use of the effective two-dimensional spectral density (2.59)

$$F_\varepsilon^{\text{eff}}(\kappa, z' - z'') = 2\pi\Phi_\varepsilon(\kappa, 0)\delta(z' - z'') \quad ,$$

(2.131) will take the form

$$F_S(\kappa, z) = \frac{\pi k^2}{2}\Phi_\varepsilon(\kappa, 0) \int_0^z \cos^2\left[\kappa^2(z - z')/2k\right]dz' \quad .$$

Taking the integral gives the expression (2.60).

2.6.4 Find the covariances of the level and phase fluctuations for observation points separated both transversely and longitudinally.
Solution. Starting from (1.54), we get

$$\psi_\chi(\varrho_1, z_1; \varrho_2, z_2) = \int d^2\kappa_1 \int d^2\kappa_2 \exp\left[i(\kappa_1 \cdot \varrho_1 + \kappa_2 \cdot \varrho_2)\right]$$

$$\times \langle \chi(\kappa_1, z_1)\chi(\kappa_2, z_2) \rangle \quad . \tag{2.132}$$

Using (2.44, 48, 59), we further find

$$\langle \chi(\kappa_1, z_1)\chi(\kappa_2, z_2) \rangle = \frac{k^2}{4} \int_0^{z_1} dz' \int_0^{z_2} dz'' \sin\left[\kappa_1^2(z_1 - z')/2k\right]$$

$$\times \sin\left[\kappa_2^2(z_2 - z'')/2k\right]\delta(\kappa_1 + \kappa_2)2\pi\Phi_\varepsilon(\kappa_1, 0)\delta(z' - z'')$$

$$= \frac{\pi k^2}{2}\delta(\kappa_1 + \kappa_2)\Phi_\varepsilon(\kappa_1, 0) \int_0^{z_<} \sin\left[\kappa_1^2(z_1 - z')/2k\right]$$

$$\times \sin\left[\kappa_1^2(z_2 - z')/2k\right]dz' \quad , \tag{2.133}$$

where $z_< = \min(z_1, z_2)$. Substituting (2.133) into (2.132) gives

$$\psi_\chi(\varrho_1, z_1; \varrho_2, z_2) = \frac{\pi k^2}{2} \int d^2\kappa\,\Phi_\varepsilon(\kappa, 0)\exp\left[i\kappa \cdot (\varrho_1 - \varrho_2)\right]$$

$$\times \int_0^{z_<} dz' \sin\left[\kappa^2(z_1 - z')/2k\right] \sin\left[\kappa^2(z_2 - z')/2k\right] \quad . \tag{2.134}$$

The expression for the phase covariance ψ_S differs from (2.134) only in that here cosines are used instead of sines

$$\psi_S(\varrho_1, z_1; \varrho_2, z_2) = \frac{\pi k^2}{2} \int d^2\kappa\,\Phi_\varepsilon(\kappa, 0)\exp\left[i\kappa \cdot (\varrho_1 - \varrho_2)\right]$$

$$\times \int_0^{z_<} dz' \, \cos \left[\kappa^2 (z_1 - z')/2k \right] \cos \left[\kappa^2 (z_2 - z')/2k \right] \quad .$$

$$\tag{2.135}$$

If we carry out the integration with respect to z' and use the relation $(z_1 + z_2)/2 - z_< = |z_1 - z_2|/2$, we obtain

$$\psi_{\chi,S}(\varrho_1, z_1; \varrho_2, z_2) = \frac{\pi k^2 z_<}{4} \int d^2 \kappa \, \Phi_\varepsilon(\kappa, 0) \exp \left[i\kappa \cdot (\varrho_1 - \varrho_2) \right]$$

$$\times \left[\cos \left[\kappa^2 (z_1 - z_2)/2k \right] \mp \frac{k}{\kappa^2 z_<} \sin \left[\kappa^2 (z_1 + z_2)/2k \right] \right.$$

$$\left. \pm \frac{k}{\kappa^2 z_<} \sin \left[\kappa^2 |z_1 - z_2|/2k \right] \right] \quad , \tag{2.136}$$

where the upper sign refers to ψ_χ and the lower ones to ψ_S.

At $z_1 = z_2 = z$, (2.135) changes into (2.58) and (2.60). Note that the covariances defined by (2.136) are only dependent on $\varrho_1 - \varrho_2$, i.e., they are invariant under simultaneous equal shifts of both observation points in the planes $z = z_1$ and $z = z_2$. At the same time, both $(z_1 - z_2)$ and $(z_1 + z_2)$ enter into (2.136). The level and phase fluctuations are thus statistically homogeneous along the transverse coordinates, but not along the longitudinal coordinate.

2.6.5 Derive (2.71) from (2.70), and (2.73) from (2.72).
Solution. The Fourier expansion

$$\psi_\varepsilon(\varrho, z) = \int_{-\infty}^\infty d^2 \kappa \int_{-\infty}^\infty dp \, \Phi_\varepsilon(\kappa, p) \exp \left[i(\kappa \cdot \varrho + pz) \right]$$

is integrated with respect to z in the limits $(-\infty, \infty)$

$$\int_{-\infty}^\infty \psi_\varepsilon(\varrho, z) dz = \int_{-\infty}^\infty d^2 \kappa \int_{-\infty}^\infty dp \, \Phi_\varepsilon(\kappa, p) \exp \left(i\kappa \cdot \varrho \right) 2\pi \delta(p)$$

$$= 2\pi \int_{-\infty}^\infty d^2 \kappa \, \Phi_\varepsilon(\kappa, 0) \exp \left(i\kappa \cdot \varrho \right) \quad ,$$

or, taking into account that $\psi_\varepsilon(\varrho, z)$ is even in z,

$$\int_{-\infty}^\infty \Phi_\varepsilon(\kappa, 0) \exp \left(i\kappa \cdot \varrho \right) d^2 \kappa = \frac{1}{\pi} \int_0^\infty \psi_\varepsilon(\varrho, z) dz \quad . \tag{2.137}$$

Operating with $\Delta_\perp^2$ on this equality and considering that $\Delta_\perp^2 \exp \left(i\kappa \cdot \varrho \right) = \kappa^4 \exp \left(i\kappa \cdot \varrho \right)$, we obtain

$$\int_{-\infty}^\infty \kappa^4 \Phi_\varepsilon(\kappa, 0) \exp \left(i\kappa \cdot \varrho \right) d^2 \kappa = \frac{1}{\pi} \int_0^\infty \Delta_\perp^2 \psi_\varepsilon(\varrho, z) dz \quad .$$

We put $\varrho = 0$, and then

$$\int_{-\infty}^{\infty} \kappa^4 \Phi_\varepsilon(\kappa, 0) d^2\kappa = \frac{1}{\pi} \int_0^{\infty} [\Delta_\perp^2 \psi_\varepsilon(\varrho, z)]_{\varrho=0} dz \quad .$$

Substituting this into (2.70) gives (2.71).

If we now put $\varrho = 0$ in (2.137), we will get

$$\int_{-\infty}^{\infty} \Phi_\varepsilon(\kappa, 0) d^2\kappa = \frac{1}{\pi} \int_0^{\infty} \psi_\varepsilon(0, z) dz \quad .$$

Substituting into (2.72) gives (2.73).

2.6.6 Find the mean squares of the level and phase fluctuations for the special case of the Gaussian covariance of the electric permittivity fluctuations

$$\psi_\varepsilon(r) = \sigma_\varepsilon^2 \exp(-r^2/2a^2) \quad . \tag{2.138}$$

Solution. Let us find $\overline{\chi^2} = \psi_\chi(0, z)$. Letting $\varrho = 0$ in (2.62), we obtain

$$\overline{\chi^2} = \frac{\pi^2 k^2 z}{2} \int_0^{\infty} \left(1 - \frac{k}{\kappa^2 z} \sin \frac{\kappa^2 z}{k}\right) \Phi_\varepsilon(\kappa, 0) \kappa \, d\kappa \quad , \tag{2.139}$$

where $\kappa^2 = \kappa_x^2 + \kappa_y^2$. Putting $\kappa_z = 0$ in (2.79), and substituting $\Phi_\varepsilon(\kappa, 0)$ into (2.139), we get

$$\overline{\chi^2} = \frac{\sqrt{2\pi} k^2 z \sigma_\varepsilon^2 a^3}{8} \int_0^{\infty} \left(1 - \frac{k}{\kappa^2 z} \sin \frac{\kappa^2 z}{k}\right) \exp(-\kappa^2 a^2/2) \kappa \, d\kappa \quad .$$

By setting $\kappa^2 a^2/2 = t$, we will change the integration variable. We will then have

$$\overline{\chi^2} = \frac{\sqrt{2\pi}}{8} k^2 \sigma_\varepsilon^2 a z \int_0^{\infty} \left[1 - \frac{\sin Dt}{Dt}\right] e^{-t} dt$$

$$= \frac{\sqrt{2\pi}}{8} k^2 \sigma_\varepsilon^2 a z \left[1 - \frac{1}{D} \int_0^{\infty} \frac{\sin Dt}{t} e^{-t} dt\right] \quad ,$$

where we introduced the wave parameter $D = 2z/ka^2$.

Consider the integral

$$f(D) = \int_0^{\infty} \frac{\sin Dt}{t} e^{-t} dt$$

in the above expression. At $D = 0$, we clearly have $f(0) = 0$. Further,

$$f'(D) = \int_0^{\infty} \cos Dt\, e^{-t} dt = \operatorname{Re}\left\{\int_0^{\infty} e^{-(1+iD)t} dt\right\} = \operatorname{Re}\left\{\frac{1}{1+iD}\right\} = \frac{1}{1+D^2} \quad ,$$

and hence

$$f(D) = \int_0^D \frac{dx}{1 + x^2} = \arctan D \quad .$$

Consequently,

$$\overline{\chi^2} = \frac{\sqrt{2\pi}}{8}\sigma_\varepsilon^2 k^2 az\left(1 - \frac{\arctan D}{D}\right) = \frac{\sqrt{2\pi}}{16}\sigma_\varepsilon^2 (ka)^3(D - \arctan D) \quad .$$

The mean square of the phase fluctuations follows from this by simply changing the sign at the second term.

2.6.7 Find the asymptotic expressions for the structure function of the phase (2.95) for the limiting cases $k\varrho^2 \ll z$ and $k\varrho^2 \gg z$.
Solution. The major contribution to (2.95) comes from the region $t \sim 1$, since for $t \gg 1$ the factor $t^{-8/3}$ is small, and for $t \ll 1$ the factor $|1 - J_0(t)|$ is small. If the relation $k\varrho^2/z \ll 1$ holds at $t \sim 1$, then

$$\left|\frac{k\varrho^2}{t^2 z}\sin\frac{t^2 z}{k\varrho^2}\right| \ll 1 \quad , \quad \text{and}$$

$$\int_0^\infty [1 - J_0(t)]\left(1 + \frac{k\varrho^2}{t^2 z}\sin\frac{t^2 z}{k\varrho^2}\right)t^{-8/3}\, dt \approx \int_0^\infty [1 - J_0(t)]t^{-8/3}\, dt = M \quad ,$$

where M is a constant. If $k\varrho^2/z \gg 1$, then at $t \sim 1$ we have

$$\left|\sin\left(\frac{t^2 z}{k\varrho^2}\right)\Big/\frac{t^2 z}{k\varrho^2}\right| \approx 1 \quad ,$$

and then

$$\int_0^\infty [1 - J_0(t)]\left(1 + \frac{k\varrho^2}{t^2 z}\sin\frac{t^2 z}{k\varrho^2}\right)t^{-8/3}\, dt \approx 2\int_0^\infty [1 - J_0(t)]t^{-8/3}\, dt = 2M \quad .$$

That is why the coefficients in the expressions for $D_S(\varrho, z)$ for the cases $k\varrho^2 \ll z$ and $k\varrho^2 \gg z$ differ by a factor of two.

2.6.8 For the statistically isotropic fluctuations $\tilde\varepsilon$, find the functions $\psi_\chi(\varrho, z)$ and $D_S(\varrho, z)$ for $\varrho \ll l_0$, where l_0 is the size of the smallest inhomogeneities of the electric permittivity.
Solution. We will start with equations (2.62) and (2.64)

$$\psi_\chi(\varrho, z) = \psi_\chi(\varrho, z) = \frac{\pi^2 k^2 z}{2}\int_0^\infty J_0(\kappa\varrho)\left(1 - \frac{k}{\kappa^2 z}\sin\frac{\kappa^2 z}{k}\right)\Phi_\varepsilon(\kappa, 0)\kappa\, d\kappa \quad ,$$

$$(2.140)$$

$$D_S(\varrho, z) = D_{\bar{S}}(\varrho, z)$$

$$= \pi^2 k^2 z \int\limits_0^\infty [1 - J_0(\kappa\varrho)] \left(1 + \frac{k}{\kappa^2 z} \sin \frac{\kappa^2 z}{k}\right) \Phi_\varepsilon(\kappa, 0) \kappa \, d\kappa \quad .$$

$$(2.141)$$

The function $\Phi_\varepsilon(\kappa, 0)$ being negligible when $\kappa > \kappa_m \sim 2\pi/l_0$, the major contributor to (2.140, 141) is the region $\kappa < 2\pi/l_0$. If $\varrho \ll l_0$, then in this region $\kappa\varrho < 2\pi\varrho/l_0 \ll 1$. We can therefore make use of the first terms of the exponential expansion $J_0(\kappa\varrho) = 1 - \kappa^2\varrho^2/4 + \ldots$. Substituting this expansion into (2.140, 141) leads to (2.98, 99).

3. Wave Propagation in Random Media. The Markovian Approximation

This chapter discusses a fairly effective method of describing wave propagation through random media that differs markedly from those considered above. It is based on the approximation of the wave field by a *Markov* random field. Section 3.1 deals with the background and content of that approximation. The main assumption here is that the wave field is statistically independent of the medium inhomogeneities that the wave has not yet passed through. However, this approximation is only possible within the framework of the parabolic equation approximation discussed in Sect. 2.1.

In the Markovian approximation, we can obtain *closed* equations for statistical moments of the wave field. These equations also appear to be valid for those regions of space where the relative fluctuations of the field are no longer small (Sect. 3.2). It turns out that the equations for the mean (i.e., coherent) field and the mutual coherence function of the second order (covariance) can be solved in the general form (Sect. 3.3). As for the equation for the fourth moment, which describes the *intensity fluctuations,* it is only amenable to examination using asymptotic and numerical methods (Sect. 3.4).

Section 3.5 is concerned with a regular procedure that allows derivations from the Markovian approximation to be obtained, and its region of validity to be refined.

3.1 General

All the approximate techniques for describing wave propagation in random media, discussed in Chap. III.4 and Chaps. 1 and 2, used the assumption that the electric permittivity fluctuations were small. It was either inherent to the technique (the method of small perturbations for the *exact* wave equation), or was introduced, because otherwise no progress could be made in solving the *approximate* equations (GOM, MSP). It was only by making this assumption that these methods could yield, in an explicitly approximate form, the wave field in a random medium, or its amplitude and phase in terms of $\tilde{\varepsilon}$. In order to derive the stochastic behavior of the various parameters of the wave, it was only necessary to average the expressions obtained, or their combinations. Clearly, the use of some form of perturbation theory for $\tilde{\varepsilon}$ imposes rather severe constraints on the applicability of the methods. For example, none of the above methods of solving the stochastic wave equation allows an adequate description of strong fluctuations of the wave field.

In Volumes I and II [3.1, 2], *Markov* processes were used to handle problems in physics involving a system of ordinary differential equations with random coefficients. It was then possible, in a number of cases, to work out an equation *directly* for the distributions or *averaged quantities* (moments, etc.). In order for the mathematics of Markovian processes to be applied to dynamic systems subject to random parametric actions (see Exercise III.1.8.22), the following conditions had to be fulfilled.

First, the dynamic causality principle had to be obeyed: the solution at any instant of time had to be only functionally dependent on the values of the random coefficients at *preceding* instants of time.

Second, the correlation time of random *actions* (i.e., random functions that enter into the equations) had to be *short* as compared to the shortest characteristic *response* time of the dynamic system. In that case, the covariances of the random actions could be approximated by delta-functions of time.

If both conditions were met, it was possible to obtain a *closed* equation for the probability density for the states of the dynamic system. For the *Gaussian* delta-correlated coefficients this was the Einstein-Fokker-Planck equation. If, in addition, the dynamic system was *linear,* we were also able to obtain closed equations for the moments (see Sect. I.6.4., 5 and Exercises I.6.7.7 and III.1.8.22).

Unlike perturbation theory, the approximation by Markovian random process uses *another* small parameter – the ratio of the correlation time of the actions, τ_0, to the correlation time of the response, τ. It is to a zeroth approximation in this parameter that the Markovian approximation corresponds. For this approximation to be valid, we may, clearly, need to impose some constraints on the intensity of the parameter fluctuations, but the resulting inequality also contains the ratio τ_0/τ, so that these constraints will appear less stringent.

Can we apply the theory of Markov processes to the problem of wave propagation through random media, i.e., to the problem of random *fields*.

Above all, the very notion of the Markov process presupposes the presence of an *ordered* variable (analogous to time), without which it would be impossible to formulate the principal property of these processes – the possibility of representing the multipoint probability density as the product of the transition probabilities. It is clear that the ordered variable can only be introduced with respect to *one* coordinate. Accordingly, it is hoped that wave propagation can be described as a Markovian random process either in a *unidimensional* problem (e.g., with a stratified random medium), or in the case where one of the coordinates is physically separated from the others (e.g., with *plane waves* or a *narrow beam* of radiation).

Further, if it is possible to separate from among the three spatial variables one that plays the role of time in the above sense, then along this coordinate the condition of dynamic *causality* must be met, i.e., the wave field under consideration must be only functionally dependent on the *preceding* (along this coordinate)[1] values of the random variable. In the general case, the wave field *does not fulfil* this requirement, since a random medium contains both forward and backscattered waves (Sect. 2.1), the latter being caused by the inhomogeneities lying *beyond* the observation point.

Nevertheless, for the *unidimensional* Helmholtz equation

$$u''(z) + k^2[1 + \tilde{\varepsilon}(z)]u = 0$$

---

[1] We will use the term "preceding" not only in relation to time, but also to any other coordinate that plays the role of time in the above sense.

describing the propagation of the scalar wave through a stratified medium, we can introduce the function

$$R(z) = \frac{iku(z) - u'(z)}{iku(z) + u'(z)}$$

which obeys the first-order equation

$$\frac{dR}{dz} = -2ikR - \frac{ik}{2}\bar{\varepsilon}(z)(1 + R)^2$$

and the "initial" condition (i.e., the boundary condition, say, at $z = 0$). In that case, the values of $R(z)$ are only functionally dependent on $\bar{\varepsilon}(\zeta)$ when $0 \leq \zeta \leq z$, so that the causality condition for R is met and this function can be approximated by a Markov process, if the correlation radius for $\bar{\varepsilon}(z)$ is sufficiently small (see, e.g. [3.3]).

However, for waves in a medium containing *three-dimensional* inhomogeneities, it is not possible to introduce an analogous function obeying the causality principle. The Markovian approximation is only possible here if *backscattered waves may be ignored*. As was found in Sect. 2.1, the latter case is covered by the parabolic equation approximation. Besides, the PEM uses a physically distinguished coordinate — that along the direction of the incident wave. Thus, the PEM generally allows a transition to the Markovian approximation, but it requires a preliminary analysis of the relationship between the longitudinal scales of the fluctuations $\bar{\varepsilon}$ and the fluctuations of the wave field v.

Mathematically, the multidimensional problem differs markedly from those treated in Volume I [3.1] in that the dynamic equation is now a *partial* differential equation. Here, instead of a random variable (or a random vector), we have a two-dimensional random *field* $v(\varrho, z)$ for each fixed value of z. But, the distribution of a random field is known (Sect. III.1.7) to be completely specified by the *characteristic functional*. In the cases of interest, therefore, the Einstein-Fokker-Planck equation must define a *functional*, rather than a function. Accordingly, the equation is more complex than that for dynamic systems with a *finite* number of degrees of freedom, since it appears to be a functional equation, not a conventional partial differential equation. To make our life easier, we will restrict ourselves to the derivation of *equations for the moments* of the field $v(\varrho, z)$.

The example of dynamic systems with a finite number of degrees of freedom tells us that the *closed* equations for the moments can only be derived from the Einstein-Fokker-Planck equation for *linear* systems. The parabolic equation (2.4) being linear, we can here too hope to obtain closed equations for the moments. Unlike (2.4), the initial equation (2.32) in the MSP is not linear and, therefore, it is not possible to derive equations for the moments of the complex phase Φ from the functional Einstein-Fokker-Planck equation [despite the fact that the solution to equation (2.32) for Φ satisfies the causality principle].

We will now turn to the longitudinal correlation radii for the fluctuations of the various field parameters. In so doing, we will proceed from the results obtained in Chap. 2 using the MSP.

The longitudinal correlation radius for the fluctuations of the phase and intensity (or level) can be estimated following the line of argument used at the end of Sect. 2.4. We saw there that the wave phase was controlled by the inhomogeneities encountered by the ray on its path to the observation point. By way of estimate, we can consider that different inhomogeneities make *independent* contributions ΔS_k to the phase. The phase progression along the ray that has encountered n inhomogeneities is $S = \sum_{k=1}^{n} \Delta S_k$. For another observation point lying on the same ray, the phase shift will be $S' = \sum_{k=1}^{n'} \Delta S_k$. If $n' > n$, then $S' = S + \sum_{k=n+1}^{n'} \Delta S_k$, and hence

$$\overline{SS'} = \left\langle S\left(S + \sum_{k=n+1}^{n'} \Delta S_k\right)\right\rangle = \overline{S^2} \quad ,$$

since $\overline{S\Delta S_k} = 0$ for $k > n$. The correlation coefficient will, therefore, be

$$K_S = \frac{\overline{SS'}}{\sqrt{\overline{S^2}\,\overline{S'^2}}} = \frac{\overline{S^2}}{\sqrt{\overline{S^2}\,\overline{S'^2}}} = \sqrt{\frac{\overline{S^2}}{\overline{S'^2}}} \quad .$$

But we have seen [see (2.81)] that $\overline{S^2} \sim z$, where z is the distance covered by the wave in an inhomogeneous medium. It follows that

$$K_S = \sqrt{z/z'} = \sqrt{z_</z_>} \quad . \tag{3.1}$$

This relationship can also be obtained in a more rigorous way by the GOM or MSP [see, e.g., (1.33) and Exercise 2.6.4]. If we fix $z_< = z$ and put $z_> = z + \zeta$, then, by (3.1) we obtain

$$K_S(z, \zeta) = \frac{1}{\sqrt{1 + (\zeta/z)}} \quad . \tag{3.2}$$

The longitudinal radius of the phase correlation is thus of order z, i.e., it is many times larger than the correlation radius of the electric permittivity fluctuations. But, as we saw earlier, this is precisely the necessary condition that makes the Markovian approximation possible.

Let us now proceed to estimate the correlation radius of the level fluctuations. At the end of Sect. 2.4, we estimated the focal distance for a characteristic inhomogeneity of size l_ε and deviation $\tilde{\varepsilon}$ from the mean electric permittivity. It was

$$F \sim \frac{l_\varepsilon}{\tilde{\varepsilon}} \quad .$$

If $\tilde{\varepsilon} \ll 1$, then $F \gg l_\varepsilon$. In the region of validity of the MSP, all the "lenses" may be considered to be weak, i.e., $F \gg z$. Here, we can argue along the same lines as in the case of the longitudinal radius of the phase correlation just considered. Should the condition $F \gg z$ be violated, the observation point may fall within the region of radiation focusing, where the intensity fluctuations are not small. But, when $\tilde{\varepsilon} \ll 1$, we also have $F \gg l_\varepsilon$, so that the size of the "region of influence" of each inhomogeneity along the z-axis is much larger than the size of the inhomogeneity itself.

When dealing with amplitude fluctuations, we should also consider the case where inhomogeneities are small as compared to the radius of the first Fresnel zone

$(l_\varepsilon \ll \sqrt{\lambda z})$. Geometrical optics is no longer applicable here, and we should thus rely on diffraction theory. It is well known that diffraction at inhomogeneities of size l_ε begins to show up at distances of about $z_{\mathrm{dif}} = l_\varepsilon^2/\lambda$. The focusing action of inhomogeneities is thus only possible at distances not larger than z_{dif}. Accordingly, the "region of influence" of the inhomogeneities has the longitudinal scale z_{dif}, and the condition $z_{\mathrm{dif}} \gg l_\varepsilon$ at which we can use the Markovian approximation takes the form $(l_\varepsilon^2/\lambda) \gg l_\varepsilon$, i.e., $\lambda \ll l_\varepsilon$.

We see that with *weak* fluctuations and *large-scale* inhomogeneities, the longitudinal correlation radius of the amplitude fluctuations in all the cases considered above appears to be *larger* than the size of the inhomogeneities, precisely what is required for the Markovian approximation.

Of course, this qualitative reasoning can in no way be regarded as a rigorous derivation of the Markovian approximation, and so its region of validity will be treated more consistently in Sect. 3.5. However, it should be emphasized at this point that nowhere along this line of reasoning did we have to make the assumption that the *amplitude* fluctuations of the wave were small. It is hoped, therefore, that the Markovian approximation will prove suitable for strong fluctuations of the field as well, if only we can ignore backscattering.

Approximating the covariance by a delta-function we used in Chap. 2 the MSP to compute the phase and level fluctuations. In Sect. 2.3, we applied the expression (2.59):

$$F_\varepsilon(\kappa, \zeta) \to F_\varepsilon^{\mathrm{ef}}(\kappa, \zeta) = 2\pi \Phi_\varepsilon(\kappa, 0)\delta(\zeta) \quad , \tag{3.3}$$

and showed that substituting F^{ef} for F in calculating the spectra of the amplitude and phase fluctuations yields correct results, if the following conditions are met:

$$z \gg \varrho \quad , \quad z \gg \lambda \quad .$$

As the covariance $\psi_\varepsilon(\varrho, \zeta)$ is related to the two-dimensional spectral density $F_\varepsilon(\kappa, \zeta)$ by

$$\psi_\varepsilon(\varrho, \zeta) = \int F_\varepsilon(\kappa, \zeta) \exp(i\kappa \cdot \varrho) d^2\kappa \quad ,$$

it is easily found that the change (3.3) is equivalent to the following change of covariance:

$$\psi_\varepsilon(\varrho, \zeta) \to \psi_\varepsilon^{\mathrm{ef}}(\varrho, \zeta) = A(\varrho)\delta(\zeta) \quad , \quad \text{where} \tag{3.4}$$

$$A(\varrho) = 2\pi \int \Phi_\varepsilon(\kappa, 0) \exp(\pm i\kappa \cdot \varrho) d^2\kappa \quad . \tag{3.5}$$

[$A(\varrho)$ being even, we may use any sign in the exponent.]

Consider the integral of $\psi_\varepsilon(\varrho, z)$ with respect to the longitudinal coordinate ζ. Using the three-dimensional spectral expansion

$$\psi_\varepsilon(\varrho, \zeta) = \int \Phi_\varepsilon(\kappa, p) \exp[i(\kappa \cdot \varrho + p\zeta)] d^2\kappa \, dp$$

and integrating with respect to ζ within infinite limits, we get

$$\int_{-\infty}^{\infty} \psi_\varepsilon(\varrho,\zeta)d\zeta = \int \Phi_\varepsilon(\boldsymbol{\kappa},p)\exp(i\boldsymbol{\kappa}\cdot\varrho)d^2\kappa\,dp \int_{-\infty}^{\infty} e^{ip\zeta}\,d\zeta$$

$$= \int \Phi_\varepsilon(\boldsymbol{\kappa},p)\exp(i\boldsymbol{\kappa}\cdot\varrho)\cdot 2\pi\delta(p)d^2\kappa\,dp$$

$$= 2\pi \int \Phi_\varepsilon(\boldsymbol{\kappa},0)\exp(i\boldsymbol{\kappa}\cdot\varrho)d^2\kappa = A(\varrho) \quad .$$

On the other hand, the integral with respect to ζ of the approximating covariance $\psi_\varepsilon^{\mathrm{ef}}(\varrho,\zeta) = A(\varrho)\delta(\zeta)$ also gives the function $A(\varrho)$, so that we have

$$\int_{-\infty}^{\infty} \psi_\varepsilon(\varrho,\zeta)d\zeta = \int_{-\infty}^{\infty} \psi_\varepsilon^{\mathrm{ef}}(\varrho,\zeta)d\zeta = A(\varrho) \quad . \tag{3.6}$$

In the following we will often make use of the combination

$$H(\varrho) = \frac{1}{\pi}[A(0) - A(\varrho)] = 2\int[1 - \cos\boldsymbol{\kappa}\cdot\varrho]\Phi_\varepsilon(\boldsymbol{\kappa},0)d^2\kappa \quad , \tag{3.7}$$

which only varies with the two-dimensional vector ϱ and is expressed, using the two-dimensional Fourier transform, in terms of the three-dimensional spectral density $\Phi_\varepsilon(\boldsymbol{\kappa},0)$.

If the random field $\tilde{\varepsilon}(\varrho,z)$ is *Gaussian,* to give its complete statistical description only requires the covariance of $\tilde{\varepsilon}$ and, in particular, the effective covariance of the form (3.4). However, if the field $\tilde{\varepsilon}(\varrho,z)$ is not supposed to be normal, then higher-order moments will be necessary

$$\langle \tilde{\varepsilon}(\varrho_1,z_1)\dots\tilde{\varepsilon}(\varrho_n,z_n)\rangle \quad .$$

It follows from the fact that the *Gaussian* field $\tilde{\varepsilon}(\varrho,z)$ is delta-correlated in z that, at any $z_1 \neq z_2$, the random variables $\tilde{\varepsilon}(\varrho_1,z_1)$ and $\tilde{\varepsilon}(\varrho_2,z_2)$ are *statistically independent.* But, with non-*Gaussian* fields, the fact that they are uncorrelated does not necessarily result in their being independent. It turns out that for the non-Gaussian fields $\tilde{\varepsilon}(\varrho,z)$, a condition similar to (3.4), under which closed equations can be derived for the moments of the random wave field $v(\varrho,z)$ is formalized as follows. Suppose that z_i $(i = 1,\dots,n)$ and z'_j $(j = 1,\dots,m)$ at any $i,\,j$ are subject to

$$z_i < z_0 \quad , \quad z'_j > z_0 \quad .$$

The sets of the random variables $\tilde{\varepsilon}(\varrho_i,z_i)$ and $\tilde{\varepsilon}(\varrho'_j,z'_j)$ must then be statistically independent. The analog of delta-correlation here is the fact that, given an arbitrarily small "gap" between the members of both groups of variables, we immediately have their complete statistical independence.

The joint cumulants of several random variables are known to vanish if they include at least one that is statistically independent of the others. Therefore, the joint cumulants for $\tilde{\varepsilon}(\varrho_i,z_i)$ and $\tilde{\varepsilon}(\varrho'_j,z'_j)$ must vanish. On the other hand, for non-Gaussian random variables, the higher cumulants must be nonzero. It follows that the cumulants for the random variables $\tilde{\varepsilon}(\varrho_i,z_i)$ must take the form of delta-functions in z_i:

$$\psi_n(\varrho_1, z_1; \varrho_2, z_2; \ldots; \varrho_n, z_n)$$
$$= A(\varrho_1, \ldots, \varrho_n, z_1)\delta(z_2 - z_1)\delta(z_3 - z_2)\ldots\delta(z_n - z_{n-1}) \quad . \tag{3.8}$$

We will refer to the random functions $\tilde{\varepsilon}(\varrho, z)$ meeting this condition as *delta-correlated* in z.

3.2 Equations for Statistical Moments of Wave Fields in the Markovian Approximation

Following the line of reasoning described above, we will now consider the parabolic equation approximation

$$2ik\frac{\partial v(\varrho, z)}{\partial z} + \Delta_\perp v(\varrho, z) + k^2\tilde{\varepsilon}(\varrho, z)v(\varrho, z) = 0 \quad ,$$
$$v(\varrho, 0) = v_0(\varrho) \quad , \tag{3.9}$$

treating $\tilde{\varepsilon}(\varrho, z)$ as a random function delta-correlated in z.

First, we will derive the equation for $\overline{v}(\varrho, z)$ [3.4]. To this end, we will use the identity transformation

$$2ik\frac{\partial v}{\partial z} + k^2\tilde{\varepsilon}(\varrho, z)v(\varrho, z)$$
$$\equiv 2ik\exp\left[-\frac{k^2}{2ik}\int_0^z \tilde{\varepsilon}(\varrho, \zeta)d\zeta\right]\frac{\partial}{\partial z}\left[\exp\left(\frac{k^2}{2ik}\int_0^z \tilde{\varepsilon}(\varrho, \zeta)d\zeta\right)v(\varrho, z)\right] \quad ,$$

which enables equation (3.9) to be written as

$$2ik\frac{\partial}{\partial z}\left[\exp\left(-\frac{ik}{2}\int_0^z \varepsilon(\varrho, \zeta)d\zeta\right)v(\varrho, z)\right]$$
$$= -\exp\left(-\frac{ik}{2}\int_0^z \tilde{\varepsilon}(\varrho, \zeta)d\zeta\right)\Delta_\perp v(\varrho, z) \quad . \tag{3.10}$$

We will integrate this equation from 0 to z, having denoted the integration variable on the right-hand side by z', and then multiply both sides by

$$\exp\left(\frac{ik}{2}\int_0^z \tilde{\varepsilon}(\varrho, \zeta)d\zeta\right) \quad .$$

Making use of

$$\int_0^z \tilde{\varepsilon}d\zeta - \int_0^{z'} \tilde{\varepsilon}d\zeta = \int_{z'}^z \tilde{\varepsilon}\, d\zeta \quad ,$$

we will obtain

$$
2ikv(\varrho, z) - 2ik \exp\left(\frac{ik}{2}\int_0^z \tilde{\varepsilon}(\varrho, \zeta)d\zeta\right) v_0(\varrho)
$$

$$
= -\int_0^z dz' \exp\left(\frac{ik}{2}\int_{z'}^z \tilde{\varepsilon}(\varrho, \zeta)d\zeta\right) \Delta_\perp v(\varrho, z') \quad . \tag{3.11}
$$

On the right-hand side of (3.11), the integrand contains the product of the two random variables $\Delta_\perp v(\varrho, z')$ and $\exp\left(ik\int_{z'}^z \tilde{\varepsilon}(\varrho, \zeta)d\zeta/2\right)$. Since $v(\varrho, z')$ is only dependent on the *preceding* (in z) values of $\tilde{\varepsilon}(\varrho, \zeta')$, where $0 \le \zeta' \le z'$, and the exponent only contains the *subsequent* values of $\tilde{\varepsilon}(\varrho, \zeta)$ for $\zeta \ge z'$, for the delta-correlated (in z) random functions $\tilde{\varepsilon}(\varrho, z)$ these two factors are statistically *independent*. Therefore, averaging (3.11) gives the relationship

$$
2ik\overline{v}(\varrho, z) - 2ik\left\langle \exp\left(\frac{ik}{2}\int_0^z \tilde{\varepsilon}(\varrho, \zeta)d\zeta\right)\right\rangle v_0(\varrho)
$$

$$
= -\int_0^z dz'\left\langle \exp\left(\frac{ik}{2}\int_{z'}^z \tilde{\varepsilon}(\varrho, \zeta)d\zeta\right)\right\rangle \Delta_\perp \overline{v}(\varrho, z') \quad , \tag{3.12}
$$

which is a *closed equation* in $\overline{v}(\varrho, z)$.

Equation (3.12) also contains the function

$$
P(z, z'; \varrho) = \left\langle \exp\left(\frac{ik}{2}\int_{z'}^z \tilde{\varepsilon}(\varrho, \zeta)d\zeta\right)\right\rangle \quad . \tag{3.13}
$$

It is *known* if the stochastic behavior of the random function $\tilde{\varepsilon}$ is specified.

Let us consider a special case where $\tilde{\varepsilon}(\varrho, z)$ is a *Gaussian* random field. The quantity $a = \int_{z'}^z \tilde{\varepsilon}(\varrho, \zeta)d\zeta$ is then *Gaussian* with a zero mean. Hence,

$$
P = \left\langle \exp\left(\frac{ika}{2}\right)\right\rangle = \exp\left[-\frac{1}{2}\left(\frac{k^2}{4}\overline{a^2}\right)\right] \quad . \tag{3.14}
$$

Using (3.4), we find

$$
\overline{a^2} = \int_{z'}^z d\zeta_1 \int_{z'}^z d\zeta_2 \langle \tilde{\varepsilon}(\varrho, \zeta_1)\tilde{\varepsilon}(\varrho, \zeta_2)\rangle
$$

$$
= A(0)\int_{z'}^z d\zeta_1 \int_{z'}^z d\zeta_2 \delta(\zeta_1 - \zeta_2) = (z - z')A(0)
$$

Thus, we arrive at

$$
P(z, z') = \exp\left[-\frac{k^2 A_0}{8}(z - z')\right] \quad , \tag{3.15}
$$

where $A_0 \equiv A(0)$.

Substituting (3.15) into (3.12) gives

$$2ik\overline{v}(\varrho, z) - 2ik \exp\left[-\frac{k^2 A_0}{8}z\right] v_0(\varrho)$$

$$= -\int\limits_0^z dz' \exp\left[-\frac{k^2 A_0}{8}(z - z')\right] \Delta_\perp \overline{v}(\varrho, z') \quad . \tag{3.16}$$

If we multiply (3.16) by $\exp(k^2 A_0 z/8)$ and differentiate with respect to z, we obtain

$$2ik \exp\left(\frac{k^2 A_0}{8}z\right)\frac{\partial \overline{v}}{\partial z} + 2ik\frac{k^2 A_0}{8}\exp\left(\frac{k^2 A_0}{8}z\right)\overline{v} = -\exp\left(\frac{k^2 A_0}{8}z\right)\Delta_\perp \overline{v}(\varrho, z) \quad ,$$

and finally

$$\frac{\partial \overline{v}}{\partial z} + \frac{k^2 A_0}{8}\overline{v} - \frac{i}{2k}\Delta_\perp \overline{v}(\varrho, z) = 0 \quad . \tag{3.17}$$

We will leave this equation for the time being and turn to the analogous equations for the moments of any order

$$\Gamma_{n,m} \equiv \langle v(\varrho_1', z)\ldots v(\varrho_n', z)v^*(\varrho_1'', z)\ldots v^*(\varrho_m'', z)\rangle \quad . \tag{3.18}$$

We will begin by deriving the differential equation for the *random* function

$$\gamma = v(\varrho_1', z)\ldots v(\varrho_n', z)v^*(\varrho_1'', z)\ldots v^*(\varrho_m'', z) \quad . \tag{3.19}$$

To this end, we will write equation (3.8) for $v(\varrho_1', z)$, denoting by Δ_1' the transverse Laplacian $\Delta_\perp'$ in the coordinates x_1, y_1 of the point ϱ_1'

$$2ik\frac{\partial v(\varrho_1', z)}{\partial z} + \Delta_1' v(\varrho_1', z) + k^2 \tilde{\varepsilon}(\varrho_1', z)v(\varrho_1', z) = 0 \quad .$$

Let us next multiply this equation by $v(\varrho_2', z)v(\varrho_3', z)\ldots v(\varrho_n', z)v^*(\varrho_1'', z)\ldots$ $v^*(\varrho_m'', z)$. Since this factor is independent of ϱ_1', it can be brought under the sign Δ_1' to yield

$$2ik\frac{\partial v(\varrho_1', z)}{\partial z}v(\varrho_2', z)\ldots v^*(\varrho_m'', z) + \Delta_1'\gamma + k^2\tilde{\varepsilon}(\varrho_1', z)\gamma = 0 \quad . \tag{3.20}$$

We then write equation (3.8) for $v(\varrho_2', z)$ and multiply it further by $v(\varrho_1', z)v(\varrho_3', z)\ldots$ $v^*(\varrho_m'', z)$ to get

$$2ikv(\varrho_1', z)\frac{\partial v(\varrho_2', z)}{\partial z}v(\varrho_3', z)\ldots v^*(\varrho_m'', z) + \Delta_2'\gamma + k^2\tilde{\varepsilon}(\varrho_2', z)\gamma = 0 \quad . \tag{3.21}$$

The equations for $v(\varrho_3', z)$, $\ldots$ $v(\varrho_n', z)$ are similar.

We will now take the equation that is complex conjugate to (3.9)

$$2ik\frac{\partial v^*}{\partial z} - \Delta_\perp v^* - k^2\tilde{\varepsilon}v^* = 0$$

and write it for $v^*(\varrho_1'', z)$. Multiplying by $v(\varrho_1', z)\ldots v(\varrho_n', z)v^*(\varrho_2'', z)\ldots v^*(\varrho_m'', z)$ gives

$$2ikv(\varrho'_1, z)\ldots v(\varrho'_n, z)\frac{\partial v^*(\varrho''_1, z)}{\partial z}v^*(\varrho''_2, z)\ldots v^*(\varrho''_m, z)$$
$$-\Delta''_1\gamma - k^2\tilde{\varepsilon}(\varrho''_1, z)\gamma = 0 \quad . \tag{3.22}$$

We can also derive similar equations for $v^*(\varrho''_2, z)\ldots, v^*(\varrho''_m, z)$.

If we now add up all the equations, beginning with (3.20), we will obtain the equation for the random function γ

$$2ik\frac{\partial\gamma}{\partial z} + \hat{L}\gamma + k^2\mu(z)\gamma = 0 \quad , \quad \text{where} \tag{3.23}$$

$$\hat{L} = \Delta'_1 + \ldots + \Delta'_n - \Delta''_1 - \ldots - \Delta''_m \quad ,$$

$$\mu(z) = \tilde{\varepsilon}(\varrho'_1, z) + \ldots + \tilde{\varepsilon}(\varrho'_n, z) - \tilde{\varepsilon}(\varrho''_1, z) - \ldots - \tilde{\varepsilon}(\varrho''_m, z) \quad . \tag{3.24}$$

Equation (3.23) only differs from (3.9) by the changes $v \to \gamma$, $\Delta_\perp \to \hat{L}$, $\tilde{\varepsilon} \to \mu$. Therefore, if we carry out the transformation as in going over from (3.9) to (3.10–12), we will end up with the equation for $\overline{\gamma} = \Gamma_{n,m}$, which is similar to (3.12),

$$2ik\Gamma_{n,m}(\varrho'_\alpha, \varrho''_\beta, z) - 2ik\left\langle \exp\left(\frac{ik}{2}\int_0^z \mu(\zeta)d\zeta\right)\right\rangle \Gamma_{n,m}(\varrho'_\alpha, \varrho''_\beta, 0)$$

$$= -\int_0^z dz'\left\langle \exp\left(\frac{ik}{2}\int_{z'}^z \mu(\zeta)d\zeta\right)\right\rangle \hat{L}\Gamma_{n,m}(\varrho'_\alpha, \varrho''_\beta, z') \quad . \tag{3.25}$$

This equation is *closed*, and the function

$$P_1(z, z') = \left\langle \exp\left(\frac{ik}{2}\int_{z'}^z \mu(\varrho'_\alpha, \varrho''_\beta, \zeta)d\zeta\right)\right\rangle \tag{3.26}$$

that enters into it is known, if the statistical behavior of $\tilde{\varepsilon}(\varrho, z)$ is given.

As in the case of $\overline{v}(\varrho, z)$, we will consider the *Gaussian* random field $\tilde{\varepsilon}$. Then, $P_1(z, z')$ is given by

$$P_1(z, z') = \exp\left[-k^2\int_{z'}^z d\zeta_1\int_{z'}^z d\zeta_2\langle\mu(\varrho'_\alpha, \varrho''_\beta, \zeta_1)\mu(\varrho'_\alpha, \varrho''_\beta, \zeta_2)\rangle/8\right] \quad . \tag{3.27}$$

Considering that

$$\mu(\varrho'_\alpha, \varrho''_\beta, \zeta) = \sum_{\alpha=1}^n \tilde{\varepsilon}(\varrho'_\alpha, \zeta) - \sum_{\beta=1}^m \tilde{\varepsilon}(\varrho''_\beta, \zeta) \quad ,$$

we get

$$\langle\mu(\varrho'_\alpha, \varrho''_\beta, \zeta_1)\mu(\varrho'_\alpha, \varrho''_\beta, \zeta_2)\rangle$$

$$= \sum_{\alpha=1}^n\sum_{\beta=1}^n \langle\tilde{\varepsilon}(\varrho'_\alpha, \zeta_1)\tilde{\varepsilon}(\varrho'_\beta, \zeta_2)\rangle - 2\sum_{\alpha=1}^n\sum_{\beta=1}^m \langle\tilde{\varepsilon}(\varrho'_\alpha, \zeta_1)\tilde{\varepsilon}(\varrho''_\beta, \zeta_2)\rangle$$

$$+ \sum_{\alpha=1}^m\sum_{\beta=1}^m \langle\tilde{\varepsilon}(\varrho''_\alpha, \zeta_1)\tilde{\varepsilon}(\varrho''_\beta, \zeta_2)\rangle \quad .$$

From (3.4),

$$\langle\tilde{\varepsilon}(\varrho_\alpha,\zeta_1)\tilde{\varepsilon}(\varrho_\beta,\zeta_2)\rangle = A(\varrho_\alpha - \varrho_\beta)\delta(\zeta_1 - \zeta_2) \quad,$$

so that

$$\int\limits_{z'}^{z} d\zeta_1 \int\limits_{z'}^{z} d\zeta_2 \langle\mu(\zeta_1)\mu(\zeta_2)\rangle = (z - z')Q_{n,m} \quad, \quad \text{where}$$

$$Q_{n,m} = \sum_{\alpha=1}^{n}\sum_{\beta=1}^{n} A(\varrho'_\alpha - \varrho'_\beta)$$

$$- 2\sum_{\alpha=1}^{n}\sum_{\beta=1}^{m} A(\varrho'_\alpha - \varrho''_\beta) + \sum_{\alpha=1}^{m}\sum_{\beta=1}^{m} A(\varrho''_\alpha - \varrho''_\beta) \quad. \tag{3.28}$$

Hence

$$P_1(z,z') = \exp\left(-k^2 Q_{n,m}(z-z')/8\right) \quad. \tag{3.29}$$

We now substitute (3.29) into (3.25), multiply both sides by $\exp(k^2 Q_{n,m}z/8)$ and differentiate the result with respect to z. We will have

$$2ik\frac{\partial\Gamma_{n,m}}{\partial z}\exp(k^2 Q_{n,m}z/8) + 2ik^3 Q_{n,m}\Gamma_{n,m}\exp(k^2 Q_{n,m}z/8)/4$$

$$= \exp(k^2 Q_{n,m}z/8)\hat{L}\Gamma_{n,m}(z) \quad,$$

and using the expression (3.24) for $\hat{L}$, we arrive at

$$\frac{\partial\Gamma_{n,m}}{\partial z} - \frac{i}{2k}(\Delta'_1 + \ldots + \Delta'_n - \Delta''_1 - \ldots - \Delta''_m)\Gamma_{n,m}$$

$$+ \frac{k^2}{8}Q_{n,m}\Gamma_{n,m} = 0 \quad. \tag{3.30}$$

The "initial" condition for (3.30) has the form

$$\Gamma_{n,m}(\varrho'_\alpha, \varrho'_\beta, 0) = v_0(\varrho'_1)\ldots v_0(\varrho'_n)v_0^*(\varrho''_1)\ldots v_0^*(\varrho''_m) \quad.$$

At $n = 1$, $m = 0$, it follows from (3.28) that

$$Q_{1,0} = A(0) = A_0 \quad, \tag{3.31}$$

and for $\Gamma_{1,0} \equiv \overline{v}(\varrho, z)$ we obtain, by (3.30), the earlier equation (3.17).

Additionally, we will write the equation for $\Gamma_{1,1} = \langle v(\varrho'_1, z)v^*(\varrho''_1, z)\rangle$, i.e., for the transverse coherence function. Letting $n = m = 1$ in (3.28), we get

$$Q_{1,1} = 2A(0) - 2A(\varrho'_1 - \varrho''_1) = 2\pi H(\varrho'_1 - \varrho''_2) \quad,$$

where we have made use of the function (3.7). Equation (3.30) at $n = m = 1$ will be

$$\frac{\partial\Gamma_{1,1}(\varrho', \varrho'', z)}{\partial z} - \frac{i}{2k}(\Delta' - \Delta'')\Gamma_{1,1} + \frac{\pi k^2}{4}H(\varrho' - \varrho'')\Gamma_{1,1} = 0 \quad. \tag{3.32}$$

In the next section, we will discuss the solution of this equation in some detail. The mutual coherence function $\Gamma_{1,1}$ is of importance in the statistical analysis of radiation behavior. Specifically, at $\varrho'' = \varrho'$, it goes over into the mean intensity of the wave:

$$\Gamma_{1,1}(\varrho', \varrho', z) = \langle |v(\varrho', z)|^2 \rangle = \langle I(\varrho', z) \rangle \quad .$$

Lastly, we will write the equation for the fourth-order moment

$$\Gamma_{2,2} = \langle v(\varrho_1', z)v(\varrho_2', z)v^*(\varrho_1'', z)v^*(\varrho_2'', z) \rangle \quad .$$

Setting $n = m = 2$ in (3.28) and considering that $A(\varrho)$ is an even function, we obtain

$$\begin{aligned}
Q_{2,2} &= 4A_0 + 2A(\varrho_1' - \varrho_2') + 2A(\varrho_1'' - \varrho_2'') - 2A(\varrho_1' - \varrho_1'') \\
&\quad - 2A(\varrho_1' - \varrho_2'') - 2A(\varrho_2' - \varrho_1'') - 2A(\varrho_2' - \varrho_2'') \\
&= 2\pi[H(\varrho_1' - \varrho_1'') + H(\varrho_1' - \varrho_2'') + H(\varrho_2' - \varrho_1'') \\
&\quad + H(\varrho_2' - \varrho_2'') - H(\varrho_1' - \varrho_2') - H(\varrho_1'' - \varrho_2'')] \quad .
\end{aligned}$$

In that case, (3.30) takes the form

$$\begin{aligned}
\frac{\partial \Gamma_{2,2}}{\partial z} &- \frac{\mathrm{i}}{2k}(\Delta_1' + \Delta_2' - \Delta_1'' - \Delta_2'')\Gamma_{2,2} + \frac{\pi k^2}{4}[H(\varrho_1' - \varrho_1'') + H(\varrho_1' - \varrho_2'') \\
&+ H(\varrho_2' - \varrho_1'') + H(\varrho_2' - \varrho_2'') - H(\varrho_1' - \varrho_2') - H(\varrho_1'' - \varrho_2'')]\Gamma_{2,2} = 0 \quad .
\end{aligned}$$

$$(3.33)$$

The moment $\Gamma_{2,2}$ is related to the wave *intensity fluctuations*. We shall be looking at their properties in Sect. 3.4.

The set of equations (3.30) is equivalent to one functional equation for the *characteristic functional* of the random fields $v(\varrho, z)$ and $v^*(\varrho, z)$ (see [3.5]). It is this equation that is the counterpart of the Einstein-Fokker-Planck equation for the problem under consideration.

3.3 Mean Field and Second-Order Coherence Function

1) Mean Field   The mean field $\bar{v}(\varrho, z)$ obeys equation (3.17)

$$\frac{\partial \bar{v}}{\partial z} + \frac{k^2 A_0}{8}\bar{v} - \frac{\mathrm{i}}{2k}\Delta_\perp \bar{v} = 0 \quad , \qquad \bar{v}(\varrho, 0) = v_0(\varrho) \quad . \tag{3.34}$$

Solving this equation is a straighforward exercise. Note that

$$\frac{\partial \bar{v}}{\partial z} + \frac{k^2 A_0}{8}\bar{v} = \exp(-k^2 A_0 z/8)\frac{\partial}{\partial z}[\exp(k^2 A_0 z/8)\bar{v}] \quad .$$

We will then write (3.34) in the following form:

$$\frac{\partial}{\partial z}[\exp(k^2 A_0 z/8)\bar{v}] - \frac{\mathrm{i}}{2k}\Delta_\perp[\exp(k^2 A_0 z/8)\bar{v}] = 0 \quad . \tag{3.35}$$

Consider the field $w(\varrho, z)$ in a *homogeneous* medium. If we set $A_0 = 0$, from (3.34) we will have

$$\frac{\partial w}{\partial z} - \frac{i}{2k}\Delta_\perp w = 0 \quad , \quad w(\varrho, 0) = v_0(\varrho) \quad . \tag{3.36}$$

A comparison of (3.36) with (3.35) reveals that these equations (and their initial conditions) coincide, as do the functions $\exp(k^2 A_0 z/8)\overline{v}(\varrho, z)$ and $w(\varrho, z)$ which obey these equations, i.e.,

$$\overline{v}(\varrho, z) = \exp(-k^2 A_0 z/8)w(\varrho, z) \quad . \tag{3.37}$$

This expression relates the *mean field* $\overline{v}(\varrho, z)$ to the field $w(\varrho, z)$ produced by the *same sources* in the plane $z = 0$, but only in the *homogeneous* medium. The influence of random inhomogeneities of the medium thus manifests itself in that the mean field decays exponentially.

This result can easily be understood if we compare the extinction coefficient $k^2 A_0/8$ in (3.37) with the value derived in the Born approximation. The latter is given by [see (III.4.79)]

$$\sigma_0 = \frac{\pi k^4}{2} \int\limits_0^{2\pi} d\varphi \int\limits_0^{\pi} \sin\theta\, d\theta \Phi_\varepsilon\left(2k\sin\frac{\theta}{2}\right) \quad .$$

This expression is written here explicitly for the statistically *isotropic* fluctuations of the electric permittivity. Integrating with respect to φ, and introducing the new integration variable $\kappa = 2k\sin(\theta/2)$ instead of θ, we will reduce the expression for σ_0 to

$$\sigma_0 = \pi^2 k^2 \int\limits_0^{2k} \Phi_\varepsilon(\kappa)\kappa\, d\kappa$$

However, in the interesting case of large-scale inhomogeneities, the $\Phi_\varepsilon(\kappa)$ is negligible in the region $\kappa > 2k$, whereby we may place the upper limit at infinity without essentially changing the value of the integral

$$\sigma_0 = \pi^2 k^2 \int\limits_0^\infty \Phi_\varepsilon(\kappa)\kappa\, d\kappa \tag{3.38}$$

(this expression was derived in Exercise III.4.9.1).

Compare this equation with the quantity $k^2 A_0/8$, where, by (3.38),

$$A_0 = 2\pi \int \Phi_\varepsilon(\kappa)d^2\kappa = 4\pi^2 \int\limits_0^\infty \Phi_\varepsilon(\kappa)\kappa\, d\kappa \quad .$$

We find that

$$\frac{k^2 A_0}{8} = \frac{\pi^2 k^2}{2} \int\limits_0^\infty \Phi_\varepsilon(\kappa)\kappa\, d\kappa = \frac{1}{2}\sigma_0 \quad . \tag{3.39}$$

According to (3.37, 39),

$$|\overline{v}|^2 = |w(\varrho, z)|^2 \exp(-k^2 A_0 z/4) = |w(\varrho, z)|^2 e^{-\sigma_0 z} \quad . \tag{3.40}$$

It follows that the decrease of the mean field can be attributed solely to the energy transfer from the ordered part of the field to its unordered (fluctuational) part. For example, for the plane incident wave, when $v_0 = $ const and $w(\varrho, z) = $ const, (3.40) takes the form

$$|\overline{v}(\varrho, z)|^2 = |v_0|^2 e^{-\sigma_0 z} \quad . \tag{3.41}$$

At the same time, as was found in Sect. 2.2, the mean field intensity here is constant

$$\langle |v^2| \rangle = |v_0|^2 = \text{const} \quad . \tag{3.42}$$

Combining (3.41) and (3.42), we obtain

$$\langle |\tilde{v}^2| \rangle = \langle |v^2| \rangle - |\overline{v}|^2 = |v_0|^2 (1 - e^{-\sigma_0 z}) \quad . \tag{3.43}$$

We see that the intensity of the fluctuational part of the field grows with the penetration of the wave into the inhomogeneous medium, and for distances $z \gg 1/\sigma_0$, nearly all the intensity of the wave is associated with its random part.

2) Second-Order Coherence Function   In the previous section, we obtained an equation for the second-order coherence function $\Gamma_{1,1}$ [see (3.30)]. We shall now transform it.

We will introduce in the plane $z = $ const the coordinate ϱ_+ of the center of mass of the points ϱ', ϱ'', and the relative coordinate ϱ:

$$\varrho_+ = \tfrac{1}{2}(\varrho' + \varrho'') \quad , \quad \varrho = \varrho' - \varrho'' \quad .$$

Then

$$\frac{\partial}{\partial \varrho'} = \frac{1}{2}\frac{\partial}{\partial \varrho_+} + \frac{\partial}{\partial \varrho} \quad , \quad \frac{\partial}{\partial \varrho''} = \frac{1}{2}\frac{\partial}{\partial \varrho_+} - \frac{\partial}{\partial \varrho} \quad ,$$

$$\Delta' - \Delta'' = 2\frac{\partial}{\partial \varrho}\frac{\partial}{\partial \varrho_+} \quad .$$

If we now set $\Gamma_{1,1}(\varrho', \varrho'', z) = \Gamma_{1,1}(\varrho_+ + \varrho/2, \varrho_+ - \varrho/2, z) = \Gamma(\varrho, \varrho_+, z)$, we will have, by (3.32),

$$\frac{\partial \Gamma(\varrho, \varrho_+, z)}{\partial z} - \frac{i}{k}\frac{\partial^2 \Gamma}{\partial \varrho \partial \varrho_+} + \frac{\pi k^2}{4}H(\varrho)\Gamma = 0 \quad . \tag{3.44}$$

Using the Fourier transformation in ϱ_+

$$\Gamma(\varrho, \varrho_+, z) = \int \gamma(\varrho, \mathbf{p}, z) \exp(i\mathbf{p} \cdot \varrho_+) d^2 p \quad , \tag{3.45}$$

and substituting into (3.45), gives

$$\frac{\partial \gamma}{\partial z} + \frac{\mathbf{p}}{k}\frac{\partial \gamma}{\partial \varrho} + \frac{\pi k^2}{4}H(\varrho)\gamma = 0 \tag{3.46}$$

We will now use the operator representation of the Taylor expansion[2]

$$f(\varrho + \varrho_0) = \exp\left(\varrho_0 \frac{\partial}{\partial \varrho}\right) f(\varrho) \quad , \tag{3.47}$$

and transform the sum of the first two terms in (3.46)

$$\frac{\partial \gamma}{\partial z} + \frac{p}{k} \frac{\partial \gamma}{\partial \varrho} = \exp\left(-\frac{pz}{k} \frac{\partial}{\partial \varrho}\right) \frac{\partial}{\partial z}\left[\exp\left(\frac{pz}{k} \frac{\partial}{\partial \varrho}\right) \gamma(\varrho, p, z)\right] \quad .$$

Equation (3.46) can therefore be written in the form

$$\exp\left(-\frac{pz}{k} \frac{\partial}{\partial \varrho}\right) \frac{\partial}{\partial z}\left[\exp\left(\frac{pz}{k} \frac{\partial}{\partial \varrho}\right) \gamma(\varrho, p, z)\right] = -\frac{\pi k^2}{4} H(\varrho)\gamma(\varrho, p, z) \quad ,$$

or, if we multiply on the left-hand side by $\exp(pz/k\,\partial/\partial\varrho)$, in the form

$$\frac{\partial}{\partial z}\left[\exp\left(\frac{pz}{k} \frac{\partial}{\partial \varrho}\right) \gamma(\varrho, p, z)\right] = -\frac{\pi k^2}{4}\exp\left(\frac{pz}{k} \frac{\partial}{\partial \varrho}\right)[H(\varrho)\gamma(\varrho, p, z)] \quad . \tag{3.48}$$

If we set $\varrho_0 = pz/k$ in (3.47), we can also write (3.48) as

$$\frac{\partial}{\partial z}\gamma\left(\varrho + \frac{pz}{k}, p, z\right) = -\frac{\pi k^2}{4} H\left(\varrho + \frac{pz}{k}\right)\gamma\left(\varrho + \frac{pz}{k}, p, z\right) \quad . \tag{3.49}$$

The solution of this equation is

$$\gamma\left(\varrho + \frac{pz}{k}, p, z\right) = \gamma^{(0)}(\varrho, p)\exp\left[-\frac{\pi k^2}{4}\int_0^z H\left(\varrho + \frac{p\zeta}{k}\right)d\zeta\right] \quad ,$$

where $\gamma^{(0)}(\varrho, p) \equiv \gamma(\varrho, p, 0)$ is the value of γ at $z = 0$. If we substitute $\varrho - pz/k$ for ϱ, we will have

$$\gamma(\varrho, p, z) = \gamma^{(0)}\left(\varrho - \frac{pz}{k}, p\right)\exp\left[-\frac{\pi k^2}{4}\int_0^z H\left(\varrho - \frac{p}{k}(z - \zeta)\right)d\zeta\right] \quad . \tag{3.50}$$

Substitution into (3.45) gives

$$\Gamma(\varrho, \varrho_+, z)$$

$$= \int \gamma^{(0)}\left(\varrho - \frac{pz}{k}, p\right)\exp\left[i p \cdot \varrho_+ - \frac{\pi k^2}{4}\int_0^z H\left(\varrho - \frac{p}{k}(z - \zeta)\right)d\zeta\right]d^2p \quad .$$

$$\tag{3.51}$$

Using (3.45) we express the Fourier transform $\gamma^{(0)}$ through $\Gamma^{(0)}(\varrho, \varrho_+) = \Gamma(\varrho, \varrho_+, 0)$:

$$\gamma^{(0)}(\varrho, p) = \frac{1}{4\pi^2}\int \exp(-i p \cdot \varrho'_+)\Gamma^{(0)}(\varrho, \varrho'_+)d^2\varrho'_+ \quad .$$

Substituting this into (3.51) gives

$$\Gamma(\varrho, \varrho_+, z) = \frac{1}{4\pi^2} \int d^2 p \, d^2 \varrho'_+ \, \Gamma^{(0)}\left(\varrho - \frac{pz}{k}, \varrho'_+\right)$$

$$\times \exp\left[i p \cdot (\varrho_+ - \varrho'_+) - \frac{\pi k^2}{4} \int_0^z H\left(\varrho - \frac{p}{k}(z - \zeta)\right) d\zeta\right] \quad .$$

To make the right-hand side of this expression more symmetric, in place of p we will introduce the new integration variable $\varrho' = \varrho - (pz/k)$ that obeys the relationship $d^2 p = k^2 d^2 \varrho'/z^2$. We will then arrive at

$$\Gamma(\varrho, \varrho_+, z) = \frac{k^2}{4\pi^2 z^2} \int d^2 \varrho' \, d^2 \varrho'_+ \, \Gamma^{(0)}(\varrho', \varrho'_+) \exp\left[\frac{ik}{z}(\varrho - \varrho')(\varrho_+ - \varrho'_+)\right.$$

$$\left. - \frac{\pi k^2}{4} \int_0^z H\left(\varrho \frac{\zeta}{z} + \varrho'\left(1 - \frac{\zeta}{z}\right)\right) d\zeta\right] \quad . \tag{3.52}$$

We have thus obtained the solution to (3.44) in the most general case: for arbitrary $H(\varrho)$ and arbitrary $\Gamma^{(0)}(\varrho, \varrho_+)$ at the boundary [3.6–8].

Let us now consider the special case of a *plane* wave falling on an inhomogeneous medium. Then $\Gamma^{(0)}(\varrho, \varrho_+) = \text{const} = I_0$ and, in (3.53), we can carry out the integration with respect to ϱ'_+, which gives rise to the factor $(2\pi z/k)^2 \delta(\varrho - \varrho')$. If we then integrate with respect to ϱ', we obtain

$$\Gamma(\varrho, \varrho_+, z) = I_0 \exp\left[-\frac{\pi k^2 z}{4} H(\varrho)\right] = \Gamma(\varrho, z) \quad . \tag{3.53}$$

For a plane wave propagating through a statistically homogeneous medium, Γ is clearly independent of the coordinate of the center of mass ϱ_+. Letting $\varrho = 0$ in (3.53) and considering that $H(0) = 0$, we get

$$\Gamma(0, \varrho_+, z) = \langle v(\varrho_+, z) v^*(\varrho_+, z)\rangle = \overline{I}(\varrho_+, z) = I_0 \quad .$$

The latter result was obtained for the plane wave already in Sect. 2.2.

The function $H(\varrho)$ was found to be increasing. Therefore, Γ decreases with ϱ. We can find some characteristic scale ϱ_k for which $\Gamma(\varrho_k, z)$ is already small as compared to $\Gamma(0, z)$, taking this scale to be, say, the solution to the equation

$$\frac{\pi k^2 z}{4} H(\varrho_k) = 1 \quad . \tag{3.54}$$

If the separation ϱ between the observation points is small in comparison with ϱ_k, then $\Gamma(\varrho, z)/\Gamma(0, z) \approx 1$ and the values of the fields at these points are strongly correlated. Then, if $\varrho \gg \varrho_k$, $\Gamma \approx 0$ and the values of the fields will appear to be spatially uncorrelated. The scale ϱ_k is thus the spatial correlation radius, or the *coherence radius*, for the fields. It is clear that ϱ_k decreases with z. If, for example $H(\varrho) \sim C_\varepsilon^2 \varrho^{5/3}$, as is the case for a turbulent medium, then

$$\varrho_k \sim (C_\varepsilon^2 k^2 z)^{-3/5} \quad .$$

This result was obtained in Sect. 1.6, using the GOM.

We will now turn to the general case where $\Gamma^{(0)}$ is not a constant (e.g., $\Gamma^{(0)}$ can only be nonzero in a certain region of ϱ'_+, which corresponds to the bounded wave beam). Let us look at the integral of Γ with respect to ϱ_+. Integrating (3.52) gives the factor $(2\pi z/k)^2 \delta(\varrho - \varrho')$, in the integrands with respect to ϱ' and ϱ'_+, thus enabling the integration with respect to ϱ' to be carried out. We will have

$$\int \Gamma(\varrho, \varrho_+, z) d^2 \varrho_+ = \exp\left[-\frac{\pi k^2}{4} z H(\varrho)\right] \int \Gamma^{(0)}(\varrho, \varrho'_+) d^2 \varrho'_+ \quad . \tag{3.55}$$

The second-order coherence function averaged over the beam cross-section thus behaves in the same way as the coherence function of the plane wave.

Letting $\varrho = 0$ in (3.55) and taking into account that

$$H(0) = 0 \quad , \quad \Gamma(0, \varrho_+, z) = I(\varrho_+, 0) \quad ,$$

we obtain the conservation law

$$\int \overline{I}(\varrho_+, z) d^2 \varrho_+ = \int I_0(\varrho_+) d^2 \varrho_+ \quad . \tag{3.56}$$

This suggests that the electric permittivity fluctuations only bring about a redistribution of the mean intensity in the plane $z = $ const, whereas the integral of this intensity over the plane $z = $ const is conserved.

Equation (3.52) allows us to work out the distributions of the mean intensity and degree of coherence of wave beams in random media (see Exercise 1).

3) Relation to the Radiative Transfer Equation We will now discuss the relation of the results obtained to the so-called radiative transfer equation (RTE) used to calculate the energy-related characteristics of the radiation (including thermal radiation) in scattering media, and widely employed in many problems of astrophysics and geophysics. The RTE is generally derived phenomenologically.

Consider the magnitude of the energy flux density $|d\vec{S}|$ with the solid angle do, with the vertex at a point R and the axis directed along the unit vector n

$$|d\vec{S}| = \mathcal{J}(R, n) do \quad , \tag{3.57}$$

where $\mathcal{J}$ is the *brightness* or *radiant intensity*, which we discussed in Sect. III.2.2 [see (III.2.31)]. The value of $\mathcal{J}$ at the point $R' = R + ndl$ differs from that at R due to two factors. First, the radiation over the path dl is *absorbed* and *scattered* in other directions ($\Delta \mathcal{J}^{(1)} = -\alpha \mathcal{J} dl$). The attenuation (extinction) coefficient α is the sum of the coefficients of scattering and absorption. Second, the energy flux in the direction n grows due to the scattering of the energy fluxes initially directed in other directions. We have

$$\Delta \mathcal{J}^{(2)} = \oint \mathcal{J}(R, n') f(n - n') do(n') dl \quad ,$$

where $f(n - n')$ is the scattering cross-section per unit volume for the change in the direction from n' to n. Then $\Delta \mathcal{J} = \Delta \mathcal{J}^{(1)} + \Delta \mathcal{J}^{(2)}$, whereby we obtain

$$\frac{d\mathcal{J}(R, n)}{dl} = -\alpha \mathcal{J}(R, n) + \oint \mathcal{J}(R, n') f(n - n') do(n') \quad ,$$

where $d\mathcal{J}/dl$ is the derivative in the direction n, which can be written as $d\mathcal{J}/dl = nd\mathcal{J}/dR$. As a result, we obtain the RTE

$$n\frac{d\mathcal{J}(R,n)}{dR} + \alpha\mathcal{J}(R,n) = \oint \mathcal{J}(R,n')f(n-n')do(n') \quad . \tag{3.58}$$

We will now move on to the case of large-scale inhomogeneities, where the scattering indicatrix is known to be strongly extended in the forward direction, i.e., the function $f(n-n')$ is only markedly distinct from zero at $n' \approx n$. We will take the initial beam to be narrow, so that $\mathcal{I}(R,n)$ is only markedly nonzero within the narrow cone of directions about the z-axis, and consequently

$$n_z = \sqrt{1-n_x^2-n_y^2} \approx 1 - n_\perp^2/2 \approx 1 \quad \text{for} \quad |n_\perp| \ll 1 \quad .$$

The same applies to n' ($n'_z \approx 1$, $|n'_\perp| \ll 1$), since the scattering only occurs at small angles. The integral in (3.58) is over the unit sphere, but the major contributor is the small region near the z-axis. This enables us to replace integration over the sphere by integration over the tangent plane normal to the z-axis

$$\oint \mathcal{J}(R;1,n'_\perp)f(n_\perp - n'_\perp)do(n') \approx \iint\limits_{-\infty}^{\infty} \mathcal{J}(R;1,n'_\perp)f(n_\perp - n'_\perp)d^2n'_\perp \quad .$$

This approximate relation holds because $\mathcal{I}(R;1,n'_\perp)$ is nonzero only for $|n'_\perp| \ll 1$, and hence the parts of the sphere far away from the polar axis, like the distant parts of the tangent plane, make no noticeable contribution to the integral.

Further,

$$n\frac{d\mathcal{J}}{dR} = n_z\frac{\partial\mathcal{J}}{\partial z} + n_\perp\nabla_\perp\mathcal{J} \approx \frac{\partial\mathcal{J}}{\partial z} + n_\perp\frac{\partial\mathcal{J}}{\partial\varrho_+} \quad , \quad \text{where}$$

$$R = \{\varrho_+, z\} \quad , \quad \nabla_\perp = \frac{\partial}{\partial R_\perp} = \frac{\partial}{\partial\varrho_+} \quad .$$

Therefore, in the case of large-scale inhomogeneities and narrow beams, the RTE can be written in the so-called *small-angle approximation* [3.6,7]

$$\frac{\partial\mathcal{J}}{\partial z} + n_\perp\frac{\partial\mathcal{J}}{\partial\varrho_+} + \alpha\mathcal{J} = \int\limits_{-\infty}^{\infty} \mathcal{J}(\varrho_+,z;n'_\perp)f(n_\perp - n'_\perp)d^2n'_\perp \quad . \tag{3.59}$$

We will now show that equation (3.59), derived phenomenologically, is closely related to equation (3.44) for $\Gamma(\varrho,\varrho_+,z)$.

When dealing with (3.44), we introduced the Fourier transform of $\Gamma(\varrho,\varrho_+,z)$ in ϱ_+. However, here we will make use of the Fourier transform in ϱ

$$\Gamma(\varrho,\varrho_+,z) = \int \exp(i\kappa\cdot\varrho)F(\varrho_+,\kappa,z)d^2\kappa \quad . \tag{3.60}$$

Multiplying (3.44) by $\exp(i\kappa\cdot\varrho)$ and integrating with respect to ϱ, we obtain using (3.7)

96

$$\frac{\partial F}{\partial z} + \frac{\kappa}{k}\frac{\partial F}{\partial \varrho_+} + \frac{k^2}{4}\int [A(0) - A(\varrho)]\exp\,(i\kappa \cdot \varrho)\Gamma(\varrho, \varrho_+, z)d^2\varrho = 0 \quad . \quad (3.61)$$

Let us now use (3.5) to transform the term containing $A(\varrho)$

$$\frac{k^2}{4}\int \exp\,(i\kappa \cdot \varrho)A(\varrho)\Gamma(\varrho, \varrho_+, z)d^2\varrho$$

$$= \frac{\pi k^2}{2}\int \exp\,[i(\kappa - \kappa') \cdot \varrho]\Phi_\varepsilon(\kappa', 0)\Gamma(\varrho, \varrho_+, z)d^2\varrho\, d^2\kappa'$$

$$= \frac{\pi k^2}{2}\int \Phi_\varepsilon(\kappa', 0)F(\varrho_+, \kappa - \kappa', z)d^2\kappa'$$

$$= \int \frac{\pi k^2}{2}\Phi_\varepsilon(\kappa - \kappa', 0)F(\varrho_+, \kappa', z)d^2\kappa' \quad .$$

As a result, (3.61) will take the form

$$\frac{\partial F}{\partial z} + \frac{\kappa}{k}\frac{\partial F}{\partial \varrho_+} + \frac{k^2 A_0}{4}F = \int\limits_{-\infty}^{+\infty}\frac{\pi k^2}{2}\Phi_\varepsilon(\kappa - \kappa', 0)F(\varrho_+, \kappa', z)d^2\kappa' \quad . \quad (3.62)$$

A comparison of (3.62) with (3.59) shows that if we put $\kappa/k = n_\perp$, $F(\varrho_+, \kappa, z) = \mathcal{J}(\varrho_+, n_\perp, z)$ and $d^2\kappa' = k^2 d^2 n'_\perp$, then (3.62) will coincide with equation (3.59), where α and f assume the values

$$\alpha = \frac{k^2 A_0}{4} \quad , \quad f(n_\perp - n'_\perp) = \frac{\pi k^4}{2}\Phi_\varepsilon[k(n_\perp - n'_\perp), 0] \quad . \quad (3.63)$$

It can easily be shown that the integral of f taken over all the directions of scattering is α. This implies that there is no true absorption and that the above extinction coefficient coincides with the scattering coefficient, i.e., with the complete effective cross-section of scattering from a unit volume.

Turning to (3.39), we see that $\alpha = \sigma_0$, i.e., the effective scattering cross-section per unit volume, α, in all directions is precisely the quantity σ_0 that entered into the expression for the extinction of the coherent part of the field. We recognize in this quantity, as for $f(n_\perp - n'_\perp)$, the effective cross-section of scattering into a unit solid angle from a unit volume [see (III.4.75) and (III.4.77)].

Thus, the equation for Γ derived from the *parabolic* equation for the field described in the *Markovian* approximation appears to be equivalent to the small-angle approximation of the RTE. This relation was first obtained in [3.9]. According to (3.60), Γ is related to $\mathcal{J}$ by

$$\Gamma(\varrho, \varrho_+, z) = k^2 \int \exp\,(ikn_\perp \cdot \varrho)\mathcal{J}(\varrho_+, n_\perp, z)d^2 n_\perp \quad . \quad (3.64)$$

In Chap. 4, we will derive the complete transfer equation (3.58) from that for the coherence function of the field in a more general case, starting from the Helmholtz rather than from the parabolic equation.

Note that, by virtue of (3.64), the solution (3.52) to the equation for Γ enables the solution of the RTE in the small-angle approximation to be simultaneously found in the analytic form. It was with precisely this problem in mind that it was initially obtained in [3.6, 7].

3.4 Fourth-Order Coherence Function and Intensity Fluctuations

Let us again write equation (3.33) fo the fourth-order moment

$$\Gamma_{2,2}(\varrho_1', \varrho_2'; \varrho_1'', \varrho_2'', z) = \langle v(\varrho_1', z)v(\varrho_2', z)v^*(\varrho_1'', z)v^*(\varrho_2'', z)\rangle \quad .$$

It has the form

$$\frac{\partial \Gamma_{2,2}}{\partial z} - \frac{i}{2k}(\Delta_1' + \Delta_2' - \Delta_1'' - \Delta_2'')\Gamma_{2,2}$$
$$+ \frac{\pi k^2}{4}[H(\varrho_1' - \varrho_1'') + H(\varrho_1' - \varrho_2'') + H(\varrho_2' - \varrho_1'')$$
$$+ H(\varrho_2' - \varrho_2'') - H(\varrho_1' - \varrho_2') - H(\varrho_1'' - \varrho_2'')]\Gamma_{2,2} = 0 \quad . \tag{3.65}$$

As we have already seen, $\Gamma_{2,2}$ is related to wave intensity fluctuations. In fact, if we bring the point ϱ_1'' into coincidence with ϱ_1', and the point ϱ_2'' with ϱ_2', then

$$\Gamma_{2,2}(\varrho_1', \varrho_2'; \varrho_1', \varrho_2', z) = \langle v(\varrho_1', z)v(\varrho_2', z)v^*(\varrho_1', z)v^*(\varrho_2', z)\rangle$$
$$= \langle I(\varrho_1', z)I(\varrho_2', z)\rangle$$

The covariance of the intensity fluctuations can therefore be expressed in terms of $\Gamma_{2,2}$ and $\Gamma_{1,1}$

$$\psi_I(\varrho_1, \varrho_2) = \langle I(\varrho_1)I(\varrho_2)\rangle - \overline{I}(\varrho_1)\overline{I}(\varrho_2)$$
$$= \Gamma_{2,2}(\varrho_1, \varrho_2; \varrho_1, \varrho_2, z) - \Gamma_{1,1}(\varrho_1, \varrho_1, z)\Gamma_{1,1}(\varrho_2, \varrho_2, z) \quad .$$

If we are only interested in intensity fluctuations, it might appear possible to dispense with the fourth moment $\Gamma_{2,2}$ and confine our discussion to the special case of the two-point moment, which is derived from $\Gamma_{2,2}$ by pairwise merging of the points ϱ_1' with ϱ_1'' and ϱ_2' with ϱ_2''. It is, however, impossible to effect such merging in (3.65), since, apart from $\Gamma_{2,2}(\varrho_1', \varrho_2'; \varrho_1', \varrho_2', z)$, it will contain new unknown functions $[\Delta_{1,2}'\Gamma_{2,2}(\varrho_1', \varrho_2'; \varrho_1', \varrho_2', z)]_{\varrho_i''=\varrho_i'}$. Therefore, even when dealing with intensity fluctuations, it is necessary to consider the complete equation (3.65).

We introduce the new variables

$$\varrho_+ = (\varrho_1' + \varrho_2' + \varrho_1'' + \varrho_2'')/4 \quad ,$$

$$\varrho_1 = (\varrho_1' + \varrho_1'')/2 - (\varrho_2' + \varrho_2'')/2 \quad ,$$

$$\varrho_2 = (\varrho_1' + \varrho_2'')/2 - (\varrho_2' + \varrho_1'')/2 \quad ,$$

$$\varrho = \varrho_1' - \varrho_1'' + \varrho_2' - \varrho_2'' \quad ,$$

$$\varrho_1' = \varrho_+ + (\varrho_1 + \varrho_2)/2 + \varrho/4 \quad ,$$

$$\varrho_2' = \varrho_+ - (\varrho_1 + \varrho_2)/2 + \varrho/4 \quad ,$$

$$\varrho_1'' = \varrho_+ + (\varrho_1 - \varrho_2)/2 - \varrho/4 \quad ,$$

$$\varrho_2'' = \varrho_+ - (\varrho_1 - \varrho_2)/2 - \varrho/4 \quad . \tag{3.66}$$

We easily find that here,

$$\Delta_1' + \Delta_2' - \Delta_1'' - \Delta_2'' = 2(\nabla_{\varrho_+}\nabla_\varrho + \nabla_{\varrho_1}\nabla_\varrho) \quad .$$

The moment $\Gamma_{2,2}$, expressed in terms of $\varrho_+, \varrho, \varrho_1, \varrho_2, z$, will be denoted by $\Gamma_4(\varrho_+, \varrho, \varrho_1, \varrho_2, z)$. Equation (3.65) will now be

$$\frac{\partial \Gamma_4}{\partial z} = \frac{i}{k}(\nabla_{\varrho_+}\nabla_\varrho + \nabla_{\varrho_1}\nabla_{\varrho_2})\Gamma_4 - \frac{\pi k^2}{4}F(\varrho_1, \varrho_2, \varrho)\Gamma_4 \quad . \tag{3.67}$$

Taking into account the fact that $H(\varrho)$ is even, we write F as

$$\begin{aligned}
F(\varrho_1, \varrho_2, \varrho) = \; & H(\varrho_1 + \varrho/2) + H(\varrho_1 - \varrho/2) \\
& + H(\varrho_2 + \varrho/2) + H(\varrho_2 - \varrho/2) \\
& - H(\varrho_1 + \varrho_2) - H(\varrho_1 - \varrho_2) \quad .
\end{aligned} \tag{3.68}$$

First, we will consider the simplest case where an inhomogeneous medium is illuminated by a *plane* wave. Here, we can put $\Gamma_4(\varrho_+, \varrho, \varrho_1, \varrho_2, 0) = 1$. Since we are dealing with statistically homogeneous fluctuations $\tilde{\varepsilon}$, it is clear that Γ_4 cannot depend on ϱ_+, since a simultaneous translation of all the four observation points by the same amount results in a physically identical situation. Thus, $\nabla_{\varrho_+}\Gamma_4 = 0$, and (3.68) is simplified to yield

$$\frac{\partial \Gamma_4}{\partial z} = \frac{i}{k}\nabla_{\varrho_1}\nabla_{\varrho_2}\Gamma_4 - \frac{\pi k^2}{4}F(\varrho_1, \varrho_2, \varrho)\Gamma_4 \quad ,$$

$$\Gamma_4(\varrho_1, \varrho_2, \varrho, 0) = 1 \quad . \tag{3.69}$$

Furthermore, this expression contains no derivatives with respect to ϱ, so that these variables can be viewed as a parameter whose value can be chosen arbitrarily. If we put $\varrho = 0$, then, by (3.68, 69),

$$\begin{aligned}
F(\varrho_1, \varrho_2, 0) \; & = F(\varrho_1, \varrho_2) \\
& = 2H(\varrho_1) + 2H(\varrho_2) - H(\varrho_1 + \varrho_2) - H(\varrho_1 - \varrho_2) \quad ,
\end{aligned} \tag{3.70}$$

$$\frac{\partial \Gamma_4}{\partial z} = \frac{i}{k}\nabla_{\varrho_1}\nabla_{\varrho_2}\Gamma_4 - \frac{\pi k^2}{4}F(\varrho_1, \varrho_2)\Gamma_4(\varrho_1, \varrho_2, z) \quad ,$$

$$\Gamma_4(\varrho_1, \varrho_2, 0) = 1 \quad . \tag{3.71}$$

Since $\varrho = \varrho_1' + \varrho_2' - \varrho_1'' - \varrho_2''$, it follows from $\varrho = 0$ that $\varrho_1' - \varrho_1'' = \varrho_2'' - \varrho_2'$, i.e., the points $\varrho_1', \varrho_2', \varrho_1'', \varrho_2''$ lie at the corners of a parallelogram whose sides are ϱ_1, ϱ_2 (Fig. 3.1). The points ϱ_1' and ϱ_2' merge with ϱ_1'' and ϱ_2'', respectively, when ϱ_2 in Γ_4 becomes zero.

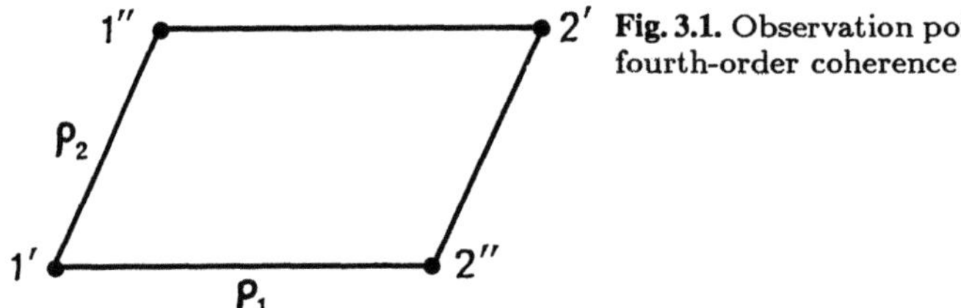

Fig. 3.1. Observation points for the calculation of the fourth-order coherence function

It is clear that the function

$$\Gamma_4(\varrho_1, \varrho_2, z) = \langle v(\varrho_+ + (\varrho_1 + \varrho_2)/2, z)v(\varrho_+ - (\varrho_1 + \varrho_2)/2, z)$$
$$\times v^*(\varrho_+ + (\varrho_1 - \varrho_2)/2, z)v^*(\varrho_+ - (\varrho_1 - \varrho_2)/2, z)\rangle$$

possesses the following symmetry properties:

$$\Gamma_4(\varrho_1, \varrho_2, z) = \Gamma_4(\varrho_2, \varrho_1, z) \quad ,$$

$$\Gamma_4(\varrho_1, -\varrho_2, z) = \Gamma_4^*(\varrho_1, \varrho_2, z) \quad . \tag{3.72}$$

Let us now consider the limiting case of $|\varrho_1| \to \infty$, $\varrho_2 = \text{const}$. This implies that both pairs of points $(\varrho_1', \varrho_1'')$ and $(\varrho_2', \varrho_2'')$ become infinitely separated, with the same separations within each pair. In that case, the correlations between the fields at the points $(\varrho_1', \varrho_1'')$ and $(\varrho_2', \varrho_2'')$ must clearly vanish, i.e.,

$$\lim_{\varrho_1 \to \infty} \Gamma_4(\varrho_1, \varrho_2, z) = \langle v(\varrho_1', z)v^*(\varrho_1'', z)\rangle \langle v(\varrho_2', z)v^*(\varrho_2'', z)\rangle$$

$$= \Gamma(\varrho_2, (\varrho_1' + \varrho_1'')/2, z)\Gamma(-\varrho_2, (\varrho_2' + \varrho_2'')/2, z). \tag{3.73}$$

By virtue of the symmetry under the permutation $\varrho_1 \rightleftarrows \varrho_2$, the function Γ_4 behaves in the same way as $|\varrho_2| \to \infty$. In the special case of the plane wave, Γ is independent of $(\varrho_1' + \varrho_1'')/2$ and $(\varrho_2' + \varrho_2'')/2$, and (3.73) takes the form

$$\lim_{\varrho_1 \to \infty} \Gamma_4(\varrho_1, \varrho_2, z) = [\Gamma(\varrho_2, z)]^2 \quad ,$$

$$\lim_{\varrho_2 \to \infty} \Gamma_4(\varrho_1, \varrho_2, z) = [\Gamma(\varrho_1, z)]^2 \quad . \tag{3.74}$$

Let us now look at those consequences of the energy conservation law which give rise to certain constraints on the form of Γ_4. As we saw earlier [see (2.25)], an exact corollary of the parabolic equation is the energy conservation law

$$\frac{\partial I(\varrho, z)}{\partial z} + \nabla_\perp \left\{ \frac{1}{2ik} [v^*(\varrho, z)\nabla_\perp v(\varrho, z) - v(\varrho, z)\nabla_\perp v^*(\varrho, z)] \right\} = 0 \quad , \tag{3.75}$$

where $I(\varrho, z) = |v(\varrho, z)|^2$. Consider the propagation of the bounded wave beam, for which $|v(\varrho, z)| \to 0$ as $|\varrho| \to \infty$. Integrating (3.75) and using the Gauss theorem for the two-dimensional case, we obtain

$$\frac{d}{dz} \int I(\varrho, z) d^2\varrho + \oint \left. S_n(\varrho, z) \right|_{\varrho=\infty} dl = 0 \quad ,$$

where $S_n|_{\varrho=\infty}$ is the component of the energy flux density along $n = \varrho/\varrho$ on the circumference of infinite radius. However, because the beam is limited, $v \to 0$ as $\varrho \to \infty$, so that $S_n|_{\varrho=\infty} = 0$, and we arrive at the conservation law

$$\int_{-\infty}^{\infty} I(\varrho, z) d^2\varrho = \text{const} = \int I(\varrho, 0) d^2\varrho \quad . \tag{3.76}$$

The left-hand side of this relationship contains the *random* intensity $I(\varrho, z) = \overline{I}(\varrho, z) + \tilde{I}(\varrho, z)$. Averaging (3.76), we get the equation

$$\int \overline{I}(\varrho, z) d^2\varrho = \int I(\varrho, 0) d^2\varrho \quad , \tag{3.77}$$

which was obtained above from the equation for Γ [see (3.56)]. Subtracting (3.77) from (3.76) gives

$$\int \tilde{I}(\varrho, z) d^2\varrho = 0 \quad . \tag{3.78}$$

The physical implication of this relationship is fairly straightforward since any random deviations of intensity from its mean have unlike signs and, therefore, always cancel out, the fluctuations only cause a redistribution of the intensity over the transverse cross-section of the beam. It will be noted that the relationship (3.78) was used in Chap. 1 [see (1.104)] to account for the fact that the integral of the level covariance was zero.

Squaring (3.78) and averaging the result gives

$$\int \psi_I(\varrho', \varrho'', z) d^2\varrho' \, d^2\varrho'' = 0 \quad .$$

If we then change the variables $\varrho = \varrho' - \varrho''$ and $\varrho_+ = (\varrho' + \varrho'')/2$, we obtain

$$\int \psi_I(\varrho_+, \varrho, z) d^2\varrho_+ \, d^2\varrho = 0 \quad . \tag{3.79}$$

It follows that the covariance of the intensity fluctuations will have a negative region at all means.

For the plane wave, we have an equality similar to (3.79)

$$\int \psi_I(\varrho, z) d^2\varrho = 0 \quad . \tag{3.80}$$

It is, nevertheless, more difficult to derive, since the field $v(\varrho, z)$ does not decrease as $\varrho \to \infty$. We can, however, derive (3.80) directly from (3.71). With this end in view, we will introduce the function

$$K(\varrho_1, \varrho_2, z) = \Gamma_4(\varrho_1, \varrho_2, z) - \Gamma^2(\varrho_1, z) \quad . \tag{3.81}$$

If $\varrho_1 = 0$, then $\Gamma_4(0, \varrho_2, z) = \langle I(\varrho_1', z) I(\varrho_2', z) \rangle$, $\Gamma^2(0, z) = \overline{I}^2$, and $K(0, \varrho_2, z) = \psi_I(\varrho_2, z)$ is the intensity covariance. Moreover, according to (3.74), K vanishes as $\varrho_2 \to \infty$. Using (3.44) [for the plane wave in this equation $\partial^2 \Gamma / \partial\varrho \partial\varrho_+ = 0$], we can easily see that K obeys the equation

$$\frac{\partial K}{\partial z} = \frac{i}{k} \nabla_{\varrho_1} \nabla_{\varrho_2} K - \frac{\pi k^2}{4} F(\varrho_1, \varrho_2) K$$

$$- \frac{\pi k^2}{4} \Gamma^2(\varrho_1, z)[F(\varrho_1, \varrho_2) - 2H(\varrho_1)] \quad . \tag{3.82}$$

If we integrate this equation with respect to ϱ_2 within infinite integration limits, then, by virtue of the above limiting relationship $\lim\limits_{\varrho_2 \to \infty} K(\varrho_1, \varrho_2, z) = 0$, the integral of the first term on the right-hand side of (3.82) vanishes, i.e., we get

$$\frac{\partial}{\partial z} \int K(\varrho_1, \varrho_2, z) d^2 \varrho_2 = -\frac{\pi k^2}{4} \int F(\varrho_1, \varrho_2) K(\varrho_1, \varrho_2, z) d^2 \varrho_2$$

$$- \frac{\pi k^2}{4} \Gamma^2(\varrho_1, z) \int [F(\varrho_1, \varrho_2) - 2H(\varrho_1)] d^2 \varrho_2 \quad . \tag{3.83}$$

Since $F(\varrho_1, \varrho_2) - 2H(\varrho_1) \to 0$ as $\varrho_2 \to \infty$, the integral of the second term on the right-hand of (3.82) converges. We now put $\varrho_1 = 0$ in (3.83). From the relationships $F(0, \varrho_2) = 0$ and $H(0) = 0$, we obtain

$$\frac{d}{dz} \int K(0, \varrho_2, z) d^2 \varrho_2 = \frac{d}{dz} \int \psi_I(\varrho_2, z) d^2 \varrho_2 = 0.$$

The solution to this equation, with the initial condition

$$\int \psi_I(\varrho_2, 0) d^2 \varrho_2 = 0$$

(at $z = 0$ there are no intensity fluctuations), is given by (3.80).

Note that we can derive expressions for the moments of an arbitrary order similar to (3.79) from equation (3.78), which holds for bounded beams.

There is no way of obtaining a solution (3.71) for Γ_4. Numerical solutions have already been obtained for two models of electric permittivity fluctuations in a medium: with spectral densities of the form $\Phi_\varepsilon(\kappa) \sim \exp(-\kappa^2 l^2/2)$ [3.10] and $\Phi_\varepsilon(\kappa) \sim \kappa^{-11/3}$ [3.11–13]. In both cases, the mean square of the intensity fluctuations $\beta^2 = (\overline{I^2} - \overline{I}^2)/\overline{I}^2$ saturates with z. The numerical results obtained for the exponential spectrum of ε show reasonable agreement with experiment.

For the region of strong fluctuations ($\beta^2 \approx 1$), asymptotical solutions to (3.71) were derived [3.14–17]. Not wishing to subject the reader to the tedium of the computation, we will only provide the basic results cited from the above work. For a more detailed discussion the reader is referred to [3.24].

The formal solution to (3.71) can be written as the limit when $N \to \infty$ of the $4N$-fold integral that is obtained if we replace the layer $(0, z)$ by N phase screens, a spacing between them being z/N. In the region of strong fluctuations, this infinite-fold (so-called *continual*) integral [3.17–19] will contain the large parameter $\sqrt{\lambda z}/\varrho_k$, which is equal to the ratio of the first Fresnel zone to the coherence radius ϱ_k defined by (3.54). This parameter has a clear physical implication. In the case where $\varrho_k \ll \sqrt{\lambda z}$, only those fields that are scattered by inhomogeneities separated by less than ϱ_k add up in a coherent way. Therefore, for $\varrho_k \ll \sqrt{\lambda z}$, the coherence radius plays the same role as the radius of the first Fresnel zone for $\sqrt{\lambda z} \ll \varrho_k$.

For $\sqrt{\lambda z}/\varrho_k \gg 1$, the asymptotic expression for $\beta^2(z)$ has the form

$$\beta^2(z) \approx 1 + \pi \int_0^z (z - z')^2 dz' \int d^2\kappa \, \kappa^4 \Phi_\varepsilon(\kappa, 0)$$

$$\times \exp\left[-\frac{\pi k^2 z'}{2} H\left(\frac{\kappa(z - z')}{k} \right) \right.$$

$$\left. -\frac{\pi k^2 (z - z')}{2} \int_0^1 H\left(\frac{\kappa(z - z')}{k} \xi \right) d\xi \right] \quad . \tag{3.84}$$

The covariance $\psi_I(\varrho, z)$ has two characteristic scales: ϱ_k and $r_0 = z/k\varrho_k$. The first scale decreases with increasing z, while the second increases. In a first approximation,

$$\psi_I(\varrho, z) = \exp\left[-\frac{\pi k^2 z}{2} H(\varrho) \right] + \dots \quad . \tag{3.85}$$

For the turbulent medium, where $\Phi_\varepsilon(\kappa) = AC_\varepsilon^2 \kappa^{-11/3}$, equation (3.84) takes the form

$$\beta^2(z) = 1 + 0.861(\beta_0^2)^{-2/5} + \dots \quad , \tag{3.86}$$

where $\beta_0^2 = 4\overline{\chi^2} = 0.307 C_\varepsilon^2 k^{7/6} z^{11/6}$ is the mean square of the relative intensity fluctuations derived using the MSP [see (2.91,92) and (2.118)]. The covariance (3.85) is then

$$\psi_I(\varrho, z) = \exp\left[-0.729 C_\varepsilon^2 k^2 z \varrho^{5/3} \right] + \dots \quad . \tag{3.87}$$

and the scales ϱ_k and r_0 will be

$$\varrho_k \sim (C_\varepsilon^2 k^2 z)^{-3/5} \quad , \quad r_0 \sim (C_\varepsilon^2 k^{1/3} z^{8/3})^{3/5} \quad .$$

It should be noted that (3.85) does not satisfy the condition (3.80). This is because the negative region of ψ_I for $\beta_0^2 \gg 1$ lies where $\varrho \gg r_0$. Since, for the statistically isotropic fluctuations, (3.80) is

$$\int_0^\infty \psi_I(\varrho, z) \varrho \, d\varrho = 0 \quad ,$$

it is clear that $\psi_I(\varrho, z)$ enters into the integral at large ϱ, with the large weight ϱ. For (3.80) to hold, it is therefore sufficient to have negative ψ_I with very small absolute values. However, these values of ψ_I are not the first terms of the asymptotic expansion, so that for (3.80) to be satisfied, it is necessary to include higher-order terms in (3.85).

To conclude this section, we compare the numerical results [3.13] with the above asymptotic expressions for the exponential spectrum $\Phi_\varepsilon(\kappa) = AC_\varepsilon^2 k^{-11/3}$. Figure 3.2 gives this calculated curve $\beta = f(\beta_0)$ (curve 1). It also provides a plot of (3.86) (curve 2), as well as the averaged experimental data (curve 3).

Figure 3.3 shows the points obtained from numerical calculations of $\psi_I(\varrho, z)$, together with the curves plotted using the asymptotic relationship (3.87). It is seen from the figure that the agreement improves as the parameter β_0 increases.

Additional recent data on the Markovian approximation can be found in [3.25–29].

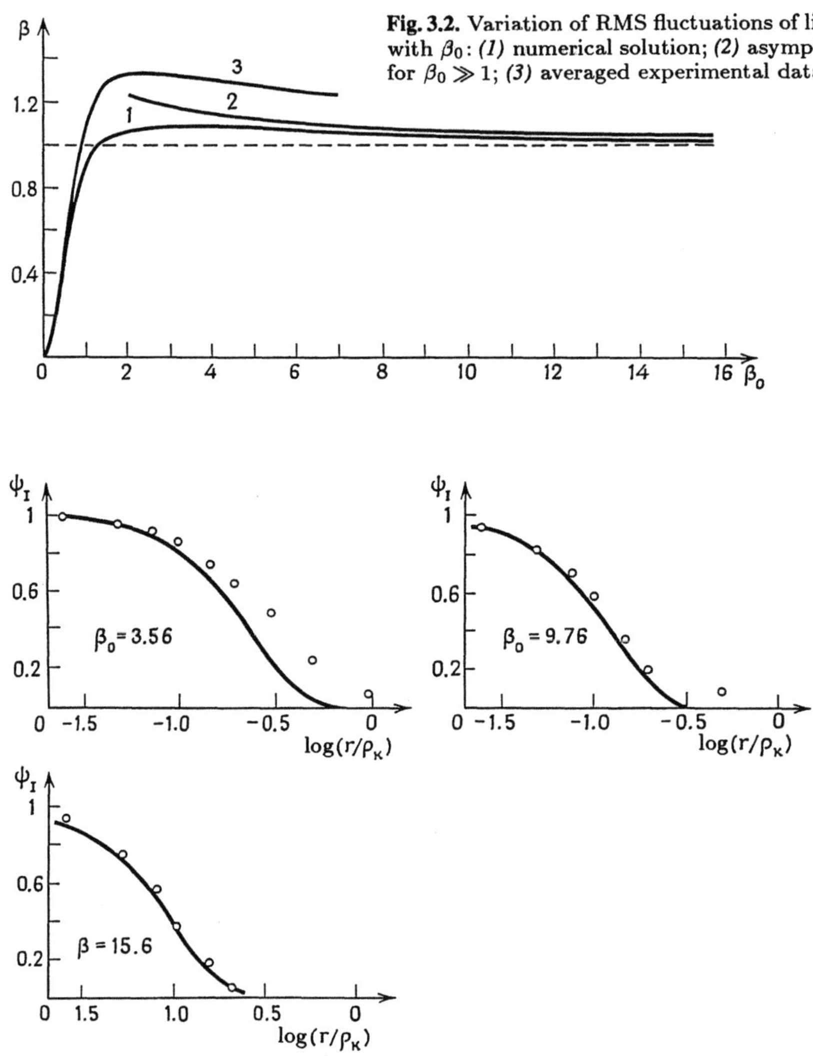

Fig. 3.2. Variation of RMS fluctuations of light intensity with β_0: *(1)* numerical solution; *(2)* asymptotic formula for $\beta_0 \gg 1$; *(3)* averaged experimental data

Fig. 3.3. Variation of ψ_I in a turbulent atmosphere at various values of β_0: open points – numerical calculation; solid curves – plots of (3.87). As β_0 increases, the asymptotic formula becomes more exact

3.5 Finite Longitudinal Correlation Radius of Fluctuations of ε. Region of Validity of the Markovian Approximation

Our assumption so far, based on the qualitative treatment given in Sect. 3.1, has been that the random function $\tilde{\varepsilon}(\varrho, z)$ is delta-correlated in z. However, we have not yet obtained any quantitative results that enable us to derive corrections to the results of the Markovian approximation on account of the fact that the longitudinal correlation radius $\tilde{\varepsilon}$ is finite. Let us now consider a more general derivation of the equations for the moments of the wave field, which will allow us to take into account the finite longitudinal correlation radius $\tilde{\varepsilon}$. From the very beginning, we will confine our discussion to the case where $\tilde{\varepsilon}$ is the *normal* random field.

We will proceed from the parabolic equation

$$\frac{\partial v(\varrho, z)}{\partial z} = \frac{i}{2k}\Delta_\perp v(\varrho, z) + \frac{ik}{2}\tilde{\varepsilon}(\varrho, z)v(\varrho, z) \quad ,$$

$$v(\varrho, 0) = v_0(\varrho) \quad , \tag{3.88}$$

and derive the system of equations for the mean field $\overline{v}$. Averaging (3.88) gives

$$\frac{\partial \overline{v}}{\partial z} = \frac{i}{2k}\Delta_\perp \overline{v} + \frac{ik}{2}\langle \tilde{\varepsilon}(\varrho, z)v(\varrho, z)\rangle \quad ,$$

$$\overline{v}(\varrho, 0) = v_0(\varrho) \quad . \tag{3.89}$$

The equation is not closed since, apart from the desired function $\overline{v}$, it contains a new unknown, $\langle \tilde{\varepsilon} v\rangle$. Since $\tilde{\varepsilon}$ is a Gaussian random function and v the functional, we may derive $\langle \tilde{\varepsilon} v\rangle$ using the Furutsu-Novikov equation (III.1.138), which in this case will be

$$\langle \tilde{\varepsilon}(\varrho, z)v(\varrho, z)\rangle$$

$$= \int_0^z dz' \int d^2\varrho' \langle \tilde{\varepsilon}(\varrho, z)\tilde{\varepsilon}(\varrho', z')\rangle \left\langle \frac{\delta v(\varrho, z)}{\delta \tilde{\varepsilon}(\varrho', z')}\right\rangle \quad . \tag{3.90}$$

The upper integration limit here is z since, according to the causality principle discussed in Sect. 3.1, $\delta v(\varrho, z)/\delta \tilde{\varepsilon}(\varrho', z') = 0$ at $z' > z$. Denoting

$$\left\langle \frac{\delta v(\varrho, z)}{\delta \tilde{\varepsilon}(\varrho', z')}\right\rangle = S_1(\varrho, z; \varrho', z')$$

and substituting (3.90), we obtain equation (3.89) in the form

$$\frac{\partial \overline{v}}{\partial z} = \frac{i}{2k}\Delta_\perp \overline{v}$$

$$+ \frac{ik}{2}\int_0^z dz' \int d^2\varrho' \langle \tilde{\varepsilon}(\varrho, z)\tilde{\varepsilon}(\varrho', z')\rangle S_1(\varrho, z; \varrho', z') \quad . \tag{3.91}$$

This equation, like (3.89), is not closed.

We now derive the equation for S_1. Operating with $\delta/\delta \tilde{\varepsilon}(\varrho', z')$ on (3.88), where $z' < z$, yields

$$\frac{\partial}{\partial z}\frac{\delta v(\varrho, z)}{\delta \tilde{\varepsilon}(\varrho', z')} = \frac{i}{2k}\Delta_\varrho \frac{\delta v(\varrho, z)}{\delta \tilde{\varepsilon}(\varrho', z')} + \frac{ik}{2}\tilde{\varepsilon}(\varrho, z)\frac{\delta v(\varrho, z)}{\delta \tilde{\varepsilon}(\varrho', z')} \quad \text{for} \quad z' < z \quad . \tag{3.92}$$

We have discarded the term $ik\delta(\varrho - \varrho')\delta(z - z')v(\varrho, z)/2$, which emerges when we differentiate $\tilde{\varepsilon}(\varrho, z)$, at $\delta(z - z') = 0$ for $z > z'$. Averaging (3.92) gives

$$\frac{\partial}{\partial z}S_1(\varrho, z; \varrho', z')$$

$$= \frac{i}{2k}\Delta_\varrho S_1(\varrho, z; \varrho', z') + \frac{ik}{2}\left\langle \tilde{\varepsilon}(\varrho, z)\frac{\delta v(\varrho, z)}{\delta \tilde{\varepsilon}(\varrho', z')}\right\rangle \quad \text{for} \quad z > z' \quad . \tag{3.93}$$

But, $\delta v/\delta\tilde{\varepsilon}$ is a functional of $\tilde{\varepsilon}$, whereby we can also make use of the Furutsu-Novikov formula to compute this quantity:

$$\left\langle \tilde{\varepsilon}(\varrho, z)\frac{\delta v(\varrho, z)}{\delta\tilde{\varepsilon}(\varrho', z')} \right\rangle$$

$$= \int\limits_0^z dz'' \int d^2\varrho'' \langle \tilde{\varepsilon}(\varrho, z)\tilde{\varepsilon}(\varrho'', z'')\rangle \left\langle \frac{\delta^2 v(\varrho, z)}{\delta\tilde{\varepsilon}(\varrho', z')\delta\tilde{\varepsilon}(\varrho'', z'')} \right\rangle \quad . \tag{3.94}$$

Let

$$\left\langle \frac{\delta^2 v(\varrho, z)}{\delta\tilde{\varepsilon}(\varrho', z')\delta\tilde{\varepsilon}(\varrho'', z'')} \right\rangle = S_2(\varrho, z; \varrho', z'; \varrho'', z'') \quad .$$

Note that S_2 does not change if the arguments are interchanged $(\varrho', z') \rightleftarrows (\varrho'', z'')$. Substituting (3.94) into (3.93) gives

$$\frac{\partial S_1(\varrho, z; \varrho', z')}{\partial z} = \frac{\mathrm{i}}{2k}\Delta_\varrho S_1$$

$$+ \frac{\mathrm{i}k}{2} \int\limits_0^z dz'' \int d^2\varrho'' \langle \tilde{\varepsilon}(\varrho, z)\tilde{\varepsilon}(\varrho'', z'')\rangle S_2(\varrho, z; \varrho', z'; \varrho'', z'')$$

$$\text{for} \quad z > z' \quad . \tag{3.95}$$

Clearly, this procedure can be carried through to arrive at an infinite chain of equations for $\overline{v}$ and S_n.

Let us consider the boundary conditions for (3.95), and similar equations for S_n. Since the equation itself is valid for $z > z'$, the boundary condition should be imposed at $z = z'$. In other words, we will have to find $S_1(\varrho, z'; \varrho', z')$. We now integrate the equation (3.88) with respect to z, and then operate on the result with $\delta/\delta\tilde{\varepsilon}(\varrho', z')$

$$v(\varrho, z) = v_0(\varrho) + \int\limits_0^z \left[\frac{\mathrm{i}k}{2}\tilde{\varepsilon}(\varrho, \zeta) + \frac{\mathrm{i}}{2k}\Delta_\varrho \right] v(\varrho, \zeta)d\zeta \quad ,$$

$$\frac{\delta v(\varrho, z)}{\delta\tilde{\varepsilon}(\varrho', z')} = \int\limits_0^z \left[\frac{\mathrm{i}k}{2}\tilde{\varepsilon}(\varrho, \zeta) + \frac{\mathrm{i}}{2k}\Delta_\varrho \right] \frac{\delta v(\varrho, \zeta)}{\delta\tilde{\varepsilon}(\varrho', z')}d\zeta$$

$$+ \int\limits_0^z \frac{\mathrm{i}k}{2}\delta(\varrho - \varrho')\delta(\zeta - z')v(\varrho, \zeta)d\zeta \quad .$$

Since $\delta v(\varrho, \zeta)/\delta\tilde{\varepsilon}(\varrho', z') = 0$ when $\zeta < z'$, the lower integration limit in the first integral can be replaced by z'. The second integral can be taken to yield

$$\frac{\delta v(\varrho, z)}{\delta\tilde{\varepsilon}(\varrho', z')} = \int\limits_{z'}^z \left[\frac{\mathrm{i}k}{2}\tilde{\varepsilon}(\varrho, \zeta) + \frac{\mathrm{i}}{2k}\Delta_\varrho \right] \frac{\delta v(\varrho, \zeta)}{\delta\tilde{\varepsilon}(\varrho', z')}d\zeta + \frac{\mathrm{i}k}{2}\delta(\varrho - \varrho')v(\varrho, z') \quad .$$

Putting here $z' = z$ (the integral here vanishes), we obtain

$$\frac{\delta v(\varrho, z)}{\delta\tilde{\varepsilon}(\varrho', z)} = \frac{\mathrm{i}k}{2}\delta(\varrho - \varrho')v(\varrho, z) \quad . \tag{3.96}$$

This relationship is the *exact* consequence of (3.88). Averaging (3.96) and setting $z' = z$, we obtain the boundary condition for (3.95)

$$S_1(\varrho, z'; \varrho', z') = \frac{ik}{2}\delta(\varrho - \varrho')\overline{v}(\varrho, z') \quad . \tag{3.97}$$

From (3.96), we can also obtain boundary conditions for S_2. To this end, we will apply the operator $\delta/\delta\tilde{\varepsilon}(\varrho'', z'')$, average the result and then put $z = z'$

$$\begin{aligned}
S_2(\varrho, z'; \varrho', z'; \varrho'', z'') &= S_2(\varrho, z'; \varrho'', z''; \varrho', z') \\
&= \frac{ik}{2}\delta(\varrho - \varrho')S_1(\varrho, z'; \varrho'', z'') \quad .
\end{aligned} \tag{3.98}$$

Taking the n-fold functional differential of (3.96), we can work out the boundary condition for S_{n+1} in a similar way.

Up until this point, we have made no assumption that $\tilde{\varepsilon}$ is delta-correlated. If we make this assumption in (3.91), i.e., if we put into it

$$\langle \tilde{\varepsilon}(\varrho, z)\tilde{\varepsilon}(\varrho', z')\rangle \to \psi_\varepsilon^{\mathrm{ef}}(\varrho, z; \varrho', z') = A(\varrho - \varrho')\delta(z - z') \quad , \tag{3.99}$$

then the integral with respect to z' in (3.91) can be taken to give

$$\frac{\partial \overline{v}}{\partial z} = \frac{i}{2k}\Delta_\perp \overline{v} + \frac{ik}{4}\int d^2\varrho' A(\varrho - \varrho')S_1(\varrho, z; \varrho', z) \quad ,$$

where the integration of the delta-function yields the factor 1/2. But, from (3.97), we already know $S_1(\varrho, z; \varrho', z)$ for coincident values of the longitudinal coordinate. If we substitute this into the integral, we obtain the *closed* equation for $\overline{v}$

$$\frac{\partial \overline{v}}{\partial z} = \frac{i}{2k}\Delta_\perp \overline{v} - \frac{k^2}{8}A(0)\overline{v}(\varrho, z) \quad , \tag{3.100}$$

which coincides with equation (3.17) derived in Sect. 3.2. If we thus suppose that $\tilde{\varepsilon}$ are delta-correlated in the *first* of the equations of our chain, it becomes closed so that the remaining equations appear to be superfluous.

In the first equation of our chain, we will now retain the exact covariance and only use the delta-correlation approximation in the second equation of the form (3.95). We can then take the integral with respect to z'' to obtain the equation

$$\frac{\partial S_1}{\partial z} = \frac{i}{2k}\Delta_\perp S_1 + \frac{ik}{4}\int d^2\varrho'' A(\varrho - \varrho'')S_2(\varrho, z; \varrho', z'; \varrho'', z) \quad . \tag{3.101}$$

If, in (3.98), we substitute z for z' and z' for z'' and interchange ϱ' and ϱ'', we will arrive at the desired value

$$S_2(\varrho, z; \varrho', z'; \varrho'', z) = \frac{ik}{2}\delta(\varrho - \varrho'')S_1(\varrho, z; \varrho', z') \quad .$$

Equation (3.101) then becomes closed and takes the form

$$\frac{\partial S_1}{\partial z} = \frac{i}{2k}\Delta_\perp S_1 - \frac{k^2 A(0)}{8}S_1 \quad , \quad \text{for} \quad z > z' \quad . \tag{3.102}$$

We now have the *system* of two equations, (3.91) and (3.102), and can discard the rest of the chain.

Clearly, if in the first $(n-1)$ equations we use the exact value of the covariance ψ_ε and replace (3.99) only in the nth equation, we will obtain a closed system of n equations, which will be the more exact, the larger n.

Let us now consider the case of $n = 2$ in more detail. First of all, we will have to solve (3.102) with the initial condition (3.97). It can easily be verified that the solution has the form

$$S_1(\varrho, z; \varrho', z')$$
$$= \frac{k^2}{4\pi(z - z')} \exp\left[\frac{ik(\varrho - \varrho')^2}{2(z - z')}\right] \overline{v}(\varrho', z') \exp\left[-\frac{k^2 A_0}{8}(z - z')\right] \quad . \qquad (3.103)$$

Substituting into (3.91) gives

$$\frac{\partial \overline{v}}{\partial z} = \frac{i}{2k} \Delta_\perp \overline{v} + \frac{ik^3}{8\pi} \int\limits_0^z \frac{dz'}{z - z'} \int d^2\varrho'$$
$$\times \exp\left[\frac{ik(\varrho - \varrho')^2}{2(z - z')} - \frac{k^2 A_0}{8}(z - z')\right]$$
$$\times \psi_\varepsilon(\varrho - \varrho', z - z')\overline{v}(\varrho', z') \quad . \qquad (3.104)$$

This equation can be readily solved using the Laplace transformation in z and Fourier transformation in ϱ. We will not follow this procedure, however, but will confine ourselves to the clarification of the conditions under which the solution to equation (3.104), to be referred to as the equation of the *second* Markovian approximation, will coincide with the solution to (3.100).

Equation (3.100) follows from (3.104) if the longitudinal scale $l_\parallel$ of $\psi_\varepsilon(\varrho - \varrho', z - z')$ is small in comparison with the scales in z of other factors. As we found in Sect. 3.3 [see (3.37)], the characteristic scale of $\overline{v}(\varrho', z')$ has the order of magnitude $(k^2 A_0)^{-1}$. The same scale enters into the factor $\exp(-k^2 A_0(z - z')/8)$ in (3.104). It is, therefore, necessary that the following condition be met:

$$k^2 A_0 l_\parallel \ll 1 \quad . \qquad (3.105)$$

If this is the case, we can substitute $\overline{v}(\varrho', z)$ for $\overline{v}(\varrho', z')$ in the integrand in (3.104), and consider that $\exp(-k^2 A_0(z - z')/8) \approx 1$. Then

$$\frac{\partial \overline{v}}{\partial z} = \frac{i}{2k} \Delta_\perp \overline{v} - \frac{k^2}{4} \int\limits_0^z dz' \int d^2\varrho' \frac{k}{2\pi i(z - z')} \exp\left[\frac{ik(\varrho - \varrho')^2}{2(z - z')}\right]$$
$$\times \psi_\varepsilon(\varrho - \varrho', z - z')\overline{v}(\varrho', z) \quad . \qquad (3.106)$$

In the region $|\varrho - \varrho'| \gg \sqrt{(z - z')/k}$, the function

$$f(\varrho - \varrho') = \frac{k}{2\pi i(z - z')} \exp\left[\frac{ik(\varrho - \varrho')^2}{2(z - z')}\right]$$

oscillates fast. If the characteristic scale of ψ_ε in ϱ, to be denoted by $l_\perp$, is large as compared with $\sqrt{(z - z')/k} \sim \sqrt{l_\parallel/k}$, the factor $\psi_\varepsilon(\varrho - \varrho', z - z')$ changes only slightly over the characteristic scale of the function f, and can be regarded as constant. As for $\overline{v}(\varrho', z)$, its characteristic transverse scale is about the size of the beam, a. Therefore, if the conditions

$$a \gg \sqrt{l_\parallel/k} \quad , \quad \text{and} \quad l_\perp \gg \sqrt{l_\parallel/k} \quad , \tag{3.107}$$

are met, we can replace $\psi_\varepsilon(\varrho - \varrho', z - z')\overline{v}(\varrho', z)$ by $\psi_\varepsilon(0, z - z')\overline{v}(\varrho, z)$. As a result, we can now take the integral with respect to ϱ' to obtain unity. Thus, if the conditions (3.107) are met, (3.106) is further simplified:

$$\frac{\partial \overline{v}}{\partial z} = \frac{i}{2k}\Delta_\perp \overline{v} - \frac{k^2}{4}\overline{v}(\varrho, z)\int_0^z \psi_\varepsilon(0, z - z')dz' \quad . \tag{3.108}$$

Finally, if

$$z \gg l_\parallel \quad , \tag{3.109}$$

then the upper integration limit in (3.108) can be replaced by infinity. Taking into account the fact that

$$\int_0^\infty \psi_\varepsilon(0, z')dz' = \tfrac{1}{2}A(0) \quad ,$$

we derive from (3.108) the equation of the first Markovian approximation (3.100).

Interestingly, even (3.108) gives a more exact expression for $\overline{v}$ when $z \ll l_\parallel$ than the first Markovian approximation. For example, if the incident wave is plane, (3.108) takes the form

$$\overline{v}(z) = v_0 \exp\left[-\frac{k^2}{4}\int_0^z (z - \zeta)\psi_\varepsilon(0, \zeta)d\zeta\right] \quad . \tag{3.110}$$

For $z \ll l_\parallel$ the relation $\psi_\varepsilon(0, \zeta) \approx \psi_\varepsilon(0, 0) = \sigma_\varepsilon^2$, holds over the entire integration region and (3.110) gives

$$\overline{v}(z) = v_0 \exp\left(-k^2 \sigma_\varepsilon^2 z^2/8\right) \quad . \tag{3.111}$$

But, if $z \gg l_\parallel$, we can put into (3.110)

$$\int_0^z (z - \zeta)\psi_\varepsilon(0, \zeta)d\zeta \approx z \int_0^\infty \psi_\varepsilon(0, \zeta)d\zeta = A_0 z/2 \quad ,$$

to arrive at

$$\overline{v}(z) = v_0 \exp\left(-k^2 A_0 z/8\right) \quad \text{for} \quad z \gg l_\parallel \quad . \tag{3.112}$$

Thus, even in a first Markovian approximation, we obtain (3.112) for the plane wave at *all* z.

Let us take a closer look at the conditions (3.105, 107, 109) under which the Markovian approximation can be applied to the mean field $\overline{v}$. The first of the conditions (3.107) implies that the longitudinal correlation scale $l_\parallel$ must be small as compared to the distance ka^2, over which the diffraction at the beam aperture begins to be noticeable. The second condition can also be written in the form $l_{||} \ll kl_{\perp}^2$, so that the scale $l_{||}$ must, likewise, be small as compared to the diffraction length corresponding to the transverse size of the inhomogeneities. Condition (3.105) implies that the mean field must be weakly attenuated in the scale $l_{||}$ [recall that according to (3.39), $k^2 A_0/4 = \sigma_0$ is the scattering coefficient], or, in other words, $l_{||} \ll d = \sigma_0^{-1}$, where d is the extinction length. Lastly, condition (3.109) implies that $l_{||}$ must be small in comparison with the line length. It can be said that the four conditions limit from above the longitudinal correlation scale for electric permittivity. Condition (3.105) then implies that σ_ε^2 is also limited. In fact, $A_0 \sim \sigma_\varepsilon^2 l_{||}$, so that (3.105) can be written as

$$\sigma_\varepsilon^2 k^2 l_{||}^2 \ll 1 \quad . \tag{3.105a}$$

It should be noted that the conditions for the Markovian approximation to be valid, unlike any of the methods considered earlier that use perturbation theory in $\tilde{\varepsilon}$, *do not limit from above the length of z.*

We have considered higher-order approximations for the mean field. In the same way, we can derive equations for the second and higher approximations for arbitrary moments $\Gamma_{n,m}$ [3.19]. We will not provide these equations here, but will only give certain qualitative arguments and conclusions of [3.19].

If we now turn to (3.23) at $n = m = 1$, i.e., to the equation for $v(\varrho_1', z)v^*(\varrho_1', z)$, we see that the only random coefficient contained in it is $\tilde{\varepsilon}(\varrho_1', z) - \tilde{\varepsilon}(\varrho_1'', z)$. Changing $\tilde{\varepsilon} \to \tilde{\varepsilon} + \text{const}$ does not influence this difference. It clearly follows that the large scale inhomogeneities do not affect $v(\varrho_1', z)v^*(\varrho_1'', z)$. Therefore, neither can the applicability conditions for the Markovian approximation for Γ contain such quantities as σ_ε^2, $l_{||}$, and $l_\perp$, which are only dependent on the behavior of the spectrum of electric permittivity fluctuations for small wave numbers.

These conditions have the form

$$kl_0 \gg 1 \quad , kz|\nabla_\perp H(\varrho)| \ll 1 \quad \text{for} \quad z \gg \varrho \quad , \tag{3.113}$$

where l_0 is the scale of the smallest inhomogeneities of the electric permittivity. The second of (3.113) is a constraint on the quantity that characterizes the intensity of the fluctuations ε. If we recall that the coherence radius of the field is found from the condition $k^2 z H(\varrho_k) \sim 1$ [see (3.54)], the second condition (3.113) can be written as

$$k\varrho_k \gg 1 \quad . \tag{3.114}$$

For the Markovian approximation to hold, it is thus required that the coherence radius be consistently much larger than the wavelength.

3.6 Exercises

3.6.1 Find the distribution of the mean intensity over the transverse cross-section of the radiation beam for which the field in the plane $z = 0$ has the form

$$v_0(\varrho) = v_0 \exp\left(-\frac{\varrho^2}{2a^2} + \frac{ik\varrho^2}{2F}\right) \tag{3.115}$$

(a Gaussian beam with effective radius a and separation F from the plane $z = 0$ to the radiation center) [3.8]. Consider the special case where $H(\varrho) = pC_\varepsilon^2 \varrho^{5/3}$, which corresponds to the turbulent fluctuations of electric permittivity.

Solution. The mean intensity $\overline{I}(\varrho_+, z)$ can be derived from Γ at $\varrho = 0$. Therefore, from (3.52) we have

$$\overline{I}(\varrho_+, z) = \frac{k^2}{4\pi^2 z^2} \int d^2\varrho' \int d^2\varrho'_+ \Gamma^{(0)}(\varrho', \varrho'_+)$$

$$\times \exp\left\{-\frac{ik\varrho'}{z}(\varrho_+ - \varrho'_+) - \frac{\pi k^2}{4} \int_0^z H[\varrho'(1 - \zeta/z)]d\zeta\right\} \ . \tag{3.116}$$

Let us work out $\Gamma^{(0)}$ using the initial condition (3.115) for the field

$$\Gamma^{(0)}(\varrho', \varrho'_+)$$
$$= v_0\left(\varrho'_+ + \frac{\varrho'}{2}\right)v_0^*\left(\varrho'_+ + \frac{\varrho'}{2}\right) = |v_0|^2$$
$$\times \exp\left(-\frac{\varrho'^2_+}{a^2} - \frac{\varrho'^2}{4a^2} + \frac{ik\varrho'_+ \cdot \varrho'}{F}\right) \ . \tag{3.117}$$

Substituting (3.117) into (3.116) and integrating with respect to ϱ_+ gives

$$\overline{I}(\varrho_+, z) = \frac{k^2 a^2 |v_0|^2}{4\pi z^2} \int d^2\varrho'$$

$$\times \exp\left(-\frac{ik\varrho'_+ \cdot \varrho'}{z} - \frac{\varrho'^2}{4a^2}g^2(z) - \frac{\pi k^2 z}{4}\int_0^1 H(\varrho't)dt\right) \ , \tag{3.118}$$

where

$$g(z) = \sqrt{1 + k^2 a^4(1/z + 1/F)^2} \ . \tag{3.119}$$

Equation (3.118) holds for any $H(\varrho)$. If $H(\varrho) = pC_\varepsilon^2 \varrho^{5/3}$, so that

$$\int_0^1 H(\varrho', t)dt = 3pC_\varepsilon^2 \varrho'^{5/3}/8 \ ,$$

and (3.118), in polar coordinates, will be

$$\overline{I}(\varrho_+ z) = \frac{k^2 a^2 |v_0|^2}{4\pi z^2} \int_0^\infty \varrho' \, d\varrho' \exp\left(-\frac{\varrho'^2 g^2}{4a^2} - \frac{3\pi\varrho}{32}C_\varepsilon^2 k^2 z\varrho'^{5/3}\right)$$

$$\times \int_0^{2\pi} \exp\left(\frac{ik\varrho'\varrho_+}{z}\cos\varphi\right)d\varphi \ .$$

The integral with respect to φ is $2\pi J_0(k\varrho'\varrho_+/z)$, so that we obtain

$$\bar{I}(\varrho_+, z) = \frac{k^2 a^2 |v_0|^2}{2z^2} \int\limits_0^\infty \exp\left[-\frac{g^2(z)}{4a^2}\varrho^2 - \frac{3\pi p}{32}C_\varepsilon^2 k^2 z \varrho^{5/3}\right]$$

$$\times J_0\left(\frac{k\varrho\varrho_+}{z}\right)\varrho\, d\varrho \quad . \tag{3.120}$$

Let us introduce the intensity $I_0(z)$ along the beam axis in a homogeneous medium, i.e., at $C_\varepsilon = 0$

$$I_0(z) = \frac{k^2 a^2 |v_0|^2}{2z^2} \int\limits_0^\infty \exp\left(-\frac{g^2 \varrho^2}{4a^2}\right)\varrho\, d\varrho = \frac{k^2 a^4 |v_0|^2}{z^2 g^2(z)} \quad .$$

The ratio $\bar{I}(\varrho_+, z)/I_0(z)$ can be written as

$$\frac{\bar{I}(\varrho_+, z)}{I_0(z)} = \frac{g^2(z)}{2a^2}\int\limits_0^\infty \exp\left(-\frac{g^2 \varrho^2}{4a^2} - \frac{3\pi p}{32}C_\varepsilon^2 k^2 z \varrho^{5/3}\right) J_0\left(\frac{k\varrho\varrho_+}{z}\right)\varrho\, d\varrho \tag{3.121}$$

If we now introduce the dimensionless integration variable $(g\varrho/2a)^2 = t$, (3.121) will then be

$$\frac{\bar{I}(\varrho_+, z)}{I_0(z)} = \int\limits_0^\infty \exp\left\{-t - \frac{3\pi p}{32}C_\varepsilon^2 k^2 z [2a/g(z)]^{5/3} t^{5/6}\right\}$$

$$\times J_0\left(\frac{2ka\varrho_+}{zg(z)}t^{1/2}\right)dt \quad . \tag{3.122}$$

Hence, the intensity along the beam axis is simply

$$\frac{\bar{I}(0, z)}{I_0(z)} = f(u) = \int\limits_0^\infty \exp(-t - ut^{5/6})dt \quad , \quad \text{where} \tag{3.123}$$

$$u = \frac{3\pi p}{32}C_\varepsilon^2 k^2 z [2a/g(z)]^{5/3} \quad .$$

The parameter u is proportional to the structure function of the phase at the basis $2a/g(z)$. Expanding $\exp(-t)$ or $\exp(-ut^{5/6})$, we obtain

$$f(u) = \sum_{n=0}^\infty \frac{\Gamma(1 + 5n/6)}{n!}(-u)^n = 1 - 0.94u + 0.75u^2 - 0.56u^3 + \dots \quad , \tag{3.124}$$

$$f(u) \approx \frac{6}{5u^{6/5}} \sum_{n=0}^\infty \frac{\Gamma[6(n+1)/5]}{n!}[-(u^{-6/5})]^n = \frac{1.10}{u^{6/5}} - \frac{1.49}{u^{12/5}} + \dots \quad . \tag{3.125}$$

The second of these is an asymptotic expansion of $f(u)$ at large u. The variation of $f(u)$ is represented in Fig. 3.4.

According to (3.56), the integral of $\bar{I}(\varrho_+, z)$ over the transverse cross-section of the beam is independent of z. The reduction in the mean intensity on the beam axis can be accounted for by beam broadening. If the effective beam area is defined by

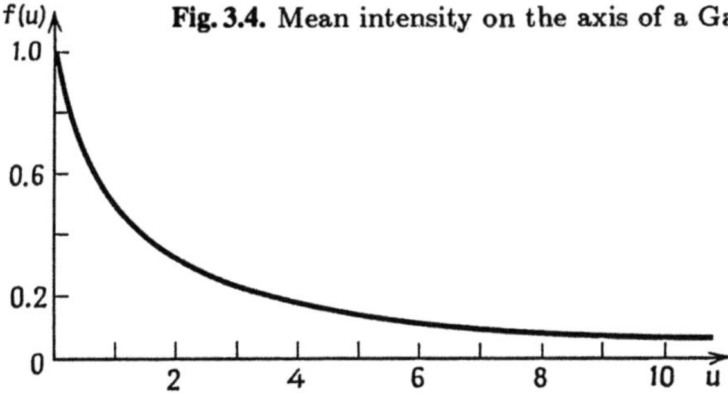

Fig. 3.4. Mean intensity on the axis of a Gaussian beam in a turbulent medium

$$\int \overline{I}(\varrho_+, z)d^2\varrho_+ = \overline{I}(0, z)S_{\mathrm{ef}}(z)$$

then, from the conservation law (3.56),

$$\overline{I}(0, z)S_{\mathrm{ef}}(z) = \overline{I}(0,0)S_{\mathrm{ef}}(0) \quad . \tag{3.126}$$

The right-hand side of this relation is independent of inhomogeneities. Therefore, putting $C_\varepsilon^2 = 0$ in (3.26) we obtain an analogous relation for the same beam in a homogeneous medium

$$I_0(z)S_{\mathrm{ef}}^{(0)}(z) = \overline{I}(0,0)S_{\mathrm{ef}}(0) \quad , \tag{3.127}$$

where $I_0(z)$ and $S_{\mathrm{ef}}^{(0)}(z)$ are the beam intensity on the axis and the beam effective cross-section in the homogeneous medium, respectively.

It follows from (3.126, 127) that

$$\frac{S_{\mathrm{ef}}(z)}{S_{\mathrm{ef}}^{(0)}(z)} = \frac{I_0(z)}{\overline{I}(0, z)} = \frac{1}{f(u)} \quad . \tag{3.128}$$

If $u \gg 1$, by (3.125), $f(u) \approx 1.1u^{-6/5}$ and

$$\frac{S_{\mathrm{ef}}(z)}{S_{\mathrm{ef}}^{(0)}(z)} \approx \frac{u^{6/5}}{1.1} \sim (C_\varepsilon^2 k^2)^{6/5} z^{6/5}[2a/g(z)]^2$$

For example, at $F = \infty$ (collimated beam), $g^2(z) = 1 + k^2a^4/z^2$. If $z \ll ka^2$ (the near zone of the source), then $g^2(z) \approx k^2a^4/z^2$, and

$$\frac{S_{\mathrm{ef}}(z)}{S_{\mathrm{ef}}^{(0)}(z)} \sim z^{16/5}$$

If now $z \gg ka^2$ (the Fraunhofer zone of the source), then $g^2 \approx 1$ and

$$\frac{S_{\mathrm{ef}}(z)}{S_{\mathrm{ef}}^{(0)}(z)} \sim z^{6/5}$$

3.6.2 The random radius vector of the "center of mass" of the intensity distribution over the beam cross-section is given by

$$\xi = \frac{\int \varrho' I(\varrho', z) d^2 \varrho'}{\int I(\varrho', z) d^2 \varrho'} \quad .$$

Express the mean square $\overline{\xi^2}$ in terms of the coherence function Γ_4.

Solution. The integral in the denominator is a constant deterministic quantity. Denoting it by P, we obtain

$$\overline{\xi^2} = P^{-2} \iint d^2 \varrho_1' d^2 \varrho_2' \varrho_1' \cdot \varrho_2' \langle I(\varrho_1', z) I(\varrho_2' z) \rangle \quad .$$

But,

$$\langle I(\varrho_1', z) I(\varrho_2' z) \rangle = \langle v(\varrho_1', z) v(\varrho_2', z) v^*(\varrho_1', z) v^*(\varrho_2', z) \rangle \quad ,$$

i.e., in the expression for Γ_4, it is necessary to put $\varrho_1'' = \varrho_1'$ and $\varrho_2'' = \varrho_2'$, which, for the new variables ϱ_+, ϱ, ϱ_1, ϱ_2 [see (3.66)], is equivalent to $\varrho_2 = 0$, $\varrho = 0$. Here, $\varrho_1' = \varrho_+ + \varrho_1/2$, $\varrho_2' = \varrho_+ - \varrho_1/2$, $\varrho_1' \varrho_2' = \varrho_+^2 - \varrho_1^2/4$ and

$$\Gamma_4(\varrho_+, \varrho, \varrho_1, \varrho_2, z) = \Gamma_4(\varrho_+, 0, \varrho_1, 0, z) \quad .$$

Hence,

$$\overline{\xi^2} = P^{-2} \iint d^2 \varrho_+ d^2 \varrho_1 (\varrho_+^2 - \varrho_1^2/4) \Gamma_4(\varrho_+, 0, \varrho_1, 0, z) \quad .$$

3.6.3 A photodetector, which is essentially a disk of radius R, measures the integral of the light intensity $I(\varrho, z)$. Find the variation of the relative fluctuations of the photocurrent J with increasing R for the case where the incident plane light wave propagates in a statistically isotropic medium.

Solution. The instantaneous value of the photocurrent is

$$J = \alpha \int_{\Sigma} I(\varrho, z) d^2 \varrho \quad , \tag{3.129}$$

where α is the sensitivity of the detector, and Σ the surface area of a circle of radius R. Averaging (3.129), we have $\overline{J} = \alpha \overline{I} \Sigma$. It also follows from (3.129) that

$$\tilde{J} = \alpha \int_{\Sigma} \tilde{I}(\varrho, z) d^2 \varrho \quad ,$$

so that the relative fluctuations $\tilde{J}/\overline{J}$ are given by

$$\frac{\tilde{J}}{\overline{J}} = \frac{1}{\Sigma} \int_{\Sigma} \frac{\tilde{I}(\varrho, z)}{\overline{I}} d^2 \varrho \quad . \tag{3.130}$$

Squaring (3.130) and averaging gives

$$\sigma^2 \equiv \overline{\left(\frac{\tilde{J}}{\overline{J}}\right)^2} = \frac{1}{\Sigma^2} \int_{\Sigma} d^2 \varrho_1 \int_{\Sigma} d^2 \varrho_2 \psi_I(\varrho_1 - \varrho_2, z) \quad , \tag{3.131}$$

where ψ_I is the covariance divided by $\bar{I}^2$. Introducing the function

$$M(\varrho) = \begin{cases} 1 & \text{for} \quad \varrho \in \Sigma \\ 0 & \text{for} \quad \varrho \notin \Sigma \end{cases} \; .$$

We can represent σ^2 in the form

$$\sigma^2 = \frac{1}{\Sigma^2} \iint\limits_{-\infty}^{\infty} d^2\varrho_1 \, d^2\varrho_2 M(\varrho_1) M(\varrho_2) \psi_I(\varrho_1 - \varrho_2, z) \quad ,$$

or, after introducing the new integration variables $\varrho = \varrho_1 - \varrho_2$ and $\varrho' = \varrho_2$,

$$\sigma^2 = \frac{1}{\Sigma^2} \int\limits_{-\infty}^{\infty} \psi_I(\varrho, z) d^2\varrho \int\limits_{-\infty}^{\infty} M(\varrho - \varrho') M(\varrho') d^2\varrho' \quad . \tag{3.132}$$

Let us denote

$$f(\varrho) = \frac{1}{\Sigma} \int\limits_{-\infty}^{\infty} M(\varrho - \varrho') M(\varrho') d^2\varrho' \quad . \tag{3.133}$$

We will then write (3.132) as

$$\sigma^2 = \frac{1}{\Sigma} \int\limits_{-\infty}^{\infty} \psi_I(\varrho, z) f(\varrho) d^2\varrho \quad . \tag{3.134}$$

It follows from the definition of M that $f\Sigma$ equals the surface area where two circles of radius R, with centers separated by ϱ, overlap. We find that

$$f(\varrho) = f(\varrho) = \begin{cases} \dfrac{2}{\pi} \left[\arccos(\varrho/2R) - \dfrac{\varrho}{2R} \sqrt{1 - (\varrho/2R)^2} \right] & \text{for} \quad \varrho < 2R \quad , \\ 0 & \text{for} \quad \varrho > 2R \quad . \end{cases}$$

Since the fluctuations are statistically isotropic (i.e., ψ_I also depends on $\varrho = |\varrho|$ only), we can write (3.134) as a single integral

$$\sigma^2 = \frac{2}{R^2} \int\limits_0^{2R} \psi_I(\varrho, z) f(\varrho) \varrho \, d\varrho \quad . \tag{3.135}$$

Consider the case where R is far larger than the correlation radius of the fluctuations. The function $\psi_I(\varrho, z)$ will then decay quickly, even at those values of ϱ at which $f(\varrho)$ has not yet changed noticeably. Let us expand $f(\varrho)$ into a Taylor expansion in $\varrho/2R$

$$f(\varrho) = 1 - \frac{4}{\pi} \frac{\varrho}{2R} + \dots$$

and substitute this expansion into (3.135):

$$\sigma^2 = \frac{2}{R^2} \int\limits_0^{2R} \psi_I(\varrho, z) \varrho \, d\varrho - \frac{4}{\pi R^3} \int\limits_0^{2R} \psi_I(\varrho, z) \varrho^2 d\varrho + \dots \quad . \tag{3.136}$$

The first integral can be written as

$$\int\limits_0^{2R} = \int\limits_0^{\infty} - \int\limits_{2R}^{\infty} \quad ,$$

which gives

$$\sigma^2 = \frac{2}{R^2} \int\limits_0^{\infty} \psi_I(\varrho, z)\varrho\, d\varrho - \frac{2}{R^2} \int\limits_{2R}^{\infty} \psi_I(\varrho, z)\varrho\, d\varrho$$

$$- \frac{4}{\pi R^3} \int\limits_0^{2R} \psi_I(\varrho, z)\varrho^2 d\varrho + \ldots \quad . \tag{3.137}$$

But, by (3.80), the integral in the first term will be zero; therefore, the term with Σ^{-1} in (3.137) vanishes. The photocurrent fluctuations thus decrease faster than R^{-2} with increasing R

$$\sigma^2 = -\frac{2}{R^2} \int\limits_{2R}^{\infty} \psi_I(\varrho, z)\varrho\, d\varrho - \frac{4}{\pi R^3} \int\limits_0^{2R} \psi_I(\varrho, z)\varrho^2 d\varrho + \ldots \quad . \tag{3.138}$$

Note that, at large ϱ, the function $\psi_I(\varrho, z)$ is negative, so that $\sigma^2 \geq 0$.

The fact that σ^2 decreases with R faster than R^{-2} is accounted for by the fact that at large R, over the detector area $\Sigma = \pi R^2$, the negative and positive intensity fluctuations cancel out almost completely.

4. Elements of Multiple Scattering

As stated in Sect. III.4.1, multiple scattering is described by an infinite perturbative series, assuming neither large-scale inhomogeneities of the medium (mostly forward scattering) nor constraints on the length of the path covered by the wave in the inhomogeneous medium. When dealing with such series, the short-hand language of Feynman diagrams is exceedingly convenient. The Feynman technique, as applied to the problem of scattering of a scalar wave in a random medium, is presented in Sect. 4.1. This technique can be translated without change to other linear problems in which random functions or quantities parametrically enter into the original stochastic equations. This section also provides equations derived using the diagram technique; they make it possible to carry out the summation of some of the infinite subsequences of perturbative series for the mean field (the Dyson equation), and for its covariance (the Bethe-Salpeter equation).

Section 4.2 takes a closer look at the Dyson equation for the mean field of a point source in a statistically homogeneous medium, as well as at various forms of that equation and some of its approximate solutions.

There is no way of solving the Bethe-Salpeter equation for the coherence functions. Therefore, the subject matter of much of Sect. 4.3 is a treatment of another issue, viz., the relation between the exact Bethe-Salpeter equation and the phenomenological radiative transfer equation (RTE) for the so-called *radiant intensity*. In other words, this section gives a consistent derivation of RTE for a random medium from the wave equation. When it is in general possible to reduce the Bethe-Salpeter equation to RTE, the derivation mentioned above enables us to relate the quantities describing the wave field to the photometric quantities that enter into the RTE. At the same time, this clarifies the question of how RTE typically reflects wave effects (diffraction).

4.1 Perturbation Theory and the Diagram Technique for the Mean Field and Covariance

This chapter deals with certain general questions of *multiple* scattering theory. We discussed a special case of this theory in Chap. 3, where an outline of the theory of wave propagation in the approximation of the Markov random process was given. This theory took into account multiple scattering, but only under conditions where the scattering was *forward*. If, however, the wavelength is not small enough as compared to the size of the inhomogeneities, scattering at any angle, and not only at small angles, becomes noticeable. If, moreover, the inhomogeneous medium is sufficiently extensive, then multiple scattering may also become important in this more general case.

We have already mentioned that multiple scattering is possible both from a multitude of *discrete* scatterers (electrons in a plasma, aerosole particles in the atmosphere, etc.), and from *continuous* inhomogeneities, e.g., from electric permittivity fluctuations in a continuous medium. Multiple scattering theory in these two cases is deduced somewhat differently, although the two forms have much in common. We will confine our discussion here to the simpler case of the theory of multiple scattering from *continuous* inhomogeneities. As for scattering theory with discrete scatterers, the reader is referred, for example, to [4.1].

Furthermore, we will consider only the simplest formulation of the problem, when the propagation of a wave is described by the *scalar* wave equation. It should be noted that with large-scale inhomogeneities, the scalar form of the problem enabled many aspects of propagation to be described for vector (electromagnetic) waves, whereas this is not the case for an arbitrary ratio of wavelength and inhomogeneity size, since polarization changes strongly in multiple scattering. We will, nevertheless, confine ourselves to the scalar wave equation, because even in this simple case the specific features of multiple scattering stand out.

Lastly, we will only concern ourselves with the problem of wave propagation in an *infinite* medium, where the fluctuations of the refractive index are a *Gaussian* random field. The last two assumptions are only made in order to simplify the theory to be discussed, and are by no means necessary.

First, consider the *mean* field of a *point* source lying at a point r_0 in an inhomogeneous medium. Random Green's function, i.e., the field of the point source in this medium obeys the wave equation

$$\hat{L}(r)G(r, r_0) \equiv \Delta G(r, r_0) + k^2[1 + \tilde{\varepsilon}(r)]G(r, r_0) = \delta(r - r_0) \tag{4.1}$$

(where Δ is the Laplacian in r), and also the radiation condition at $|r - r_0| \to \infty$.

Putting $u_0(r) = G_0(r, r_0)$ in (III.4.10), where

$$G_0(r, r_0) = -\frac{\exp(ik|r - r_0|)}{4\pi|r - r_0|} \tag{4.2}$$

is Green's function for the homogeneous medium ("free space"), and substituting (4.2) for u_0 in (III.4.10), we obtain the expansion in perturbations

$$\begin{aligned}
G(r, r_0) = G_0(r, r_0) &- k^2 \int G_0(r, r_1)\tilde{\varepsilon}(r_1)G_0(r_1, r_0)d^3r_1 \\
&+ (-k^2)^2 \int G_0(r, r_1)\tilde{\varepsilon}(r_1)G_0(r_1, r_2)\tilde{\varepsilon}(r_2) \\
&\times G_0(r_2, r_0)d^3r_1 d^3r_2 + (-k^2)^3 \int G_0(r, r_1)\tilde{\varepsilon}(r_1)G_0(r_1, r_2) \\
&\times \tilde{\varepsilon}(r_2)G_0(r_2, r_3)\tilde{\varepsilon}(r_3)G_0(r_3, r_0)d^3r_1 d^3r_2 d^3r_3 + \dots \quad . \tag{4.3}
\end{aligned}$$

Note that $G_0(r, r_0)$ is *symmetrical*, i.e., $G_0(r, r_0) = G_0(r_0, r)$. We will show that $G(r, r_0)$ meets also this condition. In fact, since G_0 is symmetrical, we can write the integral in the second term of (4.3) as

$$\int G_0(r, r_1)\tilde{\varepsilon}(r_1)G_0(r_1, r_0)d^3r_1 = \int G_0(r_1, r_0)\tilde{\varepsilon}(r_1)G_0(r, r_1)d^3r_1$$
$$= \int G_0(r_0, r_1)\tilde{\varepsilon}(r_1)G_0(r_1, r)d^3r_1 \quad .$$

We can do the same with the next term

$$\int G_0(r, r_1)\tilde{\varepsilon}(r_1)G_0(r_1, r_2)\tilde{\varepsilon}(r_2)G_0(r_2, r_0)d^3r_1 d^3r_2$$
$$= \int G_0(r_0, r_2)\tilde{\varepsilon}(r_2)G_0(r_2, r_1)\tilde{\varepsilon}(r_1)G_0(r_1, r)d^3r_1 d^3r_2$$
$$= \int G_0(r_0, r_1)\tilde{\varepsilon}(r_1)G_0(r_1, r_2)\tilde{\varepsilon}(r_2)G_0(r_2, r)d^3r_1 d^3r_2 \quad .$$

Here, we have changed the integration variables ($r_1 \rightleftharpoons r_2$). Clearly, we can transform any term in this way, thereby

$$G(r, r_0) = G(r_0, r) \quad , \tag{4.4}$$

i.e., we obtain the *reciprocity theorem* which states that, in an arbitrary linear inhomogeneous medium, the field remains unchanged if the source location and observation points are interchanged.

Let us now turn to mean Green's function. In averaging (4.3), it should be taken into account that, for the Gaussian field $\tilde{\varepsilon}$, $\langle \tilde{\varepsilon}(r_1) \dots \tilde{\varepsilon}(r_{2n-1}) \rangle = 0$, and the even moments are given by (III.1.128):

$$\langle \tilde{\varepsilon}(r_1) \dots \tilde{\varepsilon}(r_{2n}) \rangle = \sum_{\text{p.p.}} \psi_\varepsilon(r_\alpha, r_\beta) \dots \psi_\varepsilon(r_\gamma, r_\delta) \tag{4.5}$$

where the sum on the right-hand side is over all the possible pairs $(r_{\alpha_1}, r_{\alpha_2})$ of the points $r_1, \dots r_{2n}$ (the order in each pair is immaterial). For example, by (4.5),

$$\langle \tilde{\varepsilon}(r_1)\tilde{\varepsilon}(r_2)\tilde{\varepsilon}(r_3)\tilde{\varepsilon}(r_4) \rangle = \psi_\varepsilon(r_1, r_2)\psi_\varepsilon(r_3, r_4)$$
$$+ \psi_\varepsilon(r_1, r_3)\psi_\varepsilon(r_2, r_4) + \psi_\varepsilon(r_1, r_4)\psi_\varepsilon(r_2, r_3) \tag{4.6}$$

and the expression for $\langle \tilde{\varepsilon}(r_1) \dots \tilde{\varepsilon}(r_{2n}) \rangle$ will have $(2n - 1)!!$ terms. From (4.5), we average (4.3)

$$\overline{G}(r, r_0) = G_0(r, r_0)$$
$$+ k^4 \int G_0(r, r_1)G_0(r_1, r_2)G_0(r_2, r_0)\psi_\varepsilon(r_1, r_2)d^3r_1 d^3r_2$$
$$+ k^8 \int G_0(r, r_1)G_0(r_1, r_2)G_0(r_2, r_3)G_0(r_3, r_4)G_0(r_4, r_0)$$
$$\times [\psi_\varepsilon r_1, r_2)\psi_\varepsilon(r_3, r_4) + \psi_\varepsilon(r_1, r_3)\psi_\varepsilon(r_2, r_4)$$
$$+ \psi_\varepsilon(r_1, r_4)\psi_\varepsilon(r_2, r_3)]d^3r_1 \dots d^3r_4 + \dots \quad . \tag{4.7}$$

To give a graphic idea of the structure of this expansion, we will represent its elements by diagrams. These graphs (or diagrams), due to Feynman, were introduced in quantum electrodynamics and subsequently found wide use in various branches of theoretical physics. They owe their success to their compact form as compared to the analytical representation, and the fact that they simplify the treatment. Of course, the use of the "diagram technique" takes a measure of skill.

Let $G_0(r_j, r_i)$ be described by a segment of a line, r_j and r_i being the coordinates of the ends of the segment:

$$G_0(r_j, r_i) \sim \underline{\qquad\qquad} \atop r_j \qquad\quad r_i \quad .$$

The factor $k^4 \psi_\varepsilon(r_j, r_i)$ is described by a *dashed* line with the two points at the ends

$$k^4 \psi_\varepsilon(r_j, r_i) \sim \bullet\!-\!-\!-\!-\!-\!\bullet \atop r_j \qquad\quad r_i \quad .$$

The points r_j, r_i, where the lines representing G_0 and ψ_ε converge, will be referred to as the *vertices* of the diagram. Suppose that the integration is over the coordinates of all the *inner* vertices. The number of these vertices in the diagram will be called the *order* of the diagram.

These results allow us to have a corresponding Feynman diagram for any term of (4.7). So, the first term is given by

$$\underline{\qquad\qquad} \atop r \qquad\quad r_0 \quad ,$$

and the second by

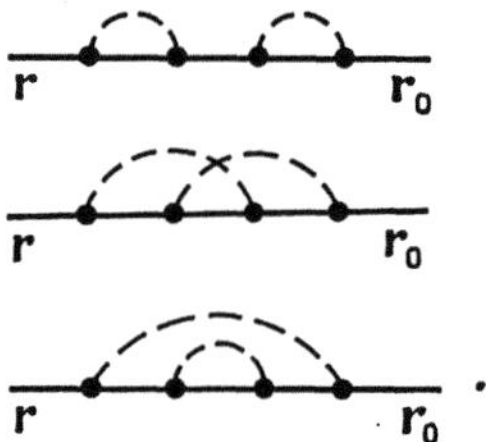

Note that since integration is over the coordinates r_1, r_2 of the inner vertices, the analytical expression described by the diagram is independent of these coordinates. Therefore, in the following, we will not indicate these coordinates on the diagrams. The third term in (4.7) is given by the three diagrams:

The unaveraged $2n$th-order term is

$$k^{4n} \int G_0(r, r_1) G_0(r_1, r_2) \dots G_0(r_{2n}, r_0) \tilde\varepsilon(r_1) \dots \tilde\varepsilon(r_{2n}) d^3 r_1 \dots d^3 r_{2n} \quad .$$

Therefore, the $2n$th-order diagrams contain $2n + 1$ lines of Green's function G_0, and $2n$ inner vertices $r_1, \dots r_{2n}$. In averaging, the factor $\langle \tilde\varepsilon(r_1) \dots \tilde\varepsilon(r_{2n}) \rangle$ gives the sum of $(2n - 1)!!$ terms, where $r_1, \dots r_{2n}$ are combined pairwise in all possible ways, and so $(2n - 1)!!$ diagrams of the $2n$th order are produced in which $2n$ vertices are connected pairwise by dashed lines in all possible ways. Finally, we will introduce a graph for the *mean* Green's function in the *inhomogeneous* medium

$$\bar{G}(r, r_0) \sim \underline{\blacksquare\blacksquare\blacksquare} \atop r \qquad r_0 \quad .$$

The expansion (4.7) can then be represented as follows:

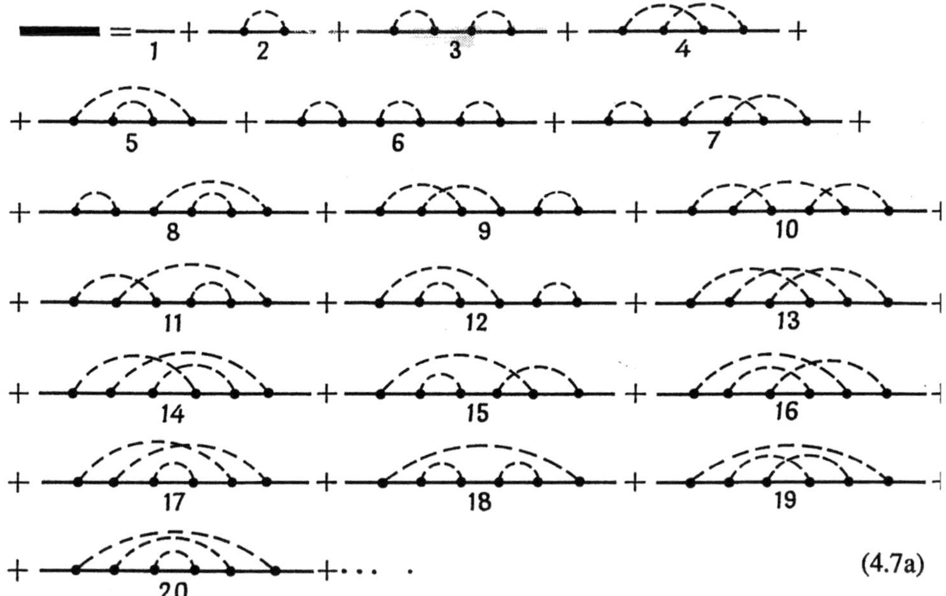

$$(4.7a)$$

Besides the terms appearing in (4.7), all 15 terms of the sixth order are represented here (six-vertex diagram).

The correspondence between the Feynman diagrams and the analytic expressions is one to one. Therefore, we can not only construct the corresponding diagram from an analytical expression but, conversely, restore the analytical representation from the diagram. For example, for diagram 19 in (4.7a) we have

$$\overline{G}^{(19)}(r, r_0) = k^{12} \int G_0(r, r_1)G_0(r_1, r_2)G_0(r_2, r_3)G_0(r_3, r_4)$$

$$\times G_0(r_4, r_5)G_0(r_5, r_6)G_0(r_6, r_0)\psi_\varepsilon(r_1, r_6)$$

$$\times \psi_\varepsilon(r_2, r_4)\psi_\varepsilon(r_3, r_5)d^3 r_1 \ldots d^3 r_6 \quad . \tag{4.8}$$

Some of the diagrams in (4.7a) contain lower-order diagrams as component parts. So, diagram 3 contains diagram 2, diagram 19 contains 4, and so on. We can make use of this to cancel analytical expressions. For example, we can write $\tilde{G}^{(19)}$ as

$$\overline{G}^{(19)}(r, r_0) = k^4 \int G_0(r, r_1)\tilde{G}^{(4)}(r_1, r_6)G_0(r_6, r_0)$$

$$\times \psi_\varepsilon(r_1, r_6)d^3 r_1 d^3 r_6 \quad , \quad \text{where} \tag{4.9}$$

$$\overline{G}^{(4)}(r_1, r_6) = k^8 \int G_0(r_1, r_1')G_0(r_1', r_2')G_0(r_2', r_3')$$

$$\times G_0(r_3', r_4')G_0(r_4', r_6)\psi_\varepsilon(r_1', r_3')$$

$$\times \psi_\varepsilon(r_2', r_4')d^3 r_1' \ldots d^3 r_4' \quad . \tag{4.10}$$

It is easily seen that substituting (4.10) into (4.9) and changing the notation of the integration variables gives (4.8).

Before we go on to develop the diagram technique, we will dwell briefly on the physical significance of the diagrams. Diagram 1 in (4.7a) describes the propagation of waves from r_0 to r without scattering (as in the homogeneous medium). Diagram 2 describes the propagation of the wave in the homogeneous medium from r_0 to r_2, where a first scattering occurs, whereupon the scattered wave comes to r_1 where it is scattered for the second time. The doubly scattered wave arrives at the observation point r. This part of its journey is described by the factors $G_0(r, r_1)G_0(r_1, r_2)G_0(r_2, r_0)$. The covariance $\psi_\varepsilon(r_1, r_2)$ in 2 indicates that both scatterers (at r_1 and r_2) are correlated, i.e., both scatterings have actually occurred at the same inhomogeneity.

Let us now consider the second-order diagrams, i.e., diagrams 3–5. They contain the same product of Green's functions

$$G_0(r, r_1)G_0(r_1, r_2)G_0(r_2, r_3)G_0(r_3, r_4)G_0(r_4, r_0) \quad .$$

This means that the wave came to r after it had been scattered at r_1; to r_1 after r_2, etc., whereas the primary wave $G_0(r_4, r_0)$ underwent a first scattering at r_4. All of these diagrams thus describe four-fold scattering. What is the difference between the processes described by these *topologically different* diagrams? In 3, the lines of the covariances connect r_1 to r_2 and r_3 to r_4. This implies that both r_1 and r_2 belong to the same inhomogeneity, and r_3 and r_4 to another. Consequently, 3 describes the process by which the wave first propagates freely from the source to the first inhomogeneity, is then scattered from it twice, and the doubly scattered wave propagates further unhindered, whereupon it is scattered twice from the second inhomogeneity. The diagram also describes four-fold scattering from two inhomogeneities, but the sequence of scatterings here is different. First, the wave is scattered from the first inhomogeneity (at r_4), then the singly scattered wave comes to the second inhomogeneity and is scattered from it (at r_3). The doubly scattered wave then returns to the first inhomogeneity to be scattered there (at point r_2 connected by dashed line with r_4), whereupon the triply scattered wave is again scattered from the second inhomogeneity (at r_1) and reaches the observation point. The sequence of scatterings from the two inhomogeneities is shown schematically in Fig. 4.1.

The diagrammatical representation (4.7a) of the solutions to (4.1) is useful not only because of its clarity, but also because it enables expansion in perturbations to be transformed using the *topological* features of the diagrams that enter into the solution. We can thus express the sum of the expansion (4.7) through the sum of a certain infinite *subsequence* of the same expansion. For this procedure to be possible, we will first have to classify the diagrams that enter into (4.7a).

The diagram in $\overline{G}$ is said to be *weakly connected* if it can be divided into two diagrams by breaking some line G_0. So, in (4.7a), diagrams 3, 6–9 and 12 are weakly connected. The remaining diagrams will be said to be *strongly connected* [2, 4, 5, 10, 11, 13–20 in (4.7a)]. Diagrams that emerge when we break the lines G_0 in a weakly connected diagram can in turn appear to be strongly or weakly connected. If there are weakly connected diagrams among the "secondary" ones, these can be broken down into simpler diagrams by breaking one solid line. Eventually, this procedure will yield a certain number of strongly connected diagrams. We will refer to the number of strongly connected diagrams derived from a weakly connected diagram

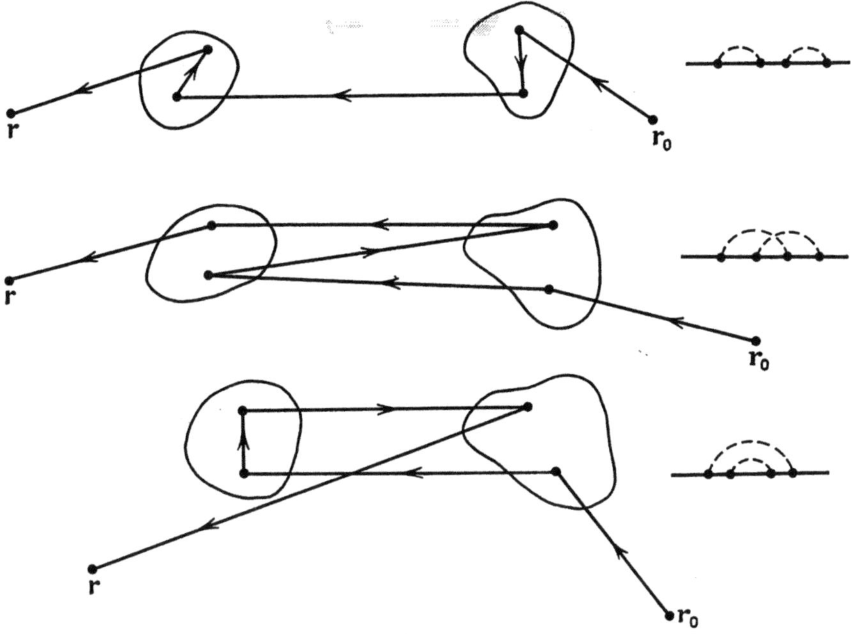

Fig. 4.1. Various scattering situations and the corresponding Feynman diagrams

as the *connection index* of the initial diagram. Returning to (4.7a), we can say that diagrams 3, 7–9 and 12 have a connection index of 2; and 6 has a connection index of 3. Strongly connected diagrams can be ascribed index 1.

We will separate all the strongly connected diagrams from (4.7a). Since each of the diagrams has the line G_0 at the beginning and end, the sum of all the strongly connected diagrams can be given by

$$\text{—}\bigcirc\text{—} \, ,\qquad\qquad (4.11)$$

where

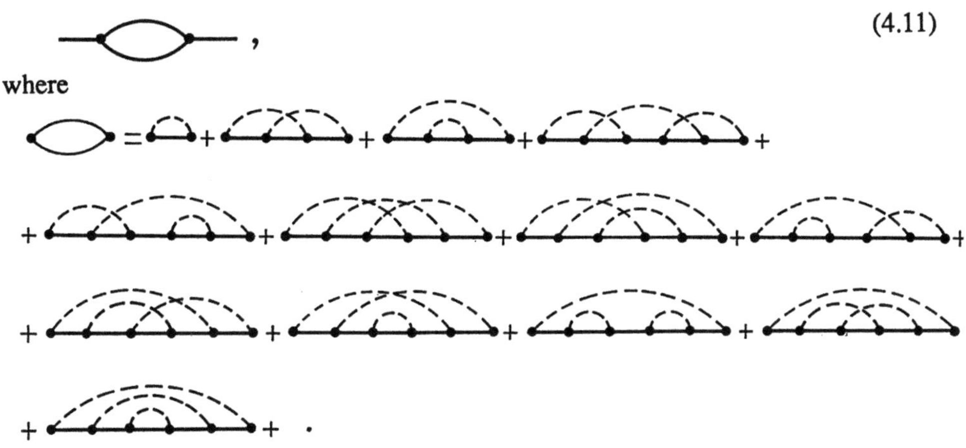

Analytically, (4.11) has the form

$$\overline{G}^{(\mathrm{str.con})}(r, r_0) = \int G_0(r, r')Q(r', r'')G_0(r'', r_0)d^3r'\, d^3r''\, , \quad \text{where} \quad (4.11a)$$

$$Q(r', r'') = k^4 G_0(r', r'')\psi_\varepsilon(r', r'') + k^8 \int G_0(r'_1, r')G_0(r_1, r_2)$$
$$\times G_0(r_2, r'')\psi_\varepsilon(r', r_2)\psi_\varepsilon(r_1, r'')d^3 r_1 d^3 r_2$$
$$+ k^8 \int G_0(r', r_1)G_0(r_1, r_2)G_0(r_2, r'')\psi_\varepsilon(r', r'')$$
$$\times \psi_\varepsilon(r_1, r_2)d^3 r_1 d^3 r_2 + \ldots \quad . \tag{4.12}$$

The function Q is called the *kernel of mass operator* (the name was borrowed from the quantum field theory).

Let us now look at the sum of all the diagrams with a connection index 2, each of which may be represented by

(4.13)

where

are some of the diagrams appearing on the right-hand side of (4.12). Since, in constructing (4.7a), we exhaust all possible ways of pairwise combination of the points, it is clear that the sum of all the possible terms of the form (4.13) will be

where

is the total sum of (4.12).

Similarly, the sum of all the diagrams with index 3 has the form

and so on. Consequently, we can represent the mean Green's function as

(4.14)

The corresponding formula only differs from the original expansion (4.7a) by the arrangement of its terms.

We will now see that (4.14) is the solution to the following equation:

(4.15)

which is called the *Dyson equation*. In the analytic form, (4.15) is

$$\overline{G}(r, r_0) = G_0(r, r_0) + \int G_0(r, r_1)Q(r_1, r_2)\overline{G}(r_2, r_0)d^3r_1 d^3r_2 \quad . \qquad (4.15a)$$

In order to show that (4.14) is the solution to (4.15), we will find this solution by successive iterations. This can be done in both analytic and graphic form. Using the latter, we will substitute the expression (4.15) for $\overline{G}$ into the right-hand side of (4.15). We get

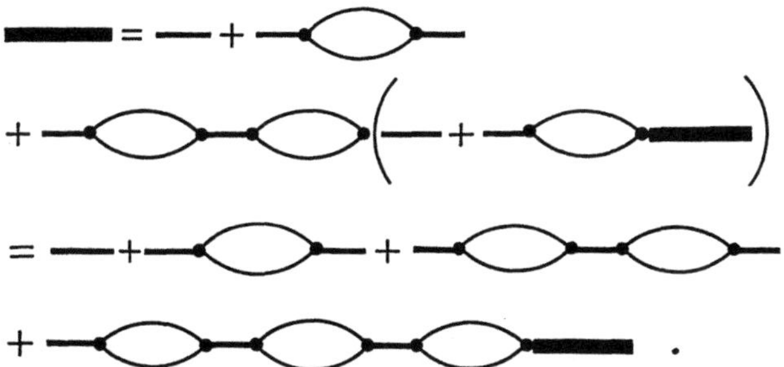

If we again substitute the right-hand side of (4.15) into the right-hand side of this equation, we will obtain

Further iteration will clearly lead to (3.14). It was mentioned above that these computations can also be made analytically, proceding from (4.15a). The result will be the expansion (4.14) written in analytic form. We will not provide the calculations here, but will leave them as an exercise for the reader, during which the power of the graphical method should become apparent.

If Q in (4.15a) is thought to be known, the equation is a linear integral equation in $\overline{G}$ which, in many cases, can be solved (see the next section). This gives an explicit expression of $\overline{G}$ in Q, i.e., the sum of (4.7a) is expressed in terms of $\overline{G}^{(\text{str.con.})}$, which is a subsequence of the same expansion.

In fact, we do not know Q exactly. It can be the sum of some of the leading terms of the expansion (4.12), or we can express Q through some new function that obeys a *nonlinear* integral equation. The latter approach is, however, too complex and will not be discussed (see [4.3]).

We will now turn to the *correlation* of two fields produced by point sources lying at r' and r''

$$\langle G(r', r_0')G^*(r'', r_0'')\rangle = \Gamma(r', r''; r_0', r_0'') \quad . \qquad (4.16)$$

To find Γ, we will have to multiply together two expansions of the form (4.3) and average the result — a more tedious problem than that considered above. It can be somewhat simplified by representing diagrammatically some quantities in (4.3) *that have not yet been averaged*. We will denote the factor $-k^2\tilde{\varepsilon}(r)$ as a cross, and $G(r, r_0)$ as a tilde. Averaging these quantities gives the elements introduced above:

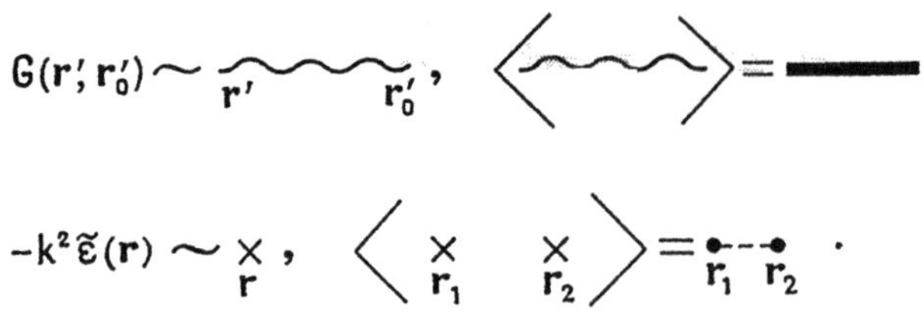

$$G(r', r_0') \sim \!\!\!\overset{}{\underset{r' \qquad r_0'}{\sim\!\!\!\sim\!\!\!\sim}}, \quad \left\langle \,\sim\!\!\!\sim\!\!\!\sim\, \right\rangle = \rule[0.4ex]{2cm}{2pt}$$

$$-k^2 \tilde{\varepsilon}(r) \sim \underset{r}{\times}, \quad \left\langle \underset{r_1}{\times} \quad \underset{r_2}{\times} \right\rangle = \underset{r_1 \;\; r_2}{\bullet\!-\!-\!\bullet}. \tag{4.17}$$

Expansion (4.3) will be given here by the infinite sum of diagrams of the following form:

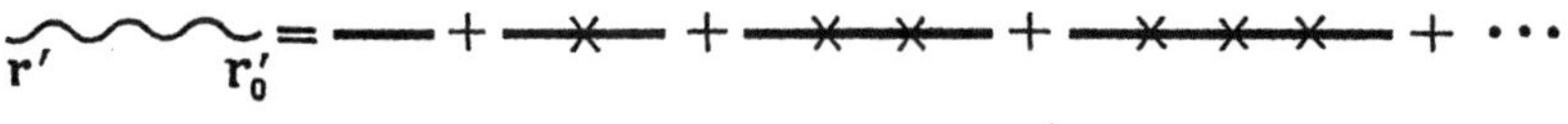

$$\underset{r' \qquad r_0'}{\sim\!\!\!\sim\!\!\!\sim} = \rule[0.4ex]{1cm}{1.5pt} + \!\!-\!\!\times\!\!-\!\! + \!\!-\!\!\times\!\!-\!\!\times\!\!-\!\! + \!\!-\!\!\times\!\!-\!\!\times\!\!-\!\!\times\!\!-\!\! + \cdots \tag{4.3a}$$

If we average (4.3a), the diagrams with an odd number of crosses will vanish and, according to the rules of averaging, we will again obtain the expansion (4.7a).

Let us write an analogous expansion for $G^*(r'', r_0'')$. It differs from (4.3a) in that here G_0^* is substituted for G_0, which will be denoted by crossing an appropriate line

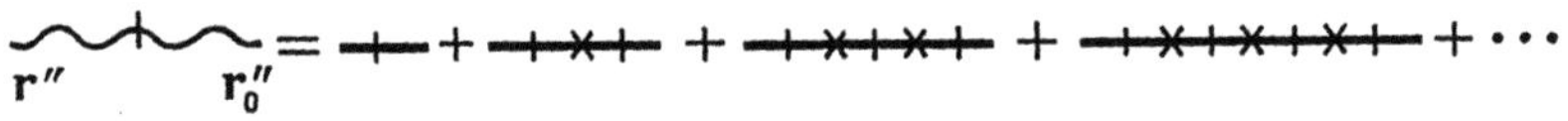

$$\underset{r'' \qquad r_0''}{\sim\!\!\!\sim\!\!\!+\!\!\!\sim} = \!-\!\!+\!\!- + \!-\!\!+\!\!\times\!\!+\!\!- + \!-\!\!+\!\!\times\!\!+\!\!\times\!\!+\!\!- + \!-\!\!+\!\!\times\!\!+\!\!\times\!\!+\!\!\times\!\!+\!\!- + \cdots \tag{4.18}$$

We will now multiply together the expansions (4.3a) and (4.18) and average the result. When multiplying together individual elements of (4.3a) and (4.18), we will place at the top elements belonging to (4.3a), and below them elements belonging to (4.18). For example, the averaged product of the third term in (4.3a) and the third term in (4.18) will take the form

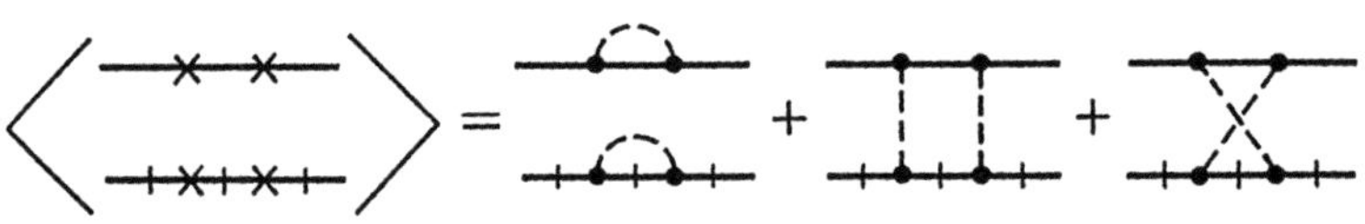

$$\left\langle \begin{array}{c} -\!\times\!-\!\times\!- \\ -\!+\!\times\!+\!\times\!+\!- \end{array} \right\rangle = \overset{\frown}{\underset{\smile}{\rule{0pt}{1em}}} + \prod + \bigtimes. \tag{4.19}$$

Let us now discuss the result. Averaging the product $\tilde{\varepsilon}(r_1)\tilde{\varepsilon}(r_2)\tilde{\varepsilon}(r_3)\tilde{\varepsilon}(r_4)$ gives, by (4.5), three terms with diagrammatical counterparts on the right-hand side of (4.19). In the first of these three diagrams, the product of the factors belonging to $G(r', r_0')$ and $G^*(r'', r_0'')$ is averaged. The resultant diagram consists of two independent pieces not connected by lines (unconnected diagram). Clearly, any unconnected diagram of this type represents one of the terms of the product $\overline{G}(r', r_0')\overline{G}_0^*(r'', r_0'')$, and the sum of all the unconnected diagrams is equal to this product.

If we introduce for $\langle G(r', r_0')G^*(r'', r_0'')\rangle$ the graphical representation

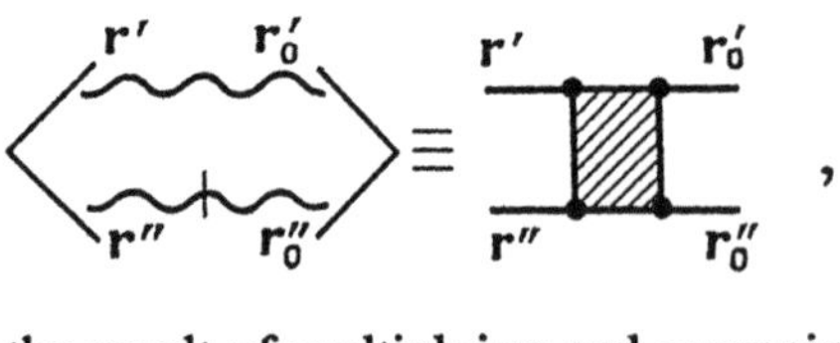

$$\left\langle \begin{array}{c} r' \qquad r_0' \\ \sim\!\!\!\sim\!\!\!\sim \\ \sim\!\!\!+\!\!\!\sim \\ r'' \qquad r_0'' \end{array} \right\rangle = \begin{array}{c} r' \qquad\quad r_0' \\ \rule{3em}{0.4pt} \\ \boxed{} \\ \rule{3em}{0.4pt} \\ r'' \qquad\quad r_0'' \end{array},$$

then the result of multiplying and averaging (4.3a) and (4.19) will be

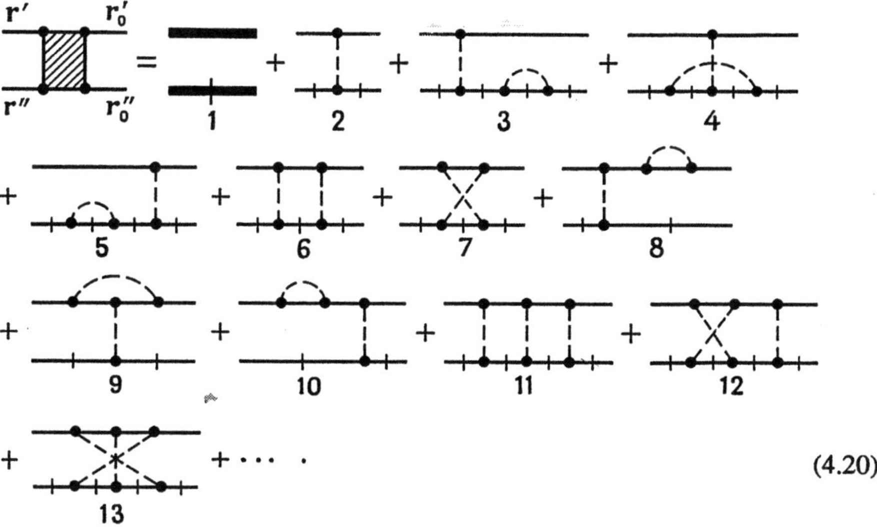

$$(4.20)$$

Here we see all the fourth-order diagrams 3–10 and only three of the sixth-order diagrams.

Let us now attach some physical interpretation to these diagrams. Diagram 1 in (4.20) describes the propagation of the wave with multiple scattering from r_0' to r' and from r_0'' to r''. In both paths, the wave is scattered from *different* inhomogeneities (Fig. 4.2).

Diagram 2 describes a process whereby both the first and second waves are singly scattered from the same inhomogeneity, etc., (two further examples are given in Fig. 4.2).

We will now classify the component diagrams of (4.20). All the diagrams, except for those belonging to 1, are connected. We will refer to the diagram for a covariance as strongly connected, if no breaking of single lines G_0 and G_0^* divides it into two independent parts, each of which will contain at least two vertices. So, in (4.20), diagrams 2, 4, 7, 9 and 13 are strongly connected. All the remaining diagrams are weakly connected, but their classification is somewhat more complex than that of the diagrams that make up $\overline{G}$.

Each of the strongly connected diagrams ends with four lines G_0 and G_0^*, i.e., has the form

$$(4.21)$$

or, in symbols,

$$\Gamma^{(\mathrm{str.con})}(r', r''; r_0', r_0'') = \int G_0(r', r_1) G_0(r_3, r_0') G_0^*(r'', r_2)$$

$$\times\, G_0^*(r_4, r_0'') K(r_1, r_2; r_3, r_4) d^3 r_1 \ldots d^3 r_4 \quad .$$

$$(4.21a)$$

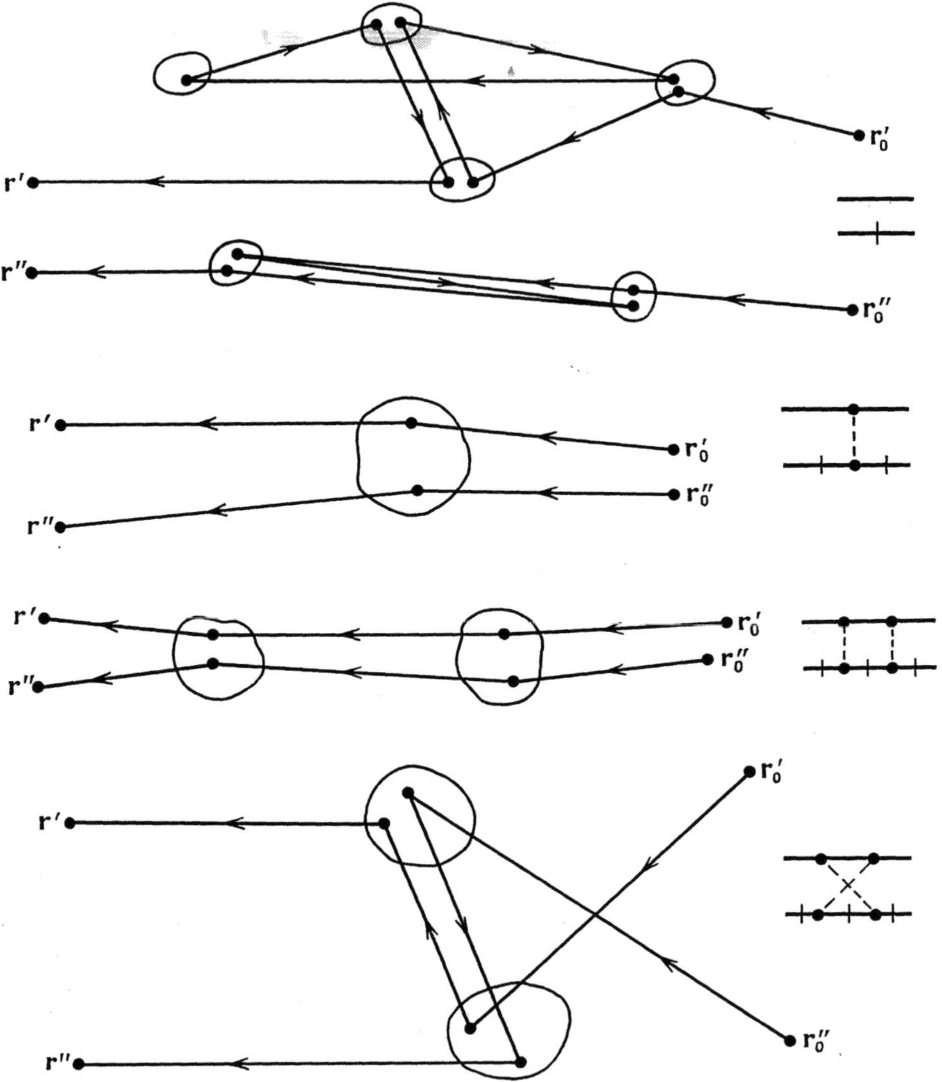

Fig. 4.2. Various scattering situations and the corresponding Feynman diagrams for the coherence function of the field

Function K is called the *kernel of intensity operator*. In diagrams it is

$$\begin{array}{c} r_1 \ \ r_3 \\ \boxed{} \\ r_2 \ \ r_4 \end{array} = \ \vdots \ + \ \cdots \ + \ \cdots \ + \ \cdots \ + \ \cdots \ + \ \cdots \ + \cdots \ . \tag{4.22}$$

In the case where the upper or lower level contains one vertex only, the analytic expression for K will contain the factor $\delta(r_1 - r_3)$ or $\delta(r_2 - r_4)$. Expansion (4.22) has the following analytic form:

$$K(r_1, r_2; r_3, r_4) = k^4 \psi_\varepsilon(r_1, r_2)\delta(r_1 - r_3)\delta(r_2 - r_4)$$

$$+ \int k^8 \psi_\varepsilon(r_1, r_5)\psi_\varepsilon(r_2, r_4)G_0^*(r_2, r_5)G_0^*(r_5, r_4)\delta(r_1 - r_3)d^3r_5$$

$$+ k^8 \psi_\varepsilon(r_1, r_4)\psi_\varepsilon(r_2, r_3)G_0(r_1, r_3)G_0^*(r_2, r_4)$$

$$+ \int k^8 \psi_\varepsilon(r_1, r_3)\psi_\varepsilon(r_5, r_2)G_0(r_1, r_5)G_0(r_5, r_3)\delta(r_2 - r_4)d^3r_5 + \ldots (4.22a)$$

(written here are the terms up to fourth order).

We will now turn to the possible types of weakly connected diagrams.

1) A weakly connected diagram can only contain *one* of the strongly connected elements that enter into K, and it will remain weakly connected due to the fact that one (or more) of the outer lines G_0 or G_0^* is replaced by elements belonging to the mean Green's function $\overline{G}$ or $\overline{G}^*$. Diagrams 3, 5, 8, and 10 in (4.20) belong to this type, and all of them belong to the multitude of diagrams of the type

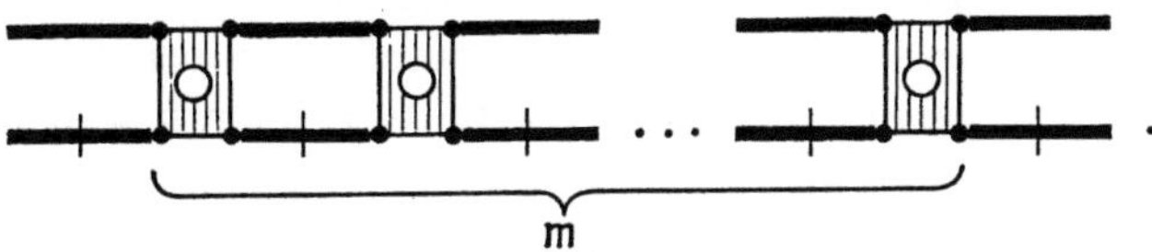

$$(4.23)$$

For example, diagram 3 in (4.20) emerges if in K we select the element $\bullet\text{---}\bullet$, in $\overline{G}^*(r_4, r_0'')$ the element $\text{---}\!\text{⌂}\!\text{---}$, and in the remaining functions $\overline{G}$ and $\overline{G}^*$, the element G_0. Clearly, the sum of all the diagrams of this type is (4.23).

2) Weakly connected diagrams with *two* strongly connected elements of K add up to

$$(4.24)$$

Diagrams 6 and 12 are included in this set of diagrams.

Similarly, diagrams containing m strongly connected elements of K add up to

It follows that (4.20) can be represented as

$$+ \cdots ,$$

$$(4.25)$$

which is similar to (4.14) for $\overline{G}$.

Just as the Dyson equation follows from (4.15), we can easily derive from (4.25) the so-called *Bethe-Salpeter* equation

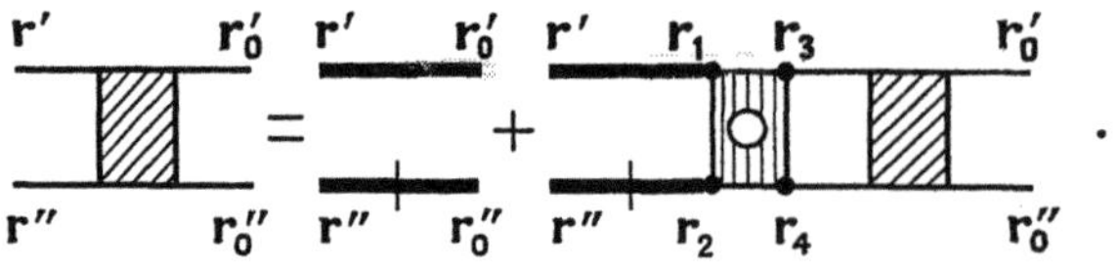

(4.26)

In fact, if (4.26) is solved by successive iterations, we arrive at (4.25)

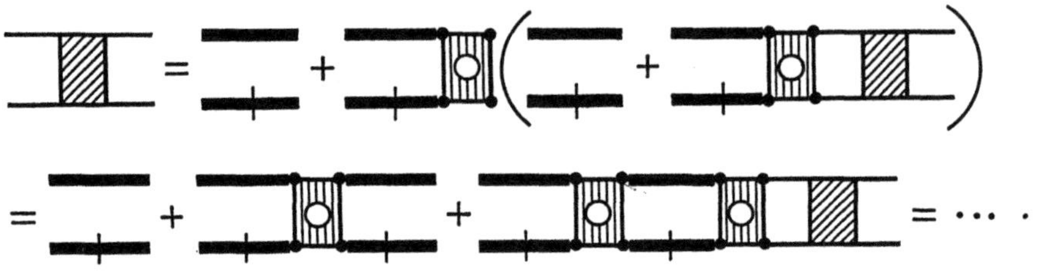

Analytically, (4.26) will be

$$\begin{aligned}\Gamma(\boldsymbol{r}',\boldsymbol{r}'';\boldsymbol{r}'_0,\boldsymbol{r}''_0) &= \overline{G}(\boldsymbol{r}',\boldsymbol{r}'_0)\overline{G}^*(\boldsymbol{r}'',\boldsymbol{r}''_0)\\ &+\int \overline{G}(\boldsymbol{r}',\boldsymbol{r}_1)\overline{G}^*(\boldsymbol{r}'',\boldsymbol{r}_2)K(\boldsymbol{r}_1,\boldsymbol{r}_2;\boldsymbol{r}_3,\boldsymbol{r}_4)\\ &\times \Gamma(\boldsymbol{r}_3,\boldsymbol{r}_4;\boldsymbol{r}'_0,\boldsymbol{r}''_0)d^3r_1\dots d^3r_4 \quad . \end{aligned} \tag{4.26a}$$

The following two sections will be concerned with an examination of the mean field and coherence function Γ. However, before we conclude this section, a general remark is in order.

Up until this point, we have made no mention of the specific form of $G_0(\boldsymbol{r}',\boldsymbol{r}'')$, and the above treatment relied on expansion (4.3). This expansion is the solution of the following integral equation:

$$G(\boldsymbol{r}',\boldsymbol{r}'') = G_0(\boldsymbol{r}',\boldsymbol{r}'') - k^2\int G_0(\boldsymbol{r}',\boldsymbol{r}_1)\tilde{\varepsilon}(\boldsymbol{r}_1)G(\boldsymbol{r}_1,\boldsymbol{r}'')d^3r'' \quad . \tag{4.27}$$

Generally speaking, in this equation an arbitrary function G_0 could be regarded as the initial one, whereby we would arrive at the same diagram representations, and the same equations (4.15) and (4.26). Moreover, the dimensionality of the space where the radius vector is $\boldsymbol{r}$ is also of no consequence. Therefore, $\boldsymbol{r}$ can also be thought of as a vector in an n-dimensional space with arbitrary n.

A wide variety of problems in physics can be reduced to an equation of the form (4.27). For example, the parametric oscillator with the random frequency $\omega = \omega_0\sqrt{1+\tilde{\varepsilon}(t)}$, where $\tilde{\varepsilon}(t)$ is a random process, is described by the stochastic Green's function $G(t,t_0)$ which obeys the differential equation

$$\ddot{G}(t,t_0) + \omega_0^2[1+\tilde{\varepsilon}(t)]G(t,t_0) = \delta(t-t_0)$$

with the initial conditions $G(0,t_0)=0$, $\dot{G}(0,t_0)=0$. This problem can be formulated as an integral equation of the form (4.27)

$$G(t,t_0) = G_0(t,t_0) - \omega_0^2\int\limits_0^\infty G_0(t,t')\tilde{\varepsilon}(t')G(t',t_0)dt' \quad , \quad \text{where}$$

$$G_0(t, t_0) = \frac{\theta(t - t_0)}{\omega_0} \sin\left[\omega_0(t - t_0)\right] \quad,$$

and θ is a step function.

Another example is the scattering of waves from a surface with a randomly distributed impedance [4.4]. For further examples, the reader is referred to [4.5, 6].

Thus, the technique discussed above covers the many problems described by linear integral equations with a Gaussian random kernel.

However, it should be noted that, along with the diagram technique, they often use the *operator* form of perturbation theory (see [4.7]), which also yields approximate equations for statistical moments of the solution to (4.27).

4.2 Mean Field of Point Sources in an Infinite Random Medium

Consider the Dyson equation (4.15a)

$$\overline{G}(r, r_0) = G_0(r, r_0) + \int G_0(r, r_1) Q(r_1, r_2) \overline{G}(r_2, r_0) d^3 r_1 d^3 r_2 \quad . \tag{4.28}$$

Operating with $(\Delta + k^2)$ and considering that

$$(\Delta + k^2) G_0(r, r_0) = \delta(r - r_0) \quad, \tag{4.29}$$

we obtain

$$\Delta \overline{G}(r, r_0) + k^2 \overline{G}(r, r_0) - \int Q(r, r') \overline{G}(r', r_0) d^3 r' = \delta(r - r_0) \quad . \tag{4.30}$$

Comparing (4.30) with (4.29), we find that, unlike G_0, $\overline{G}$ obeys an integro-differential, and not a differential equation. An equation of the form (4.30) follows, for example, for Maxwell's equations when the electric displacement D and the field strength E in the isotropic medium are related in a spacially nonlocal manner:

$$D(r) \equiv E(r) + 4\pi P(r), \quad P(r) = \int \chi(r, r') E(r') d^3 r' \quad,$$

where P is the polarization, and χ the polarizability of the medium. Indeed, from Maxwell's equations

$$\mathrm{curl}\, E = \mathrm{i} k H \quad, \quad \mathrm{curl}\, H = -\mathrm{i} k D \quad,$$

we then obtain

$$\Delta E + k^2 E + 4\pi k^2 \int \chi(r, r') E(r') d^3 r' = \mathrm{grad}\, \mathrm{div}\, E \quad .$$

Again, comparing this equation with (4.30), we see that the quantity

$$-\frac{Q(r, r')}{4\pi k^2} = \chi(r, r')$$

can be treated as the polarizability of the medium with spatial dispersion. Physically, this is obvious since the mean field at a point r is also dependent on the surrounding inhomogeneities.

Consider now a *statistically homogeneous* medium, i.e., suppose that $\psi_\varepsilon(r_1, r_2) = \psi_\varepsilon(r_1 - r_2)$. Turning to the expansion (4.12a) for $Q(r_1, r_2)$, we can easily see that in this case too, the kernel Q is only dependent on $r_1 - r_2$, i.e., $Q = Q(r_1 - r_2)$. Finally, since Green's function $G_0(r, r_0)$ also varies with the difference of its arguments, it is clear that the mean Green's function in the inhomogeneous medium will also have the form $\overline{G}(r, r_0) = \overline{G}(r - r_0)$. Under these conditions, (4.28) can be solved using the Fourier transforms

$$G_0(r - r_1) = \int g_0(\kappa)\exp\left[i\kappa \cdot (r - r_1)\right]d^3\kappa \quad,$$

$$\overline{G}(r - r_1) = \int \overline{g}(\kappa)\exp\left[i\kappa \cdot (r - r_1)\right]d^3\kappa \quad,$$

$$Q(r_1 - r_2) = \int q(\kappa)\exp\left[i\kappa \cdot (r_1 - r_2)\right]d^3\kappa \quad. \tag{4.31}$$

Then

$$g_0(\kappa) = [8\pi^3(k^2 - \kappa^2 + io]^{-1} \quad. \tag{4.32}$$

Here, we have introduced into the denominator an infinitesimal imaginary correction io, which ensures that only divergent waves are present when we take the integrals (4.31). Of course, it can be viewed as an infinitesimal absorption $(k \rightarrow k + io/2)$.

In this case, (4.28) has the form

$$\overline{G}(r - r_0) = G_0(r - r_0) + \int G_0(r - r_1)Q(r_1 - r_2)$$

$$\times \overline{G}(r_2 - r_0)d^3r_1 d^3r_2 \quad, \tag{4.33}$$

i.e., the two-fold integral is now a two-fold convolution. When subject to Fourier transformation, it gives the product of the Fourier transforms, and (4.33) thus becomes a purely algebraic equation

$$\overline{g}(\kappa) = g_0(\kappa) + (8\pi^3)^2 g_0(\kappa)q(\kappa)\overline{g}(\kappa) \quad. \tag{4.34}$$

Solving this equation for $\overline{g}$, by (4.32) we obtain

$$\overline{g}(\kappa) = \frac{g_0(\kappa)}{1 - (8\pi^3)^2 g_0(\kappa)q(\kappa)} = \frac{1}{8\pi^3[k^2 - \kappa^2 - 8\pi^3 q(\kappa) + io]} \quad. \tag{4.35}$$

Substituting (4.35) into (4.31) and expressing $q(\kappa)$ in terms of $Q(r)$, using the inverse Fourier transform, we obtain

$$\overline{G}(r - r_0) = \frac{1}{8\pi^3} \int \frac{\exp\left[i\kappa \cdot (r - r_0)\right]}{k^2 - \kappa^2 - \int Q(r')\exp\left(-i\kappa \cdot r'\right)d^3r' + io}d^3\kappa \quad. \tag{4.36}$$

This relationship explicitly expresses $\overline{G}$ in terms of Q.

Although (4.36) is an *exact* formula, the expression for Q is only known as an expansion that is generally not amenable to summation. Equation (4.36) is therefore only useful in the sense that it allows an appropriate approximation for $\overline{G}$ to be derived from one approximation or other for Q.

One of the simplest approximations, which will be considered in more detail later in this section, is the so-called *Bourret approximation*. It only keeps the *first term* of (4.12a), i.e.

$$Q \approx Q_1 = \overset{\frown}{\underset{r_1 \quad r_2}{\bullet\!-\!\!-\!\bullet}} = k^4 G_0(r_1 - r_2)\,\psi_\varepsilon(r_1 - r_2). \tag{4.37}$$

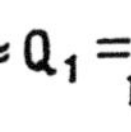

Graphically, the corresponding approximation $\overline{G}_1$ for $\overline{G}$ will be

$$\overline{G}_1 = \text{\small$\curvearrowright\!\!\curvearrowright\!\!\curvearrowright\!\!\curvearrowright\!\!\curvearrowright$} \;.$$

Substituting (4.37) for Q in (4.14), we arrive at the following diagrammatic representation of $\overline{G}_1$:

$$\text{\small$\curvearrowright\!\!\curvearrowright\!\!\curvearrowright$} = -\!-\!- + -\!\overset{\frown}{\bullet\!-\!\bullet}\!- + -\!\overset{\frown}{\bullet\!-\!\bullet}\!-\overset{\frown}{\bullet\!-\!\bullet}\!- + -\!\overset{\frown}{\bullet\!-\!\bullet}\!-\overset{\frown}{\bullet\!-\!\bullet}\!-\overset{\frown}{\bullet\!-\!\bullet}\!- + \cdots$$

$$\tag{4.38}$$

Accordingly, if we substitute (4.37) into (4.36) and take the integral, we can find $\overline{G}_1$, i.e., sum up (4.38).

If a random medium is not only statistically homogeneous, but also *statistically isotropic*, then clearly

$$Q(r') = Q(r') \quad \text{and} \quad \overline{G}(r - r_0) = \overline{G}(|r - r_0|) \quad .$$

In that case, (4.36) can be simplified further by the use of spherical coordinates (both in r' and in κ), and integration over all the angle variables. As a result, (4.36) becomes

$$\overline{G}(R) = \frac{1}{4\pi^2 R i} \int\limits_{-\infty}^{\infty} \frac{\exp(i\kappa R)\kappa \, d\kappa}{\left(k^2 - \kappa^2 - (4\pi/\kappa)\displaystyle\int\limits_0^{\infty} Q(r')\sin(\kappa r')r'\,dr'\right)} \quad ,$$

$$R = |r - r_0| \quad . \tag{4.39}$$

The Dyson equation (4.28) can be transformed so that, instead of Green's function of the unperturbed medium, G_0, it will contain a partially summed infinite subsequence of the expansion for $\overline{G}$ [4.8,9]. This transformation allows the convergence of the expansion for $\overline{G}$ to be improved. To effect this transformation, we will first write the Dyson equation in the operator form. First, we will introduce the linear operators $\hat{M}$, $\hat{M}_0$ and $\hat{Q}$ with the kernels $\overline{G}$, G_0 and Q, respectively

$$(\hat{M}f)(r) \equiv \int \overline{G}(r, r_0)f(r_0)d^3r_0 \quad ,$$

$$(\hat{M}_0 f)(r) \equiv \int G_0(r, r_0)f(r_0)d^3r_0 \quad ,$$

$$(\hat{Q}f)(r) \equiv \int Q(r, r')f(r')d^3r' \quad ,$$

where $(\hat{M}f)(r)$ signifies that f is transformed by $\hat{M}$ into $(\hat{M}f)$ at a point r.

Multiplying (4.28) by $f(r_0)$ and integrating with respect to r_0, we can write

$$\hat{M}f = \hat{M}_0 f + \hat{M}_0 \hat{Q}\hat{M}f \quad .$$

Since it is valid for any f, the relationship can also be written as an equation for $\hat{M}$

$$\hat{M} = \hat{M}_0 + \hat{M}_0\hat{Q}\hat{M} = \hat{M}_0(\hat{1} + \hat{Q}\hat{M}) \quad . \tag{4.28a}$$

It was found in the previous section that the function $G(r', r'')$ (not averaged) is symmetrical, i.e., $G(r', r'') = G(r'', r')$, and that after averaging

$$\overline{G}(r', r'') = \overline{G}(r'', r') \quad .$$

This suggestes that, in the operator form, the transposed operator $\hat{M}^t$ coincides with $\hat{M}$

$$\hat{M}^t = \hat{M} \quad , \tag{4.40}$$

and the symmetry of G_0 suggests

$$\hat{M}_0^t = \hat{M}_0 \quad . \tag{4.41}$$

Diagrammatically, (4.40) implies that each diagram in the expansion for $\overline{G}$ is either symmetric about the vertical axis through the diagram center, or, if not symmetric, enters into $\overline{G}$ in combination with another diagram produced from the original nonsymmetrical one by reflection about the vertical axis. For example, (4.7a) contains symmetrical diagrams 1–6, 10, 13 and 17–20, and pairs of nonsymmetrical diagrams (7, 9), (8, 12), (11, 15), and (14, 16).

It follows from the fact that $\overline{G}(r', r'')$ is symmetric that $Q(r', r'') = Q(r'', r')$, since the diagrams for Q show the same type of symmetry as those for $\overline{G}$. Accordingly,

$$\hat{Q}^t = \hat{Q} \quad . \tag{4.42}$$

It is well known that transposing a product of operators reverses the order in which they appear, i.e., $(\hat{A}\hat{B}\hat{C}\ldots)^t = \ldots\hat{C}^t\hat{B}^t\hat{A}^t$. We will apply this operation to (4.28a). Using (4.40–42), we get

$$\hat{M} = \hat{M}_0 + \hat{M}\hat{Q}\hat{M}_0 = (\hat{1} + \hat{M}\hat{Q})\hat{M}_0 \quad . \tag{4.28b}$$

Suppose that we know some approximate solution $\hat{M}_1$ to (4.28b) corresponding to some approximate expression $\hat{Q}_1$, i.e., we know the solution to

$$\hat{M}_1 = (\hat{1} + \hat{M}_1\hat{Q}_1)\hat{M}_0 \quad . \tag{4.43}$$

It is always possible to obtain such a solution. We will have to substitute into (4.36) some approximate expression Q_1 (e.g., the Bourret approximation) for the unknown exact expression Q. We then multiply (4.28a) on the left-hand side by $(\hat{1} + \hat{M}_1\hat{Q}_1)$

$$(\hat{1} + \hat{M}_1\hat{Q}_1)\hat{M} = (1 + \hat{M}_1\hat{Q}_1)\hat{M}_0(1 + \hat{Q}\hat{M}) \quad .$$

Substituting (4.43) into the right-hand side of this equation, we obtain

$$(\hat{1} + \hat{M}_1\hat{Q}_1)\hat{M} = \hat{M}_1(\hat{1} + \hat{Q}\hat{M}) \quad ,$$

which may be written in the form

$$\hat{M} = \hat{M}_1 + \hat{M}_1(\hat{Q} - \hat{Q}_1)\hat{M} \quad . \tag{4.44}$$

Equation (4.44) is analogous to the Dyson equation (4.28a), $\hat{M}_1$ being substituted for $\hat{M}_0$ and $(\hat{Q} - \hat{Q}_1)$ for $\hat{Q}$. Clearly, if the norm of $\hat{Q} - \hat{Q}_1$ is less than that of the original operator $\hat{Q}$, the iterative expansion of (4.44) will converge faster than the original one.

Applying (4.44) to a delta-function gives for the kernels of $\hat{M}$, $\hat{M}_1$, $\hat{Q}$ and $\hat{Q}_1$

$$\overline{G}(r, r_0) = \overline{G}_1(r, r_0) + \int \overline{G}_1(r, r_1)[Q(r_1, r_2) - Q_1(r_1, r_2)]$$
$$\times \overline{G}(r_2, r_0)d^3r_1 d^3r_2 \quad . \tag{4.44a}$$

For a statistically homogeneous medium, (4.44a) can be solved using Fourier transforms. In the spectral form, i.e., for Fourier transforms, the solution will have a form similar to (4.35)

$$\overline{g}(\kappa) = \frac{\overline{g}_1(\kappa)}{1 - (8\pi^3)^2 g_1(\kappa)[q(\kappa) - q_1(\kappa)]} \quad . \tag{4.45}$$

In order to improve further the convergence of the expansion of $\overline{G}$, we can repeat this procedure. For example, we can specify an approximate expression for $Q - Q_1 = Q_2$ and, using (4.45), find the desired expression for $\overline{G}_2$.

Let us now take a closer look at the Bourret approximation for the statistically isotropic fluctuations ε.

Substituting the approximation (4.37) for $Q(r)$ into (4.39) gives

$$\overline{G}_1(R) = \frac{1}{4\pi^2 Ri} \int_{-\infty}^{\infty} \frac{\exp(i\kappa R)\kappa\, d\kappa}{k^2 - \kappa^2 + (k^4/\kappa) \int_0^{\infty} \psi_\varepsilon(r)e^{ikr} \sin(\kappa r)dr + io} \quad . \tag{4.46}$$

The significance of this result can be clarified by considering a particular example, $\psi_\varepsilon(r) = \sigma_\varepsilon^2 \exp(-\alpha r)$, where $\alpha^{-1} = l_\varepsilon$ is the correlation radius of the fluctuations ε. Given this ψ_ε, the integral in the integrand's denominator in (4.46) is easily taken. In order that we might then carry out the integration with respect to κ, we will take κ to be complex. This enables the integration path to be closed by an infinite semi-circle in the upper half-plane κ, and the integration then reduces to summing up the residues at the poles in the region $\mathrm{Im}\{\kappa\} > 0$.

In this example, the roots of the denominator are readily obtainable:

$$\kappa_1^2 = -k^2 \frac{\alpha(\alpha - 2ik)}{2}\delta \quad , \quad \kappa_2^2 = (k + i\alpha)^2 + \frac{\alpha(\alpha - 2ik)}{2}\delta \quad , \tag{4.47}$$

$$\delta = 1 - \sqrt{1 + \frac{4k^4\sigma_\varepsilon^2}{\alpha^2(\alpha - 2ik)^2}} \quad .$$

To simplify the treatment further, we will consider sufficiently small fluctuations, where $|\delta| \ll 1$, which is the case if

$$\left| \frac{4k^4\sigma_\varepsilon^2}{\alpha^2(\alpha - 2ik)^2} \right| = \frac{4k^4\sigma_\varepsilon^2}{\alpha^2(\alpha^2 + 4k^2)} \ll 1 \quad . \tag{4.48}$$

In that case,

$$\delta \approx -2k^4\sigma_\varepsilon^2/\alpha^2(\alpha - 2ik)^2 \quad \text{for} \quad |\delta| \ll 1 \quad , \quad \text{and}$$

$$\kappa_1^2 = k^2\left[1 + \frac{k^2\sigma_\varepsilon^2}{\alpha^2 + 4k^2} + i\frac{2k^3\sigma_\varepsilon^2}{\alpha(\alpha^2 + 4k^2)} + \dots\right] \quad . \tag{4.49}$$

We will require that, along with (4.48), the conditions

$$\frac{k^2\sigma_\varepsilon^2}{\alpha^2 + 4k^2} \ll 1 \quad , \quad \frac{2k^3\sigma_\varepsilon^3}{\alpha(\alpha^2 + 4k^2)} \ll 1 \tag{4.50}$$

be met, which allow us to extract approximately the square root of (4.49) and write the root κ_1, which lies in the upper half-plane, as

$$\kappa_1 = k\left[1 + \frac{k^2\sigma_\varepsilon^2}{2(\alpha^2 + 4k^2)} + i\frac{k^3\sigma_\varepsilon^2}{\alpha(\alpha^2 + 4k^2)} + \dots\right] \quad . \tag{4.51}$$

Note that the three constraints (4.48) and (4.50) can be combined to give one constraint

$$\frac{k^2\sigma_\varepsilon^2}{\alpha^2 + 4k^2} \ll \min\left\{1, \frac{\alpha}{2k}, \frac{\alpha^2}{4k^2}\right\} \quad .$$

Using the obvious relationship

$$\frac{\alpha^2}{\alpha^2 + 4k^2} < \min\left\{1, \frac{\alpha}{2k}, \frac{\alpha^2}{4k^2}\right\} \quad ,$$

we will make the last constraint somewhat stronger by replacing it by the requirement

$$\frac{k^2\sigma_\varepsilon^2}{\alpha^2 + 4k^2} \ll \frac{\alpha^2}{\alpha^2 + 4k^2} < \min\left\{1, \frac{\alpha}{2k}, \frac{\alpha^2}{4k^2}\right\} \quad ,$$

which ensures that the three constraints be met. However, the last constraint will clearly reduce to

$$\frac{k^2}{\alpha^2}\sigma_\varepsilon^2 = k^2 l_\varepsilon^2 \sigma_\varepsilon^2 \ll 1 \quad . \tag{4.52}$$

As for the second root κ_2, for $|\delta| \ll 1$, it is

$$\kappa_2 = k + i\alpha + \dots \quad . \tag{4.53}$$

If we now take the residues of the integrand with respect to κ in (4.46) at the poles κ_1 and κ_2, then

$$\begin{aligned}
\overline{G}_1(R) &= -\frac{1 - (\delta/2)}{1 - \delta}\frac{1}{4\pi}\frac{e^{i\kappa_1 R}}{R} + \frac{\delta}{8\pi(1 - \delta)}\frac{e^{i\kappa_2 R}}{R} \\
&\approx -\frac{e^{i\kappa_1 R}}{4\pi R} + \frac{\delta}{8\pi}\frac{e^{i\kappa_2 R}}{R} \quad .
\end{aligned} \tag{4.54}$$

Let us examine this expression.

$\overline{G}_1$ describes two divergent waves, the amplitude of the second being far smaller than that of the first. Compare the attenuation coefficients of the two waves, i.e., $\text{Im}\,\{\kappa_1\}$ and $\text{Im}\,\{\kappa_2\}$

$$\frac{\text{Im}\,\{\kappa_1\}}{\text{Im}\,\{\kappa_2\}} = \frac{k^4 \sigma_\varepsilon^2}{\alpha^2(\alpha^2 + 4k^2)} \quad .$$

According to (4.48), this ratio is small, i.e.,

$$\text{Im}\,\{\kappa_1\} \ll \text{Im}\,\{\kappa_2\} \quad .$$

Thus, the second wave in (4.54) not only has a small amplitude, but is damped much faster than the principal wave, namely over a distance about the size of an inhomogeneity $\alpha^{-1} = l_\varepsilon$. Therefore, we can with sufficient accuracy put

$$\overline{G}_1(R) \approx -\frac{e^{i\kappa_1 R}}{4\pi R} \quad . \tag{4.55}$$

This expression differs from G_0 only in that a complex κ_1 is used instead of a real k (here $\text{Im}\,\{\kappa_1\} > 0$), i.e., the mean field decays exponentially. In addition, the real part of κ_1 is different from k: $\text{Re}\,\{\kappa_1\} > k$.

Both results are easily explained: the inhomogeneities cause the wave's energy to be pumped from the deterministic into the random component of the field, i.e., the mean field is decreased. Furthermore, the inhomogeneities increase the phase path of the rays, which amounts to increasing the real part of the wave number.

Note that, by (4.52), the relative changes in the wave number are small

$$\left| \frac{\kappa_1 - k}{k} \right| \ll 1 \quad . \tag{4.56}$$

This enables us to examine the Bourret approximation described by (4.46) in the general form, i.e., without specifying the covariance ψ_ε.

The poles κ_n of the integrand in (4.46) are the roots of the equation

$$k^2 - \kappa_n^2 + \frac{k^4}{\kappa_n} \int_0^\infty \psi_\varepsilon(r) e^{ikr} \sin(\kappa_n r) dr + io = 0 \quad . \tag{4.57}$$

Of especial interest here is the one that lies in the upper half-plane κ, i.e., the root κ_1 of (4.57), such that it has the smallest imaginary part, since the contributions to $\overline{G}_1$ of the other roots are small and decay fast. But, by (4.56), the role of the integral term in (4.57) is small and, if we discard it, we will get $\kappa_n = k + io$, and the desired root κ_1 will be close to this value. Accordingly, (4.57) can be solved by successive iteration: in a zeroth approximation $\kappa_1 = k$, and a first-order approximation results if we substitute $\kappa_n = k$ into the integral term (4.57)

$$k^2 - \kappa_n^2 + k^3 \int_0^\infty \psi_\varepsilon(r) e^{ikr} \sin(kr) dr = 0 \quad . \tag{4.58}$$

In order to obtain the effective wave number $k_{\text{ef}} = \kappa_1$, we solve this equation, taking the integral term to be small,

$$k_{\mathrm{ef}} = k_1 + ik_2 = k\left[1 + \frac{1}{2}k\int_0^\infty \sin\,(kr)e^{ikr}\psi_\varepsilon(r)dr\right]$$

$$= k\left[1 + \frac{k}{4}\int_0^\infty \sin\,(2kr)\psi_\varepsilon(r)dr + i\frac{k}{2}\int_0^\infty \sin^2\,(kr)\psi_\varepsilon(r)dr\right] \quad . \tag{4.59}$$

We can conveniently express k_{ef} in terms of the spectral expansion of $\psi_\varepsilon(r)$, which for statistically isotropic fluctuations is given by

$$\psi_\varepsilon(r) = \frac{4\pi}{r}\int_0^\infty \Phi_\varepsilon(\kappa')\,\sin\,(\kappa'r)\kappa'\,d\kappa' \quad . \tag{4.60}$$

Let us substitute (4.60) into (4.59) and change the order of the integration with respect to r and κ'. Taking account of the relationships

$$\int_0^\infty \frac{\sin\,ax\,\sin\,bx}{x}dx = \frac{1}{4}\ln\left(\frac{a+b}{a-b}\right)^2 \quad ,$$

$$\int_0^\infty \frac{\sin\,ax\,\cos\,bx}{x}dx = \frac{\pi}{2}\theta(a-b) = \begin{cases} \pi/2 & \text{for}\quad a>b \\ 0 & \text{for}\quad a<b \end{cases} \quad ,$$

we integrate with respect to r

$$k_{\mathrm{ef}} = k\left[1 + \frac{\pi k}{4}\int_0^\infty \kappa\ln\left(\frac{2k+\kappa}{2k-\kappa}\right)^2\Phi_\varepsilon(\kappa)d\kappa + \frac{i\pi^2 k}{2}\int_0^{2k}\Phi_\varepsilon(\kappa)\kappa\,d\kappa\right] \quad . \tag{4.61}$$

In this approximation, the averaged Green's function is, by (4.46),

$$\overline{G_1}(R) \approx \frac{1}{4i\pi^2 R}\int_{-\infty}^\infty \frac{\exp\,(i\kappa R)\kappa\,d\kappa}{k_{\mathrm{ef}}^2 - \kappa^2} = -\frac{\exp\,(ik_{\mathrm{ef}}R)}{4\pi R} \quad . \tag{4.62}$$

We have thus found that for $R \gg l_\varepsilon$, the mean Green's function had the same form as in a homogeneous medium, except that k_{ef} was substituted for k. The ratio k_{ef}/k can be treated as the effective refractive index of a random medium. From (4.61),

$$n_{\mathrm{ef}} = n_1 + in_2 \quad ,$$

$$n_1 = 1 + \frac{\pi k}{4}\int_0^\infty \ln\left(\frac{2k+\kappa}{2k-\kappa}\right)^2\Phi_\varepsilon(\kappa)\kappa\,d\kappa \quad ,$$

$$n_2 = \frac{\pi^2 k}{2}\int_0^{2k}\Phi_\varepsilon(\kappa)\kappa\,d\kappa \quad . \tag{4.63}$$

It is seen from these expressions that $n_1 > 1$ at any spectral density $\Phi_\varepsilon(\kappa)$, since $\ln\,[(2k+\kappa)/(2k-\kappa)]^2 > 0$, and n_2 is always positive. This was explained above, when discussing the case of $\psi_\varepsilon \approx \exp\,(-\alpha r)$.

According to (4.61), the power attenuation coefficient of the mean field is

$$\gamma = 2 \operatorname{Im}\{k_{\mathrm{ef}}\} = \pi^2 k^2 \int_0^{2k} \Phi_\varepsilon(\kappa)\kappa \, d\kappa \quad . \tag{4.64}$$

It is easily verified that this expression coincides with the scattering coefficient σ_0 derived in the Born approximation for the statistically isotropic fluctuations ε [see (III.4.79)].

$$\sigma_0 = \frac{\pi k^4}{2} \int_0^\pi \int_0^{2\pi} \Phi_\varepsilon\left(2k\sin\frac{\theta}{2}\right)\sin\theta \, d\theta \, d\varphi = \pi^2 k^4 \int_0^\pi \Phi_\varepsilon\left(2k\sin\frac{\theta}{2}\right)\sin\theta \, d\theta \quad .$$

If we take here $2k\sin(\theta/2) = \kappa$ to be the new integration variable, we obtain the right-hand side of (4.64), i.e., $\gamma = \sigma_0$.

In the case of large-scale inhomogeneities ($k \gg \kappa_{\mathrm{m}}$), the upper integration limit in (4.64) can be replaced by infinity, and we will thus arrive at equation (3.38) derived in the Markovian approximation. For n_1 in (4.63) we can, for $k \gg \kappa_{\mathrm{m}}$, put

$$\frac{1}{4}\ln\left(\frac{2k+\kappa}{2k-\kappa}\right)^2 \approx \frac{\kappa}{2k} \quad .$$

Hence,

$$n_1 \approx 1 + \frac{\pi}{2}\int_0^\infty \Phi_\varepsilon(\kappa)\kappa^2 d\kappa \quad \text{and by the order of magnitude}$$

$$n_1 - 1 \sim \frac{\kappa_{\mathrm{m}}}{k}n_2 \ll \gamma \quad ,$$

i.e., the real part of the wave number changes little (relative to the wave number for the homogeneous medium) as compared with its imaginary part.

Let us now consider the boundaries of validity of the Bourret approximation. Our first approximation was the kernel of mass operator Q in the approximation (4.37). Let us now take the next approximation for Q, which will include the next infinite subsequence of diagrams from the exact expansion for Q

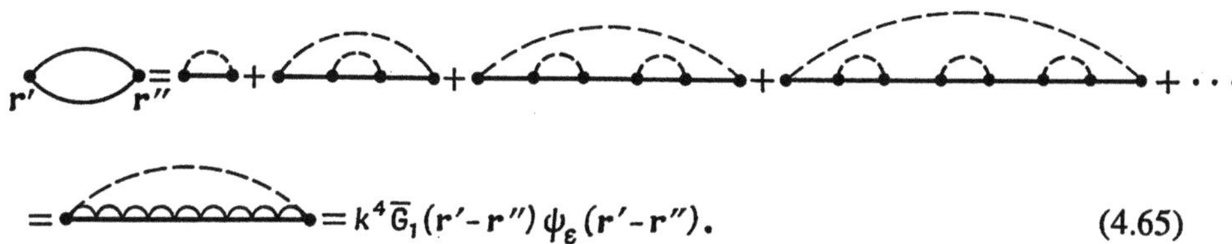

$$= k^4\overline{G}_1(\mathbf{r}'-\mathbf{r}'')\,\psi_\varepsilon(\mathbf{r}'-\mathbf{r}'') \quad . \tag{4.65}$$

Thus, Q_2 differs from Q_1 in that, instead of G_0, it contains the mean Green's function $\overline{G}_1$ found in the Bourret approximation

$$Q_2(r) = k^4\overline{G}_1(r)\psi_\varepsilon(r) \quad .$$

The poles of Green's function will then also be given by an equation similar to (4.57), but instead of $\exp(ikr)$ the integrand will contain $\exp(ik_{\mathrm{ef}}r)$

$$\kappa^2 = k^2 + \frac{k^4}{\kappa} \int_0^\infty \psi_\varepsilon(r) \exp(ik_{\mathrm{ef}}r) \sin(\kappa r)dr \quad . \tag{4.66}$$

We have found that in a first approximation, the root of this equation is $\kappa_1 = k_{\mathrm{ef}}$. We can, therefore, substitute k_{ef} for κ on the right-hand side of (4.66)

$$\kappa_2^2 = k^2 + \frac{k^4}{k_{\mathrm{ef}}} \int_0^\infty \psi_\varepsilon(r) \exp(ik_{\mathrm{ef}}r) \sin(k_{\mathrm{ef}}r)dr \quad . \tag{4.67}$$

Compare this with

$$\kappa_1^2 = k^2 + k^3 \int_0^\infty \psi_\varepsilon(r) e^{ikr} \sin(kr)dr \quad . \tag{4.68}$$

We let $k_{\mathrm{ef}} = k + \Delta k$, where, from (4.56), $|\Delta k| \ll k$. Then, $1/k_{\mathrm{ef}} \approx 1/k - \Delta k/k^2$. We will substitute this into (4.67) and subtract (4.68) from (4.67). If we replace the sine by the difference of the exponentials using the Euler formula, we obtain

$$\kappa_2^2 - \kappa_1^2 = \frac{k^3}{2i} \int_0^\infty \psi_\varepsilon(r)(e^{2ik_{\mathrm{ef}}r} - e^{2ikr})dr - \frac{k^2 \Delta k}{2i} \int_0^\infty \psi_\varepsilon(r)[e^{2ik_{\mathrm{ef}}r} - 1]dr \quad .$$

Again, to estimate this difference we will make use of the model covariance $\psi_\varepsilon(r) = \sigma^2 \exp(-\alpha r)$. The integrals here are easily taken, and the expression $\kappa_2^2 - \kappa_1^2$ takes the form

$$\kappa_2^2 - \kappa_1^2 = \Delta k \left[\frac{k^3 \sigma^2}{(\alpha - 2ik)(\alpha - 2ik_{\mathrm{ef}})} - \frac{k^2 k_{\mathrm{ef}} \sigma^2}{\alpha(\alpha - 2ik)} \right] \quad .$$

Since we estimate the difference $\kappa_2^2 - \kappa_1^2$ to first order in Δk, we can put $k_{\mathrm{ef}} \approx k$ in the brackets, and then

$$\kappa_2^2 - \kappa_1^2 \approx i\Delta k \frac{2k^4 \sigma^2}{\alpha(\alpha - 2ik)^2} \quad ,$$

or, since

$$\kappa_2^2 - \kappa_1^2 = (\kappa_2 - \kappa_1)(\kappa_2 + \kappa_1) \approx 2k(\kappa_2 - \kappa_1) \quad ,$$

we have

$$\kappa_2 - \kappa_1 \approx i\Delta k \frac{k^3 \sigma^2}{\alpha(\alpha - 2ik)^2} \quad . \tag{4.69}$$

In order that the shift Δk found in the Bourret approximation may have a significance, we will have to require that the modulus of the correction of the next approximation $\kappa_2 - \kappa_1$ be small as compared to $|\Delta k|$, i.e., that

$$\left| \frac{\kappa_2 - \kappa_1}{\Delta k} \right| = \frac{k^3 \sigma^2}{\alpha(\alpha^2 + 4k^2)} \ll 1 \quad . \tag{4.70}$$

However, this coincides with the second condition (4.50) and holds *a fortiori* if the condition (4.52), for which the principal relationships were derived in the Bourret approximation is met. We can thus say that the above relationships for $\overline{G}$ derived in the Bourret approximation hold if

$$(kl_\varepsilon)^2\sigma^2 \ll 1 \quad ,$$

where l_ε is the correlation radius of the random field $\tilde{\varepsilon}$.

It should be noted that (4.70) follows from a comparison of the Bourret approximation not with the exact solution, which is unknown, but with the next approximation [which, however, includes, as is seen from (4.65), an infinite number of diagrams for Q]. This condition cannot, therefore, be considered sufficient, but is necessary.

A number of other approximate solutions besides the Bourret approximation are known, as well as a number of more complex equations (including nonlinear ones), for the mean Green's function. For further details the reader is referred to [4.3, 10] and the extensive bibliography given in [4.11]. A good review of the works devoted to applications of multiple scattering theory to various problems of propagation of electromagnetic radiation through random media is given in [4.12].

4.3 Field Coherence Function. The Optical Theorem and Radiative Transfer Equation

We obtained above the Bethe-Salpeter equation (4.26a) for the coherence function

$$\Gamma(r', r''; r_0', r_0'') = \langle G(r', r_0')G^*(r'', r_0'') \rangle$$

of the field of two point sources:

$$\begin{aligned}
\Gamma(r', r''; r_0', r_0'') = \ &\overline{G}(r', r_0')\overline{G}^*(r'', r_0'') \\
&+ \int \overline{G}(r', r_1)\overline{G}^*(r'', r_2)K(r_1, r_2; r_3, r_4)\Gamma(r_3, r_4; r_0', r_0'') \\
&\times d^3r_1 \ldots d^3r_4 \quad .
\end{aligned} \tag{4.71}$$

Of course, it is not difficult to derive from this the equation for the coherence function of the field $u(r)$ produced in an inhomogeneous medium by an arbitrary distribution of volume sources $j(r)$. The field $u(r)$ is related to $j(r)$ by

$$u(r) = \int G(r, r_0)j(r_0)d^3r_0 \quad .$$

Hence, the mean field is

$$\overline{u}(r) = \int \overline{G}(r, r_0)j(r_0)d^3r_0 \tag{4.72}$$

and the coherence function is obtained by averaging the product $u(r')u^*(r'')$

$$\begin{aligned}
\Gamma_u(r', r'') &\equiv \langle u(r')u^*(r'') \rangle \\
&= \int \langle G(r', r_0')G(r'', r_0'') \rangle j(r_0')j^*(r_0'')d^3r_0' d^3r_0'' \\
&= \int \Gamma(r', r''; r_0', r_0'')j(r_0')j^*(r_0'')d^3r_0' d^3r_0'' \quad .
\end{aligned}$$

Therefore, if we multiply (4.71) by $j(r_0')j^*(r_0'')$ and integrate with respect to r_0' and r_0'', we will obtain

$$\Gamma_u(r', r'') = \overline{u}(r')\overline{u}^*(r'') + \int \overline{G}(r', r_1)\overline{G}^*(r'', r_2)$$

$$\times K(r_1, r_2; r_3, r_4)\Gamma_u(r_3, r_4)d^3r_1 \ldots d^3r_4 \quad . \tag{4.73}$$

This equation gives the coherence function of the field produced by an arbitrary deterministic distribution of sources $j(r)$.

Unlike the Dyson equation, the Bethe-Salpeter equation (4.71) cannot be solved for Γ.

This section will be concerned with the relation between the Bethe-Salpeter equation and the phenomenological radiative transfer equation (3.57) (RTE). In Sect. 3.3, this relation was found in the special case where the propagation of waves was described by the parabolic equation approximation, and for the radiative equation — by the small-angle approximation. We are now going to treat the general case, but to simplify the problem we will confine ourselves to the *statistically homogeneous* random medium, and the coherence function $\Gamma(r', r'')$ will be considered only in the region *free of sources*, i.e., where $j(r) = 0$.

First, we will transform (4.73) into an integro-differential equation. It will be recalled that $\overline{G}$ satisfies (4.30), which we will write in the form

$$\hat{D}(r)\overline{G}(r, r') = \delta(r - r') \quad , \tag{4.74}$$

where $\hat{D}(r)$ is the Dyson operator

$$\hat{D}(r) = \frac{\partial^2}{\partial r^2} + k^2 - \hat{Q}_r \tag{4.75}$$

and $\hat{Q}_r$ is the mass operator, i.e., an integral operator with the kernel $Q(r, r')$

$$\hat{Q}_r f(r) \equiv \int Q(r, r')f(r')d^3r' \quad .$$

If we apply $\hat{D}(r)$ to (4.72) by (4.74), we will get

$$\hat{D}(r)\overline{u}(r) = j(r) \quad .$$

Hence, in the source-free region the following equation holds:

$$\hat{D}(r)\overline{u}(r) = 0 \quad . \tag{4.76}$$

We will now apply $\hat{D}(r)$ to (4.73). From (4.76, 74) we find

$$\hat{D}(r')\Gamma_u(r', r'') = \int \overline{G}^*(r'', r_2)K(r', r_2; r_3, r_4)\Gamma_u(r_3, r_4)d^3r_2 d^3r_3 d^3r_4 \quad . \tag{4.77}$$

Similarly, if we operate with $\hat{D}^*(r'')$ on (4.73), then, using the equations that are complex conjugates to (4.74) and (4.76), we obtain

$$D^*(r'')\Gamma_u(r',r'') = \int \overline{G}(r',r_1)K(r_1,r'';r_3,r_4)\Gamma_u(r_3,r_4)d^3r_1d^3r_3d^3r_4 \quad .$$

$$(4.78)$$

Next, we subtract (4.78) from (4.77). If we denote by r_0 the integration variables r_2 in (4.77) and r_1 in (4.78), this difference will be

$$[\hat{D}(r') - \hat{D}^*(r'')]\Gamma_u(r',r'')$$
$$= \int [\overline{G}^*(r'',r_0)K(r',r_0;r_3,r_4) - \overline{G}(r',r_0)K(r_0,r'';r_3,r_4)]$$
$$\times \Gamma_u(r_3,r_4)d^3r_0d^3r_3d^3r_4 \quad .$$

$$(4.79)$$

But, by the definition (4.75),

$$[\hat{D}(r') - \hat{D}^*(r'')]\Gamma_u(r',r'')$$
$$= [\Delta' - \Delta'']\Gamma_u(r',r'') - \int d_r^3r_0[Q(r',r_0)\Gamma_u(r_0,r'')$$
$$- Q^*(r'',r_0)\Gamma_u(r',r_0)] \quad .$$

$$(4.80)$$

We will now introduce the new variables $R = (r' + r'')/2$ and $r = r' - r''$. Then,

$$\Delta' - \Delta'' = 2\nabla_R\nabla_r \quad .$$

The function Γ_u expressed in terms of R, r will now be denoted by $\Gamma(R,r)$

$$\Gamma(R,r) \equiv \Gamma_u(R + r/2, R - r/2) \quad .$$

Going over to the new variables, substituting (4.80) into (4.79) and taking into account the statistical homogeneity of the medium $[\overline{G}(r',r'') = \overline{G}(r' - r''), Q(r',r'') = Q(r' - r'')]$, we find

$$2\nabla_R\nabla_r\Gamma(R,r) - \int d^3r_0\left[Q\left(R+\frac{r}{2} - r_0\right)\Gamma\left(\frac{R}{2} + \frac{r_0}{2} - \frac{r}{4}, r_0 - R + \frac{r}{2}\right)\right.$$
$$\left. - Q^*\left(R - \frac{r}{2} - r_0\right)\Gamma\left(\frac{R}{2} + \frac{r_0}{2} + \frac{r}{4}, R + \frac{r}{2} - r_0\right)\right]$$
$$= \int\left[\overline{G}^*\left(R - \frac{r}{2} - r_0\right)K\left(R+\frac{r}{2}, r_0; r_3, r_4\right) - \overline{G}\left(R + \frac{r}{2} - r_0\right)\right.$$
$$\left. \times K\left(r_0, R - \frac{r}{2}; r_3, r_4\right)\right]\Gamma\left(\frac{r_3 + r_4}{2}, r_3 - r_4\right)d^3r_0d^3r_3d^3r_4 \quad . \quad (4.81)$$

We will now change integration variables in various terms in (4.81), taking as the independent variable each time the argument of the functions Q, Q^*, $\overline{G}^*$, or $\overline{G}$, that enters into a given term, and denoting $R' = (r_3 + r_4)/2$ and $r' = r_3 - r_4$. Equation (4.81) will then be

$$2\nabla_R\nabla_r\Gamma(R,r) - \int\left[Q(r'')\Gamma\left(R - \frac{r''}{2}, r - r''\right)\right.$$
$$\left. - Q^*(r'')\Gamma\left(R - \frac{r''}{2}, r + r''\right)\right]d^3r''$$

$$= \int \left[\overline{G}^*(r'')K\left(R+\frac{r}{2},R-\frac{r}{2}-r'';R'+\frac{r'}{2},R'-\frac{r'}{2}\right)\right.$$
$$\left. -\overline{G}(r'')K\left(R+\frac{r}{2}-r'',R-\frac{r}{2};R'+\frac{r'}{2},R'-\frac{r'}{2}\right)\right]$$
$$\times \Gamma(R',r')d^3R'\,d^3r'\,d^3r'' \quad . \tag{4.82}$$

Note that (4.82) is an *exact* corollary of the Bethe-Salpeter equation.

Consider the relation of (4.82) to the energy conservation law, which in any source-free region is written as [see (2.20)]

$$\mathrm{div}\,\vec{S}(R) = 0 \quad , \quad \text{where}$$

$$\vec{S}(R) = \frac{1}{2ik}[u^*\nabla_R u - u\nabla_R u^*] \quad .$$

Averaging these expressions gives

$$\mathrm{div}\,\langle\vec{S}(R)\rangle = 0 \quad ,$$

$$\langle\vec{S}(R)\rangle = \frac{1}{2ik}\langle u^*(R)\nabla_R u(R) - u(R)\nabla_R u^*(R)\rangle \quad . \tag{4.83}$$

The mean energy flux density $\langle\vec{S}\rangle$ can be expressed in terms of $\Gamma(R,r)$. In fact, differentiating

$$\Gamma(R,r) = \langle u(R+r/2)u^*(R-r/2)\rangle$$

with respect to r, we find

$$\nabla_r\Gamma(R,r) = \tfrac{1}{2}\langle u^*(R-r/2)\nabla_R u(R+r/2) - u(R+r/2)\nabla_R u^*(R-r/2)\rangle \quad .$$

Setting here $r = 0$, we get

$$[\nabla_r\Gamma(R,r)]_{r=0} = \tfrac{1}{2}\langle u^*(R)\nabla_R u(R) - u(R)\nabla_R u^*(R)\rangle \quad ,$$

so that

$$\langle\vec{S}(R)\rangle = (ik)^{-1}[\nabla_r\Gamma(R,r)]_{r=0} \quad . \tag{4.84}$$

Returning to (4.82), if we put $r = 0$, we will obtain, by (4.83, 84),

$$2\nabla_R\nabla_r\Gamma(R,r)|_{r=0} = 2\nabla_R(ik\langle\vec{S}(R)\rangle) = 0 \quad .$$

Therefore, at $r = 0$, (4.82) will be

$$-\int\left[Q(r'')\Gamma\left(R-\frac{r''}{2},-r''\right) - Q^*(r'')\Gamma(R-\frac{r''}{2},r'')\right]d^3r''$$
$$= \int\left[\overline{G}^*(r'')K\left(R,R-r'';R'+\frac{r'}{2},R'-\frac{r'}{2}\right)\right.$$
$$\left. -\overline{G}(r'')K\left(R-r'',R;R'+\frac{r'}{2},R'-\frac{r'}{2}\right)\right]\Gamma(R',r')d^3r'\,d^3R'\,d^3r'' \quad . \tag{4.85}$$

It should be noted that the form of Γ in (4.85) depends on the particular form of the source field $j(r)$, whereas Q, $\overline{G}$ and K are independent of the selection of $j(r)$. Since, by selecting appropriate sources, we can arbitrarily change Γ without changing Q, $\overline{G}$ and K, equation (4.85) must hold identically for Γ. If we write the left-hand side of (4.85) as a three-fold integral

$$-\int\left[Q(r'')\delta\left(R-\frac{r''}{2}-R'\right)\delta(r''+r')\right.$$
$$\left.-Q^*(r'')\delta\left(R-\frac{r''}{2}-R'\right)\delta(r''-r')\right]\Gamma(R',r')d^3r'd^3R'd^3r''\quad,$$

substitute this equation into (4.85) and equate the coefficients at $\Gamma(R',r')$ on the left and right-hand sides, we obtain

$$Q(-r')\delta\left(R-R'+\frac{r'}{2}\right)-Q^*(r')\delta\left(R-R'-\frac{r'}{2}\right)$$
$$=\int\left[\overline{G}(r'')K\left(R-r'',R;\,R'+\frac{r'}{2},R'-\frac{r'}{2}\right)\right.$$
$$\left.-\overline{G}^*(r'')K\left(R,R-r'';\,R'+\frac{r'}{2},R'-\frac{r'}{2}\right)\right]d^3r''\quad.\tag{4.86}$$

This equation, which relates the mean Green's function to the kernels of mass and intensity operators, is called the *optical theorem* [4.13].

The physical significance of the optical theorem consists in that the attenuation of the mean field is expressed in terms of $\mathrm{Im}\{k_{\mathrm{ef}}\}$, i.e., $\mathrm{Im}\{Q\}$. It is, however, caused not by absorption (the medium is transparent), but by *scattering*, which is described by the operator $\hat{K}$. Accordingly, there must be some relation between these quantities. It is this relation that is expressed by the optical theorem. Note that when we use approximate expressions for $\overline{G}$, Q, and K, these *must* satisfy (4.86), otherwise they will violate energy conservation.

Our next task will be to transform (4.82) into the radiative transfer equation. We will use the approximation (4.65) for Q

$$Q(r) = k^4\psi_\varepsilon(r)\overline{G}(r)\quad,\tag{4.87}$$

where $\overline{G}$ is the mean Green's function in the Bourret approximation. For the intensity operator, we will use the leading term of the expansion

$$K(r_1,r_2;\,r_3,r_4) = k^4\psi_\varepsilon(r_1-r_2)\delta(r_1-r_3)\delta(r_2-r_4)\quad.\tag{4.88}$$

This approximation of the intensity operator kernel is called the *ladder approximation*. The name derives from the fact that if in (4.25) we use only the leading term of the expansion (4.22) for K, then, in diagrammatic representation, Γ will resemble a ladder

Substituting (4.87, 88) into (4.82) gives

$$2\nabla_R \nabla_r \Gamma(R, r) = k^4 \int \left\{ \overline{G}(r')[\psi_\varepsilon(r') - \psi_\varepsilon(r - r')]\Gamma\left(R - \frac{r'}{2}, r - r'\right) \right.$$
$$\left. - \overline{G}^*(r')[\psi_\varepsilon(r') - \psi_\varepsilon(r + r')]\Gamma\left(R - \frac{r'}{2}, r + r'\right) \right\} d^3 r'$$

$$(4.89)$$

It is seen, above all, that at $r = 0$ the right-hand side of this equation vanishes, i.e., in the approximation under consideration, both the energy conservation law and the optical theorem are satisfied [in other words, approximations (4.87, 88) are coordinated so that the energy conservation law is not violated]. Furthermore, it is seen from (4.89) that the equation for Γ only contains the differences $\psi_\varepsilon(r') - \psi_\varepsilon(r \pm r')$, which can be expressed in terms of the structure function $D_\varepsilon(r)$. This means that large-scale inhomogeneities do not influence the field coherence function (it will be recalled that the situation was exactly the same in the approximation of the Markov random process discussed in Sect. 3.3).

It is quite obvious that a necessary condition for the *approximate* equation (4.89) to be applicable is the applicability of the Bourret approximation discussed in the previous section. In this connection, we will consider the characteristic scales of $\Gamma(R, r)$ in both arguments.

As was found in the previous section, the mean field decays exponentially with increasing distance from a source, the characteristic space scale being $1/\mathrm{Im}\,\{k_{\mathrm{ef}}\} \equiv d$. Since the attenuation of the mean field results from scattering from inhomogeneities, it is clear that the same scale d will be the characteristic longitudinal scale of Γ in R. The characteristic transverse scale of Γ in R may be either the mean diameter of the beam, or the characteristic distance over which such averaged characteristics of the medium as σ^2, or the correlation radius l_ε, vary noticeably.

As for the characteristic scale r_k of the function $\Gamma(R, r)$ in the difference variable r, this can be of the order of either the correlation radius l_ε of the inhomogeneities of the medium, or the radius of the first Fresnel zone (which was the case in the region of validity of the method of smooth perturbations), or else it can be smaller still (as with strong fluctuations of the field considered in Chap. 3). In any case, we can always believe that r_k is much smaller than the characteristic scale d of the function Γ in R.

Such a relationship between the scales of Γ can be used to further simplify equation (4.89). Indeed, the difference $|\psi_\varepsilon(r') - \psi_\varepsilon(r \pm r')|$ at $r \approx r_k$ is only noticeably different from zero when $r' < (l_\varepsilon + r) \sim (l_\varepsilon + r_k)$ (Fig. 4.3). But, as assumed, the quantity $(l_\varepsilon + r_k)$ is small in comparison with the characteristic scale d of $\Gamma(R, r)$ in R. This means that, in (4.89), we can put

$$\Gamma\left(R - \frac{r'}{2}, r \pm r'\right) \approx \Gamma(R, r \pm r') \quad .$$

We then obtain

$$2\nabla_R \nabla_r \Gamma(R, r) = k^4 \int \overline{G}(r')[\psi_\varepsilon(r') - \psi_\varepsilon(r - r')]\Gamma(R, r - r')d^3 r'$$
$$- k^4 \int \overline{G}^*(r')[\psi_\varepsilon(r') - \psi_\varepsilon(r + r')]\Gamma(R, r + r')d^3 r' \quad .$$

146

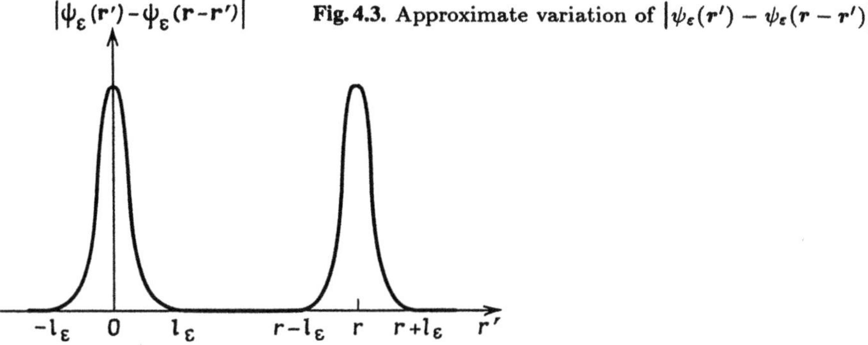

Fig. 4.3. Approximate variation of $|\psi_\varepsilon(r') - \psi_\varepsilon(r-r')|$ with r'

Let us change the integration variable $r' \to -r'$ in the second integral. Since $\overline{G}^*(-r') = \overline{G}^*(r')$ and $\psi_\varepsilon(-r') = \psi_\varepsilon(r')$, we get

$$2\nabla_R\nabla_r\Gamma(R,r) = k^4 \int [\overline{G}(r') - \overline{G}^*(r')]$$
$$\times [\psi_\varepsilon(r') - \psi_\varepsilon(r-r')]\Gamma(R, r-r')d^3r' \quad . \tag{4.90}$$

When $\overline{u}(r) = 0$ [$\Gamma(R,r)$ then coincides with the field covariance], (4.90) is equivalent to the RTE. To confirm this, we will introduce the Fourier transform of $\Gamma(R,r)$ in the difference variable[1]

$$\Gamma(R,r) = \int f(R,\kappa)\exp(i\kappa \cdot r)d^3\kappa \quad , \tag{4.91}$$

$$f(R,\kappa) = (8\pi^3)^{-1} \int \Gamma(R,r)\exp(-i\kappa \cdot r)d^3r \quad . \tag{4.92}$$

To simplify the transition to the Fourier transforms in (4.90), we will first calculate the three-dimensional spectral density of the mean Green's function $\overline{G}$, which will be taken in the approximation

$$\overline{G}(r) = -\frac{\exp(ik_{ef}r)}{4\pi r} \quad , \tag{4.93}$$

where $k_{ef} = k_1 + ik_2$ and $k_1 \gg k_2$. It is easily seen that the spectral density of $\overline{G}(r) - \overline{G}^*(r)$ is

$$z(\kappa) = z(\kappa) = \frac{1}{8\pi^3} \int [\overline{G}(r') - \overline{G}^*(r')]\exp(-i\kappa \cdot r')d^3r'$$
$$= \frac{2k_1k_2}{4\pi^3i[(k_1^2 - k_2^2 - \kappa^2)^2 + 4k_1^2k_2^2]} \quad . \tag{4.94}$$

We will now show that, by virtue of the inequality $k_2 \ll k_1$, (4.94) can be approximately replaced by a delta-function. The modulus of $z(\kappa)$ has a maximum $z(\kappa_0) =$

---

[1] If $\overline{u} \neq 0$, it is first necessary to subtract from Γ the product of the mean fields in order that the Fourier transform of the covariance may be found (see [4.14] and Exercise 1).

$1/8\pi^3 i k_1 k_2$ at $\kappa = \kappa_0 = \sqrt{k_1^2 - k_2^2}$. At the same time, at $\kappa^2 = k_1^2 - k_2^2 \mp 2k_1 k_2$, i.e., at $\kappa \approx \kappa_0 \mp k_2$, we get $z(\kappa_0 \mp k_2) = z(\kappa_0)/2$. Thus, $z(\kappa)$ drops by half as κ_0 changes slightly by $k_2 = \mathrm{Im}\,\{k_{\mathrm{ef}}\} = 1/d$. Further, it is easily verified that

$$\frac{1}{\pi} \int\limits_{-\infty}^{\infty} \frac{2k_1 k_2}{x^2 + 4k_1^2 k_2^2}\,dx = 1 \quad .$$

Therefore, if (as we have already assumed) the scale d is large as compared to the scales of $\psi_\varepsilon(r')$ and $\Gamma(R, r - r')$ in r', then the function

$$\frac{1}{\pi} \frac{2k_1 k_2}{(k_1^2 - k_2^2 - \kappa^2)^2 + 4k_1^2 k_2^2} \quad ,$$

concentrated within the narrow interval $\Delta\kappa \sim 2k_2$, can be replaced by $\delta(k_1^2 - k_2^2 - \kappa^2)$. Thus,

$$z(\kappa) \to \frac{1}{4\pi^2 i}\delta(k_1^2 - k_2^2 - \kappa^2) \quad .$$

Using the relationship

$$\delta(\kappa^2 - \kappa_0^2) = \frac{\delta(\kappa - \kappa_0) + \delta(\kappa + \kappa_0)}{2\kappa_0}$$

and considering that $\delta(\kappa + \kappa_0) = 0$ (since $\kappa > 0$ and $\kappa_0 > 0$), we can write

$$z(\kappa) \approx \frac{\delta(\kappa - \sqrt{k_1^2 - k_2^2})}{8\pi^2 i \sqrt{k_1^2 - k_2^2}} \quad ,$$

and if we take account of the smallness of k_2, then

$$z(\kappa) \approx \frac{\delta(\kappa - k_1)}{8\pi^2 i k_1} \quad . \tag{4.95}$$

We can now easily carry out the Fourier transformation of (4.90). If we substitute into it the expressions

$$\overline{G}(r') - \overline{G}^*(r') = \int \frac{\delta(\kappa' - k_1)}{8\pi^2 i k_1}\exp{(i\kappa' \cdot r')}d^3\kappa'$$

and (4.91), and perform some calculations using

$$\int \psi_\varepsilon(r')\exp{(i\kappa \cdot r')}d^3 r' = 8\pi^3 \Phi_\varepsilon(\kappa) \quad ,$$

we obtain

$$2i\nabla_R \int \exp{(i\kappa \cdot r)}\kappa f(R, \kappa)d^3\kappa$$

$$= -\frac{i\pi k^4}{k_1}\int d^3\kappa \exp{(i\kappa \cdot r)}f(R, \kappa)\int d^3\kappa'\delta(\kappa' - k_1)\Phi_\varepsilon(\kappa' - \kappa)$$

$$+\frac{i\pi k^4}{k_1}\int d^3\kappa' \exp{(i\kappa' \cdot r)}\delta(\kappa' - k_1)\int d^3\kappa f(R, \kappa)\Phi_\varepsilon(\kappa' - \kappa) \quad .$$

If we then change the integration variables $\kappa \rightleftarrows \kappa'$ in the last integral and equate the coefficients at $\exp(i\kappa r)$, we obtain

$$\kappa \nabla_R f(R, \kappa) = -\frac{\pi k^4}{2k_1} f(R, \kappa) \int d^3 \kappa' \delta(\kappa' - k_1)\Phi_\varepsilon(\kappa' - \kappa)$$

$$+ \frac{\pi k^4}{2k_1}\delta(\kappa - k_1)\int d^3\kappa' f(R, \kappa')\Phi_\varepsilon(\kappa - \kappa') \quad .$$

We will now introduce spherical coordinates on the right-hand side of this equation for the integration variable κ' by setting $\kappa' = \kappa' n'$, where n' is the unit vector. Then, $d^3\kappa' = \kappa'^2 d\kappa' do(n')$ and the equation becomes

$$\kappa \nabla_R f(R, \kappa) = -\frac{\pi k^4}{2k_1} k_1^2 f(R, \kappa) \oint \Phi_\varepsilon(k_1 n' - \kappa)do(n')$$

$$+ \frac{\pi k^4}{2k_1}\delta(\kappa - k_1)\int d^3\kappa' f(R, \kappa')\Phi_\varepsilon(\kappa - \kappa') \quad . \tag{4.96}$$

We will seek its solution in the form

$$f(R, \kappa) = \delta(\kappa - k_1)\mathcal{J}(R, n)\frac{1}{k_1^2} \quad , \tag{4.97}$$

where $\kappa = \kappa n$.

Substituting (4.97) into (4.96), we obtain for $\mathcal{J}(R, n)$

$$\kappa n \delta(\kappa - k_1)\nabla_R \mathcal{J}(R, n) = -\frac{\pi k^4 k_1}{2}\delta(\kappa - k_1)\mathcal{J}(R, n)$$

$$\times \oint \Phi_\varepsilon(k_1 n' - \kappa n)do(n') + \frac{\pi k^4}{2k_1}\delta(\kappa - k_1)\int_0^\infty \kappa'^2 d\kappa' \oint do(n')$$

$$\times \delta(\kappa' - k_1)\mathcal{J}(R, n')\Phi_\varepsilon(\kappa n - \kappa' n) \quad .$$

Since $\varphi(\kappa)\delta(\kappa - k_1) = \varphi(k_1)\delta(\kappa - k_1)$, we can substitute k_1 for κ and κ' throughout. When we then compare the coefficients at the delta-functions, we arrive at

$$n\nabla_R \mathcal{J}(R, n) = -\left[\oint \frac{\pi k^4}{2}\Phi_\varepsilon(k_1(n - n'))do(n')\right]\mathcal{J}(R, n)$$

$$+ \oint \frac{\pi k^4}{2}\Phi_\varepsilon(k_1(n - n'))\mathcal{J}(R, n')do(n') \quad , \tag{4.98}$$

where the even nature of Φ_ε is taken into account.

This expression is nothing other than the radiative transfer equation (3.57), with the attenuation coefficient equal to

$$\alpha = \oint \frac{\pi k^4}{2}\Phi_\varepsilon(k_1 n')do(n') \quad . \tag{4.99}$$

The effective scattering cross-section from a unit volume into a unit solid angle has the form

$$\sigma(n, n') = \frac{\pi k^4}{2}\Phi_\varepsilon(k_1(n - n')) \quad . \tag{4.100}$$

In fact, in this notation (4.98) assumes the standard form of the RTE [4.15]:

$$n\nabla_R \mathcal{J}(R, n) = -\alpha \mathcal{J}(R, n) + \oint \sigma(n, n')\mathcal{J}(R, n')do(n') \quad . \tag{4.101}$$

A comparison of (4.99) and (4.100) shows that

$$\alpha = \oint \sigma(n, n')do(n') \quad , \tag{4.102}$$

i.e., the attenuation of the field is due entirely to scattering. Comparing further (4.100) with the formula (III.4.79), derived in the Born approximation, we see that the value of σ defined by (4.100) differs from its value in the Born approximation in that in the latter, k_1 is substituted for k. We have seen that the difference between k_1 and k is caused by multiple scattering. However, in the approximation considered in the previous section, k_1 only differs slightly from k, and if $\Phi_\varepsilon(\kappa)$ is a sufficiently smoothly-varying function, this difference can be neglected.

The fundamental importance of the results derived resides in the following. In the phenomenological theory of radiation transport, the brightness, or radiant intensity, $\mathcal{J}$, is in no way connected with the parameters describing the *wave* field. We now have the possibility of establishing this connection. Combining (4.91) and (4.97) gives

$$\Gamma(R, r) = \frac{1}{k_1^2} \int e^{i\kappa r \cdot n}\delta(\kappa - k_1)\mathcal{J}(R, n)\kappa^2 d\kappa\, do(n) \quad ,$$

or, after integrating with respect to κ,

$$\Gamma(R, r) = \oint \mathcal{J}(R, n)\exp(ik_1 r \cdot n)do(n) \quad . \tag{4.103}$$

The radiant intensity is thus the angular spectrum of the coherence function. Putting $r = 0$ in (4.103), we get an equation relating the mean intensity $\bar{I} = \langle |u|^2 \rangle = \Gamma(R, 0)$ to the radiant intensity

$$\bar{I}(R) = \oint \mathcal{J}(R, n)do(n) \quad . \tag{4.104}$$

Further, differentiating (4.103) with respect to r and then putting $r = 0$, we get

$$\nabla_r \Gamma(R, r)|_{r=0} = ik_1 \oint n\mathcal{J}(R, n)do(n) \quad .$$

Substituting this into (4.84) and ignoring the difference between k and k_1, we arrive at the mean energy flux density

$$\langle \vec{S}(R) \rangle = \oint n\mathcal{J}(R, n)do(n) \quad . \tag{4.105}$$

We can thus express all the main characteristics of the wave field in terms of $\mathcal{J}(R, n)$: the energy density (intensity), the energy flux density and the coherence function[2].

---

[2] Note that in Sect. III.2.2, similar expressions were given for the random wave field in the *homogeneous* medium.

Of course, it is not always possible to reduce the exact Bethe-Salpeter equation to the RTE. In the general case, the coherence function $\Gamma(\boldsymbol{R}, \boldsymbol{r})$ can be represented by the Fourier integral of the form (4.91), which only contains the plane waves $\exp(i\boldsymbol{\kappa} \cdot \boldsymbol{r})$ with arbitrary values of $\boldsymbol{\kappa}$. It is only in that expansion $|\boldsymbol{\kappa}| = k$, that (4.103) holds.

Further, in deriving the RTE we used the Bourret approximation for $\overline{G}$ and kept only the leading term of the expansion of the intensity operator (ladder approximation). However, a more detailed treatment shows that the latter approximation holds simultaneously with the former (i.e., when the Bourret approximation is valid), so that both are subject to the necessary condition

$$k^2 l_\varepsilon^2 \sigma^2 \ll 1 \quad . \tag{4.106}$$

Lastly, in deriving (4.90) from (4.89), we assumed that the characteristic scale of $\Gamma(\boldsymbol{R}, \boldsymbol{r})$ in $\boldsymbol{r}$ is small as compared to its characteristic scale in $\boldsymbol{R}$. It appears, however, that for this assumption to hold, the condition (4.106) is also necessary.

Nevertheless, situations are possible where the RTE is *a fortiori* not valid. This is the case, for example, when we are interested in *back*scattering. In the RTE, the scattered waves are added up in an *incoherent* way (intensity addition), which is especially obvious in the phenomenological derivation of this equation (Sect. 3.3). But, with backscattering, a scattered wave travels through exactly the same inhomogeneities as the incident wave, and so phase relations between the waves are important. As a result, the RTE appears to be unsuitable for describing backscattering [4.16, 17]. When this type of scattering is present "cyclic" diagrams play an important role [e.g., 7 and 13 in (4.20)].

We will restrict ourselves to a fairly simple problem of the propagation of a *scalar* wave field in a *statistically homogeneous* random medium. A more general case of the *statistically inhomogeneous* medium is considered in [4.14]. In [4.18], the RTE is derived for the electromagnetic field with spatial and frequency dispersion.

Let us consider another aspect of the problem, which has not even been touched upon in the phenomenological theory of radiation transport [4.15], namely the *diffraction content* of the RTE. As we have seen, the RTE is equivalent to (4.90) for the coherence function $\Gamma(\boldsymbol{R}, \boldsymbol{r})$, which has in turn been derived from the wave equation. Therefore, if the analytic expression for $\mathcal{J}(\boldsymbol{R}, \boldsymbol{n})$ is known, using (4.103), we can then restore Γ, which must contain information about diffraction effects (see Exercise 2). This gives rise to the question of how we should formulate the boundary conditions for the RTE in order to obtain a description of diffraction effects when reverting to $\Gamma(\boldsymbol{R}, \boldsymbol{r})$. This, and a number of other questions concerning the relation between coherence theory and radiation transport theory, are treated in [4.19, 20], as well as in the review [4.21] and the monograph [4.25], this latter being devoted entirely to the statistical and wave aspects of radiation transport theory.

To summarize, the general theory of multiple scattering takes into account both large and small inhomogeneities. This theory enabled approximate closed equations for field moments to be obtained, which are actually derived using partial summation of expansions in perturbations. Significantly, the general theory of multiple scattering allows *all* the equations that were derived by the various approximate methods to be obtained and, more important, makes it possible to delineate the boundaries of validity of these methods. A substantial advance of the theory is the "statistical wave" derivation of the RTE and the establishment of the diffraction content of the equation. Other important results have also been obtained. In particular, we can now derive the RTE, including the transformation of the coherent into the incoherent part of the field (the equation is derived for scalar waves in Exercise 1, and for electromagnetic waves in [4.18]).

In terms of multiple scattering theory, we can derive approximate closed equations not only for the *continuous* random medium, but also for the moments of the wave field scattered by a multitude of *discrete* scatterers (diffraction from bodies that are randomly distributed and oriented). Here, too, it is possible to obtain the RTE (see, e.g., [4.22, 23]) and understand the microscopic significance of the phenomenological parameters that appear in this equation. The scattering cross-section of a unit volume is generally not coincident with the scattering cross-section from one particle multiplied by the particle concentration since, at high concentrations, the so-called collective effects caused by the fact that a given particle is illuminated not only by direct waves, but also by those scattered by other particles, begin to manifest themselves.

As far as future possibilities of multiple scattering theory are concerned, it may be expected that it will eventually allow us to solve the problem of wave propagation through a medium with electric permittivity fluctuations that are *not small* (which is the case, for example, in a plasma, if the frequency of the electromagnetic wave is close to the plasma frequency; or in a liquid near the critical point).

In addition, there is a wide variety of problems of practical interest that undoubtedly belong to multiple scattering theory, but are complicated by many factors. The diffraction of partially coherent fields in a random medium, the scattering of waves from rough surfaces surrounded by a random medium, the diffraction of waves in a random medium with discrete scatterers, the thermal radiation of random media, etc., all fall into this category (see the review [4.24] and the bibliography given in it).

4.4 Exercises

4.4.1 Derive the RTE from (4.89) for the case where the mean field $\overline{u}$ is nonzero. *Solution.* We will introduce the notation $F(R, r) = \overline{u}(R + r/2)\overline{u^*}(R - r/2)$ and substitute into (4.89) $\Gamma(R, r) = \psi_u(R, r) + F(R, r)$, where $\psi_u(R, r) = \langle \tilde{u}(R + r/2)\tilde{u}^*(R - r/2) \rangle$. Using (4.76), we can easily obtain

$$2\nabla_R \nabla_r F(R, r) = k^4 \int \psi_\varepsilon(r') \left[\overline{G}(r') F\left(R - \frac{r'}{2}, r - r' \right) \right.$$
$$\left. - \overline{G}^*(r') F\left(R + \frac{r'}{2}, r - r' \right) \right] d^3 r' \quad . \tag{4.107}$$

Substituting $\Gamma = \psi_u + F$ into (4.89) we find, by (4.107),

$$2\nabla_R\nabla_r\psi_u(\boldsymbol{R},\boldsymbol{r}) = k^4 \int \left\{ \overline{G}(\boldsymbol{r}')[\psi_\varepsilon(\boldsymbol{r}') - \psi_\varepsilon(\boldsymbol{r}-\boldsymbol{r}')]\psi_u\left(\boldsymbol{R}-\frac{\boldsymbol{r}'}{2},\boldsymbol{r}-\boldsymbol{r}'\right) \right.$$

$$\left. - \overline{G}^*(\boldsymbol{r}')[\psi_\varepsilon(\boldsymbol{r}') - \psi_\varepsilon(\boldsymbol{r}+\boldsymbol{r}')]\psi_u\left(\boldsymbol{R}-\frac{\boldsymbol{r}'}{2},\boldsymbol{r}+\boldsymbol{r}'\right) \right\} d^3 r'$$

$$- k^4 \int \left\{ \overline{G}(\boldsymbol{r}')\psi_\varepsilon(\boldsymbol{r}-\boldsymbol{r}')F\left(\boldsymbol{R}-\frac{\boldsymbol{r}'}{2},\boldsymbol{r}-\boldsymbol{r}'\right) \right.$$

$$\left. - \overline{G}^*(\boldsymbol{r}')\psi_\varepsilon(\boldsymbol{r}+\boldsymbol{r}')F\left(\boldsymbol{R}-\frac{\boldsymbol{r}'}{2},\boldsymbol{r}+\boldsymbol{r}'\right) \right\} d^3 r' \quad . \tag{4.108}$$

As in the transition from (4.89) to (4.90), we ignore $r'/2$ in the arguments ψ_u and F and arrive at

$$2\nabla_R\nabla_r\psi_u(\boldsymbol{R},\boldsymbol{r}) = k^4 \int [\overline{G}(\boldsymbol{r}') - \overline{G}^*(\boldsymbol{r}')][\psi_\varepsilon(\boldsymbol{r}') - \psi_\varepsilon(\boldsymbol{r}-\boldsymbol{r}')]$$

$$\times \psi_u(\boldsymbol{R},\boldsymbol{r}-\boldsymbol{r}')d^3 r' - k^4 \int [\overline{G}(\boldsymbol{r}') - \overline{G}^*(\boldsymbol{r}')]$$

$$\times \psi_\varepsilon(\boldsymbol{r}-\boldsymbol{r}')F(\boldsymbol{R},\boldsymbol{r}-\boldsymbol{r}')d^3 r' \quad , \tag{4.109}$$

which is a generalization of (4.90).

Let us introduce the Fourier transforms

$$\psi_u(\boldsymbol{R},\boldsymbol{r}) = \int f(\boldsymbol{R},\boldsymbol{\kappa})\exp(i\boldsymbol{\kappa}\cdot\boldsymbol{r})d^3\kappa \quad ,$$

$$F(\boldsymbol{R},\boldsymbol{r}) = \int f_0(\boldsymbol{R},\boldsymbol{\kappa})\exp(i\boldsymbol{\kappa}\cdot\boldsymbol{r})d^3\kappa \quad . \tag{4.110}$$

Carrying out the Fourier transformation of (4.109), we obtain, by (4.94, 95),

$$\boldsymbol{\kappa}\nabla_R f(\boldsymbol{R},\boldsymbol{\kappa}) = -k_1\frac{\pi k^4}{2}f(\boldsymbol{R},\boldsymbol{\kappa}) \oint \Phi_\varepsilon(k_1 \boldsymbol{n}' - \boldsymbol{\kappa})do(\boldsymbol{n}')$$

$$+ \frac{\pi k^4}{2k_1}\delta(\kappa - k_1)\int d^3\kappa' \Phi_\varepsilon(\boldsymbol{\kappa} - \boldsymbol{\kappa}')[f(\boldsymbol{R},\boldsymbol{\kappa}') + f_0(\boldsymbol{R},\boldsymbol{\kappa}')] \quad . \tag{4.111}$$

If we seek a solution in the form

$$f(\boldsymbol{R},\boldsymbol{\kappa}) = \frac{\delta(\kappa - k_1)}{k_1^2}\mathcal{J}(\boldsymbol{R},\boldsymbol{n}) \quad , \quad \text{for} \quad \boldsymbol{\kappa} = \kappa\boldsymbol{n} \quad ,$$

this gives

$$\boldsymbol{n}\nabla_R\mathcal{J}(\boldsymbol{R},\boldsymbol{n}) = -\alpha\mathcal{J}(\boldsymbol{R},\boldsymbol{n}) + \oint \sigma(\boldsymbol{n},\boldsymbol{n}')\mathcal{J}(\boldsymbol{R},\boldsymbol{n}')do(\boldsymbol{n}')$$

$$+ \frac{\pi k^4}{2}\int \Phi_\varepsilon(k\boldsymbol{n} - \boldsymbol{\kappa}')f_0(\boldsymbol{R},\boldsymbol{\kappa}')d^3\kappa' \tag{4.112}$$

in the same notation as in (4.99, 100).

Equation (4.112) differs from the standard RTE (4.101) in that it has an additional (last) term describing the transformation of the coherent into the incoherent part of the field. Equation (4.103) now holds not for Γ, but for the covariance

$$\psi_u(\boldsymbol{R}, \boldsymbol{r}) = \oint \mathcal{J}(\boldsymbol{R}, \boldsymbol{n}) \exp(ik_1 \boldsymbol{n} \cdot \boldsymbol{r}) do(\boldsymbol{n}) \quad . \tag{4.113}$$

4.4.2 The plane wave $u_0(r)$ falls on a volume V in which $\tilde{\varepsilon}(r) \neq 0$, and which is small as compared to the extinction length d, so that the scattering of the wave from the inhomogeneities can be described in the Born approximation. Solve the scattering problem, proceeding from the RTE derived in the previous problem and ignoring the difference between k and k_1.

Solution. Since the scattering volume is small as compared to d, the energy of the scattered field is much smaller than that of the incident wave. Therefore, in the RTE (4.112), we can ignore the terms describing the extinction and scattering of the incoherent field, i.e., on the right-hand side, only the last term remains:

$$\boldsymbol{n}\nabla_R \mathcal{J}(\boldsymbol{R}, \boldsymbol{n}) = \frac{\pi k^4}{2} \int \Phi_\varepsilon(k\boldsymbol{n} - \boldsymbol{\kappa}', \boldsymbol{R}) f_0(\boldsymbol{R}, \boldsymbol{\kappa}') d^3\kappa' \quad . \tag{4.114}$$

We have deliberately separated the dependence of Φ_ε on $\boldsymbol{R}$ here, since the fluctuations ε are not statistically homogeneous. The scattering volume being small as compared to d, we can assume that $u(r) = u_0(r) = u_0 \exp(ik\boldsymbol{n}_0 r)$, where $\boldsymbol{n}_0$ is the unit vector along the direction of propagation of the primary wave, and u_0 the amplitude. Then (in the notation of the previous problem)

$$F(\boldsymbol{R}, \boldsymbol{r}) = |u_0|^2 \exp(ik\boldsymbol{n}_0 \cdot \boldsymbol{r}) \quad ,$$

$$f_0(\boldsymbol{R}, \boldsymbol{\kappa}) = \frac{|u_0|^2}{8\pi^3} \int \exp(ik\boldsymbol{n}_0 \cdot \boldsymbol{r} - i\boldsymbol{\kappa} \cdot \boldsymbol{r}) d^3 r = |u_0|^2 \delta(k\boldsymbol{n}_0 - \boldsymbol{\kappa}) \quad . \tag{4.115}$$

Substituting (4.115) into (4.114) gives

$$\boldsymbol{n}\nabla_R \mathcal{J}(\boldsymbol{R}, \boldsymbol{n}) = \frac{\pi k^4 |u_0|^2}{2} \Phi_\varepsilon(k\boldsymbol{n} - k\boldsymbol{n}_0, \boldsymbol{R}) \quad . \tag{4.116}$$

The function Φ_ε is only distinct from zero inside the scattering volume V. Since $\boldsymbol{n}\nabla_R \mathcal{J}(\boldsymbol{R}, \boldsymbol{n}) = d\mathcal{J}/dl$, where d/dl is the derivative along the unit vector $\boldsymbol{n}$ (Fig. 4.4), it is clear that the solution to (4.116) has the form

$$\mathcal{J}(\boldsymbol{R}, \boldsymbol{n}) = \int_0^\infty \frac{\pi k^4 |u_0|^2}{2} \Phi_\varepsilon[k(\boldsymbol{n} - \boldsymbol{n}_0), \boldsymbol{R} - \boldsymbol{n}l] dl \quad . \tag{4.117}$$

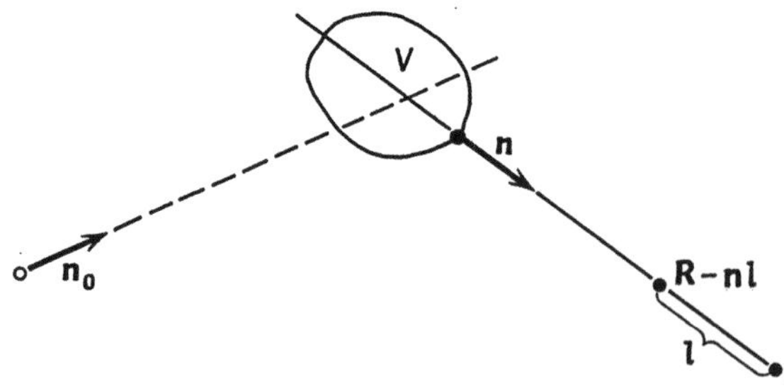

Fig. 4.4. Explanation of the notation to exercise 2

We will now find $\psi_u(R, r)$, using (4.113)

$$\psi_u(R, r) = \oint e^{ikn \cdot r} do(n) \int_0^\infty \frac{\pi k^4 |u_0|^2}{2} \Phi_\epsilon[k(n - n_0), R - nl] dl \quad . \quad (4.118)$$

Let us introduce the new integration variable $r' = -nl$, which is the radius vector of the point $R - nl$ from the point R. Then

$$d^3 r' = l^2 dl\, do(n) \quad , \quad \text{i.e.} \quad ,$$

$$dl\, do(n) = \frac{d^3 r'}{l^2} = \frac{d^3 r'}{r'^2} \quad .$$

Equation (4.118) takes the form

$$\psi_u(R, r) = \frac{\pi k^4 |u_0|^2}{2} \int_V \exp\left(-\frac{ikr \cdot r'}{r'}\right) \Phi_\epsilon\left(k(n - n_0), R + r'\right) \frac{d^3 r'}{r'^2} \quad .$$

$$(4.119)$$

In Chap. III.4, this equation was derived in the single scattering approximation. It is seen from the derivation given there that it takes due account of diffraction effects. Consequently, this problem confirms directly that the RTE describes the diffractional behavior of the field.

4.4.3 When waves are scattered by bodies in a randomly inhomogeneous medium, we observe the phenomenon of enhanced backscattering predicted in [4.17], which means that the mean field intensity of the field scattered directly back to the source $\overline{I}_s$, is on average higher than the intensity of the backscattered field in a homogeneous space, I_{s0}.

Let r_s be the radius vector of a source, r_r the reception point located near r_s, and R the radius vector of a scatterer. Show that the enhancement coefficient for backscattering $K = \overline{I}_s/I_{s0}$ may be expressed through the covariance of the relative intensity fluctuations α corresponding to a single passage of the source-scattered "leg"

$$K(r_s, r_r) = 1 + \psi_\alpha(r_s, r_r) \quad . \tag{4.120}$$

Solution. For a point source and a relatively small scatterer, the coherence radius of the incident field must be larger than the size of a body, or at least of the "flash" that forms the scattered field. The incident field $U_{\text{inc}}(r)$ and the scattered field $u_{\text{sc}}(r_r)$ will then be expressed in terms of Green's function of the turbulent medium $G(r_1, r_2)$:

$$u_{\text{inc}}(r) = AG(r, r_s) \quad , \quad u_{\text{sc}}(r_r) = fG(r_r, R)u_{\text{inc}}(R),$$

where A = const, and f is the scattering amplitude. In this notation, the mean intensity of the scattered field $\overline{I}_s = \overline{|u_s|^2}$ is

$$\overline{I}_s = |Af|^2 \langle |G(R, r_s)G(r_r, R)|^2 \rangle = |Af|^2 \langle I_G(R, r_s) I_G(r_r, R) \rangle \quad . \tag{4.121}$$

where $I_G = |G|^2$ is the intensity of the spherical wave after it has once covered the route.

When the medium is statistically homogeneous and isotropic and has large-scale inhomogeneities, $\overline{I}_G$ coincides with the intensity I_{G0} of the spherical wave in a homogeneous medium: $\overline{I}_G = I_{G0}$. Assuming that $\alpha = \Delta I_G / I_{G0} = (I_G - I_{G0})/I_{G0}$, we write (4.121) in the form

$$I_s = |Af|^2 I_{G0}^2 [1 + \psi_\alpha(\boldsymbol{r}_s, \boldsymbol{r}_r)] \quad , \quad \text{where}$$

$$\psi_\alpha(\boldsymbol{r}_s, \boldsymbol{r}_r) = \langle \alpha(\boldsymbol{R}, \boldsymbol{r}_s)\alpha(\boldsymbol{r}_r, \boldsymbol{R})\rangle \quad .$$

Considering further that, in the absence of inhomogeneities, $I_s = |Af|^2 I_{G0}^2 \equiv I_{s0}$, for the backscattering enhancement coefficient we will get (4.120).

If the receiver is superposed on the source, i.e., we have $\boldsymbol{r}_s = \boldsymbol{r}_r$, then, by the reciprocity theorem, we get $G(\boldsymbol{r}_r, \boldsymbol{R}) = G(\boldsymbol{R}, \boldsymbol{r}_s)$ and $\alpha(\boldsymbol{r}_r, \boldsymbol{R}) = \alpha(\boldsymbol{R}, \boldsymbol{r}_s)$, whereby

$$K(\boldsymbol{r}_s, \boldsymbol{r}_s) = 1 + \langle \alpha^2 \rangle \quad .$$

This quantity is always larger than unity, irrespective of the approximation we chose in order to compute $\langle \Delta I_G^2 \rangle$. As the receiver is moved away from the source, the fluctuations of Green's function G in the forward and backward "legs" of the path become independent, and so $K = 1$. From energy conservation considerations, as K near the source is increased and when $K = 1$ nearly everywhere away from the source, K slightly decreases at points that are separated from $\boldsymbol{r}_s$ by a distance of the order of the coherence radius.

The backscattering enhancement effect is in principle associated with the multiple scattering of waves in a randomly inhomogeneous medium. In a region of saturated fluctuations, where the field is normalized and $\langle \alpha^2 \rangle \to 1$, the enhancement can be as high as $K = 2$, and in the field of strong fluctuations, K is slightly higher than two. Still larger values of K are realized when there is a phase screen or a statistically uneven interface between the source and the scatterer. In [4.25, 26], we provide additional facts about the enhancement effect and a number of related phenomena are described: the far correlation effect, the partial wave front reversal, and so on.

5. Rough Surface Scattering

The chapter is devoted to the basic methods of describing the scattering of waves at rough surfaces: the perturbation technique (Sect. 5.1) and Kirchhoff's technique (Sect. 5.2). Using these methods, some of the main results are obtained: the scattering cross-section for the rough surface, the spatial covariance for the field, and other characteristics.

Section 5.3 sketches the more general problems (e.g., the case of two-scale roughness) and touches upon some more powerful techniques (the integral equation method for Green's function).

5.1 Slightly Rough Surfaces. Perturbation Technique

The surfaces of real bodies are rough to varying degrees, whereby waves are not reflected and refracted in the same way as from ideally smooth interfaces. Clearly, the degree of "smoothness" is primarily determined by the relationship between the wavelength and the geometric parameters of the surface roughness. Therefore, a surface which is "ideally" smooth for radio waves or sound may be rough for light or ultrasound.

Roughness may vary with time (high seas, thermal fluctuations of the surface shape, etc.), or may be essentially unchanged (terrain and sea bed irregularities, the surface of paper, frosted glass or any solid body in general). Finally, as with volume inhomogeneities in a medium, problems concerning the reflection and refraction of waves from rough surfaces, or diffraction and scattering from them, may be either deterministic or stochastic. The latter, as usual, occur in those cases where we are interested not in some specific, "individual" kind of rough surface, but in the characteristics of an *ensemble* of such surfaces. We deal with a statistical ensemble when we have a large number of irregularities in an area of illuminated surface (*rough* surfaces), but situations are, of course, possible where these irregularities are few. For example, we may be interested in scattering from a single body placed on a plane. The ensemble here will consist of realizations with a wide variety of bodies, the geometrical and/or physical characteristics of the body being random (subject to some stochastic distributions).

Earlier (Sect. III.2.1), we classified wave scattering from bodies with a random shape or position as belonging to problems of type (3). These problems normally consist of finding the statistical behavior of the scattered field from the known roughness statistics, but it is often necessary to solve the inverse problem and determine the characteristics of the surface from the statistics of the scattered field.

The character of scattering is determined by many factors. Apart from the height of the irregularities and the wavelength of the incident radiation, it is influenced by the size of the scattering surface, the angle of incidence, the polarization of the primary wave, the reflective and refractive behavior of matter, and so forth. Depending on the interplay of the various parameters, some approximate techniques or other are used to derive the scattered field. We will only consider here the two simplest and most common techniques: the *method of small perturbations* and the *Kirchhoff approximation*. A discussion of more general techniques which take into account multiple scattering can be found in [5.1, 2] (see also [5.3]).

We confine our discussion to the case of the surface that is *on average plane*. Let the height of the irregularities be $z = \zeta(x, y) \equiv \zeta(\varrho)$, so that $\langle \zeta \rangle = 0$. If the surface $z = \zeta(\varrho)$ is an interface of two media, then two fields u_1 and u_2 in the first and second media, respectively, should be subject to some "two-sided" boundary conditions. For example, if u_1 and u_2 are *acoustic velocity potentials,* then at the interface the normal components of velocity are equal, i.e.,

$$\frac{\partial u_1}{\partial N} = \frac{\partial u_2}{\partial N} \quad ,$$

where N is the normal to the surface. The pressures are also equal at the interface,

$$\varrho_1 u_1 = \varrho_2 u_2 \quad ,$$

where ϱ_1 and ϱ_2 are densities of the media. With electromagnetic waves, the tangential components of the fields, E_t and H_t, must be continuous at the interface. The scattered waves here propagate in both media.

For simplicity, we will consider *scalar* waves and simplify the problem still further by assuming that waves can only propagate through one medium, i.e., that the surface $z = \zeta(\varrho)$ is either "absolutely soft" ($u = 0$) or "absolutely rigid" ($\partial u / \partial N = 0$). In both cases, complete reflection occurs. With electromagnetic waves, this corresponds to the perfectly conducting surface ($E_t = 0$) and the surface of a perfect magnetic substance ($H_t = 0$).

Thus, for the absolutely soft surface, the boundary condition has the form

$$u|_S = u(\varrho, z)|_{z=\zeta(\varrho)} = 0 \quad , \tag{5.1}$$

and for the absolutely rigid one

$$\frac{\partial u}{\partial N}\bigg|_S = (N, \nabla u)\bigg|_{z=\zeta(\varrho)} = 0 \quad , \tag{5.2}$$

where the unit vector of the external normal N has the components

$$(N_\perp, N_z) = (\alpha \nabla_\perp \zeta, -\alpha) \quad \text{and} \quad \alpha = [1 + (\nabla_\perp \zeta)^2]^{-1/2} \quad . \quad \text{Here} \tag{5.3}$$

$$\nabla_\perp = \left(\frac{\partial}{\partial x}, \frac{\partial}{\partial y} \right)$$

is the transverse differential operator.

There are no rigorous methods of solving the wave equation with the random boundary conditions (5.1) or (5.2), and the problem only lends itself to an approximate solution, under certain constraints imposed on the size and shape of the irregularities. We will now consider two cases of such constraints, namely the surfaces with roughness that on the scale of wavelength λ is either *small* and *gently sloping*, or *smooth*, which allow the two techniques mentioned above to be employed. In the first case we will apply the method of small perturbations, and in the second, Kirchhoff's approximation.

The *gentleness* of roughness means that the slopes of individual irregularities are on average small, i.e.,

$$\langle (\nabla_\perp \zeta)^2 \rangle \sim \sigma_\zeta^2 / l_\zeta^2 \ll 1 \quad , \tag{5.4}$$

where $\sigma_\zeta^2 \equiv \langle \zeta^2 \rangle$ is the mean square of deviation from the unperturbed surface $z = 0$, and l_ζ is the characteristic size of the irregularities. Of course, this inequality, which is valid for the mean square, is used even before averaging in practical calculations.

The *smallness* of roughness means that the moments $\langle \zeta^m \rangle$ are small in comparison with mth powers of the wavelength, i.e., $\langle \zeta^m \rangle \ll \lambda^m$, in particular

$$\sigma_\zeta^2 \ll \lambda^2 \quad . \tag{5.5}$$

As a result, for small and gently sloping irregularities, we may use the expansion both of the boundary condition and of the solution in the parameters $\zeta/\lambda \ll 1$ and $|\nabla_\perp \zeta| \sim \sigma_\zeta / l_\zeta \ll 1$, which is precisely the idea behind the method of small perturbation.

In the case of *smooth* irregularities, the deviation ζ is not limited, and the gentleness condition is replaced by the requirement of a small curvature of roughness, which mathematically implies that the curvature radii of the surface, R_{cur}, must be large as compared to λ

$$R_{\mathrm{cur}} \gg \lambda \quad \text{or} \quad k R_{\mathrm{cur}} \gg 1 \quad . \tag{5.6}$$

Accordingly, for $\zeta \gtrsim \lambda$, the inhomogeneities must be more extended in the x- and y-directions (large-scale irregularities). Under these conditions, we apply Kirchhoff's approximation: in the neighborhood of any point on the surface, the field is expressed as the sum of the incident wave and the wave reflected from the contiguous plane at that point, using the *local* values of the "plane" coefficients of reflection (in the electromagnetic problem, the Fresnel coefficients). By "incident wave" we understand not only the primary wave, but also the waves that come to this point after reflecting from other parts of the surface. The simplest case is that without multiple reflection.

For smooth roughness, the ratio σ_ζ / l_ζ may not necessarily be small. Kirchhoff's approximation is in principle applicable to the case of *steep* irregularities ($\sigma_\zeta / l_\zeta \gtrsim 1$). It is clear, however, that as the steepness increases, the contribution of multiple reflection will grow. Kirchhoff's approximation does not allow us to take this into account in any simple way. But, we cannot neglect this contribution, even in the case of large-scale roughness, unless a measure of smoothness is present. A measure of smoothness is also required in order that the mutual shadowing of elements of rough surface may be ignored, although shadowing is taken into consideration more easily than multiple reflection. Thus, unlike the method of small perturbations, Kirchhoff's approximation does not regard the slope of irregularities $|\nabla_\perp \zeta|$ as a small parameter. Gentleness is only required to simplify the problem, i.e., in a measure that makes it possible to ignore multiple reflection and shadowing. We shall be looking at Kirchhoff's method in Sect. 5.2, and now turn to the case of small irregularities.

We have already said that the method of small perturbations is based on the expansion of the field u which we seek, and the boundary conditions in the small parameters $\zeta/\lambda \sim \sigma_\zeta/\lambda \ll 1$ and $|\nabla_\perp \zeta| \sim \sigma_\zeta/l_\zeta \ll 1$. The approach dates back to Rayleigh, who proposed it for sinusoidal irregularities. It was then used by Mandelshtam for the statistical problem concerned with light scattering from inhomogeneities caused by the thermal fluctuations of the surface of a liquid.

Let $U(\mathbf{r})$ be the primary monochromatic field incident on a rough surface [we omit the factor $\exp(-i\omega t)$ for simplicity]. We write the solution to the Helmholtz equation in the form

$$u(\mathbf{r}) = U(\mathbf{r}) + \sum_{n=0}^{\infty} u^{(n)}(\mathbf{r}) \quad , \tag{5.7}$$

where the nth term is of the order of $(\sigma_\zeta/\lambda)^n$ or $(\sigma_\zeta/l_\zeta)^n$. This is an expansion in scattering multiplicity.

For the absolutely soft surface, the boundary condition (5.1) expanded in ζ takes the form

$$u(\varrho, z)\Big|_{z=\zeta(\varrho)} = u_0 + \left(\frac{\partial u}{\partial z}\right)_0 z + \left(\frac{\partial^2 u}{\partial z^2}\right)_0 \frac{z^2}{2} + \dots \Big|_{z=\zeta(\varrho)} = 0 \quad . \tag{5.8}$$

Here, and in the remainder of the book, the subscript 0 refers to $z = 0$. Here, $\nabla_\perp \zeta$ does not appear in the boundary condition, so that the gentleness condition (5.4) for the absolutely soft surface is superfluous, and (5.7) uses only one small parameter $\sigma_\zeta/\lambda \ll 1$.

Substituting (5.7) into (5.8), we obtain the boundary conditions (now in the plane $z = 0$, which is sometimes called *mean surface*), for the successive approximations of the field

$$u_0^{(0)} + U_0 = 0 \quad ,$$

$$u_0^{(1)} + \left[\frac{\partial}{\partial z}(u^{(0)} + U)\right]_0 \zeta = 0 \quad ,$$

$$u_0^{(2)} + \left(\frac{\partial u^{(1)}}{\partial z}\right)_0 \zeta + \left[\frac{\partial^2}{\partial z^2}(u^{(0)} + U)\right]_0 \frac{\zeta^2}{2} = 0 \quad . \tag{5.9}$$

Finding the n-fold scattered field $u^{(n)}(\mathbf{r})$ thus reduces to the problem of the wave field having a predetermined value $v^{(n)}(\varrho) = u_0^{(n)} \equiv u^{(n)}(\varrho, 0)$ in the plane

$z = 0$ (Sect. III.2.2). This value is known as long as the earlier successive approximations from $u^{(0)}$ to $u^{(n-1)}$ are known. The random character of the boundary values of $u_0^{(n)}$ is due to the presence of the random function $\zeta(\varrho)$ in (5.9).

In order to find the field in an nth approximation, we may use either Green's function (III.2.16), or the Rayleigh technique [expansion in plane waves – equations (III.2.19–21)], both approaches being equivalent for the plane boundary $z = 0$. We will now give the treatment using Green's function (III.2.16), which for fields in a zeroth and first approximation, gives [here $r = \{\varrho, z\}$, $R = \sqrt{z^2 + (\varrho + \varrho')^2}$]

$$u^{(0)}(r) = \frac{1}{2\pi} \int U(\varrho', 0) \frac{\partial}{\partial z}\left(\frac{e^{ikR}}{R}\right) d^2\varrho' \quad ,$$

$$u^{(1)}(r) = \frac{1}{2\pi} \int \frac{\partial}{\partial z}[u^{(0)} + U]_0 \zeta(\varrho')\left(\frac{e^{ikR}}{R}\right) d^2\varrho' \quad . \tag{5.10}$$

When fields themselves are specified in the plane $z = 0$, it is advantageous to compute $u^{(n)}(r)$ both over a bounded area Σ of this plane, and over the entire infinite plane. If we assume that the bounded area is large as compared to the wavelength, we can make use of Kirchhoff's hypothesis that equations (5.10), written for the infinite plane, also hold over the finite area under consideration and we need only bound the integration region in (5.10) by the confines of the area.

Let us consider the plane wave incident on a rough surface at an angle θ_0 with

$$U(r) = A \exp(ik n_i \cdot r) = A \exp[ik(n_\perp^i \cdot \varrho + n_z^i z)]$$
$$= A \exp[ik(n_\perp^i \varrho - z \cos \theta_0)] \quad , \tag{5.11}$$

where $n_i = (n_\perp^i, n_z^i)$ is the unit vector in the direction of incidence; $n_z^i = -\cos \theta_0$ (Fig. 5.1). In the case of the infinite surface it is clear, without calculation, from the first of (5.10) that in a zeroth approximation, we deal with the *specular* wave

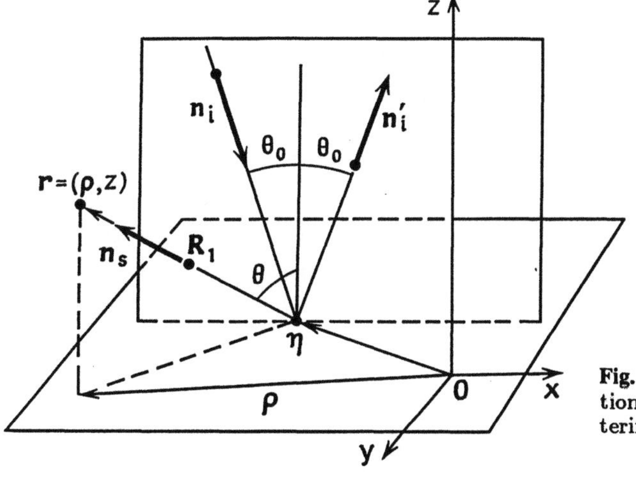

Fig. 5.1. The geometry and notation for the problem of wave scattering by a rough surface

$$u^{(0)}(\mathbf{r}) = -A\exp\left[ik(\mathbf{n}^i_\perp \cdot \boldsymbol{\varrho} - n^i_z z)\right] = -A\exp\left[ik(\mathbf{n}^i_\perp \cdot \boldsymbol{\varrho} + z\cos\theta_0)\right]. \quad (5.12)$$

If the area is finite, then according to Kirchhoff's hypothesis the field $u^{(0)}$ near it is also the specular wave (5.12). But, sufficiently far from the area, the field $u^{(0)}$ becomes a directed spherical wave with an intensity maximum in the direction of specular reflection.

The zero order field $u^{(0)}(\mathbf{r})$ is only of interest to the extent to which it determines the singly scattered field $u^{(1)}$. It follows from (5.11, 12) that

$$\left[\frac{\partial}{\partial z}(u^{(0)} + U)\right]_0 = 2\left(\frac{\partial U}{\partial z}\right)_0 = 2ikn^i_z A\exp\left(ikn^i_\perp \cdot \boldsymbol{\varrho}\right) \quad ,$$

therefore, by (5.10),

$$u^{(1)} = \frac{ikn^i_z A}{\pi}\int \exp\left(ikn^i_\perp \cdot \boldsymbol{\varrho}'\right)\zeta(\boldsymbol{\varrho}')\frac{\partial}{\partial z}\left(\frac{e^{ikR}}{R}\right)d^2\varrho'$$

$$\approx -\frac{k^2 n^i_z A z}{\pi}\int \zeta(\boldsymbol{\varrho}')\exp\left(ikn^i_\perp \cdot \boldsymbol{\varrho}' + ikR\right)\frac{d^2\varrho'}{R^2} \quad . \quad (5.13)$$

The second form of this relationship refers to the case where $kz \gg 1$, when the observation point is separated from the plane $z = 0$ by at least several wavelengths.

The mean of $u^{(1)}$ is zero, and its mean intensity $\overline{I}_1 = \langle|u^{(1)}|^2\rangle$ is calculated in much the same way as $\overline{I}_1$ in the case of scattering from volume inhomogeneities. Given the assumption that the fluctuations $\zeta(\boldsymbol{\varrho})$ are statistically homogeneous, using (5.13), we find

$$\overline{I}_1 = \frac{k^4 (n^i_z)^2 z^2}{\pi^2}|A|^2 \iint\limits_{-\infty}^{\infty} \psi_\zeta(\boldsymbol{\varrho}' - \boldsymbol{\varrho}'')\exp\left[ikn^i_\perp \cdot (\boldsymbol{\varrho}' - \boldsymbol{\varrho}'')\right.$$

$$\left. + ik(R' - R'')\right]\frac{d^2\varrho'\,d^2\varrho''}{R'^2 R''^2} \quad (5.14)$$

where $\psi_\zeta(\boldsymbol{\varrho}' - \boldsymbol{\varrho}'') = \langle\zeta(\boldsymbol{\varrho}')\zeta(\boldsymbol{\varrho}'')\rangle$ is the covariance of roughness, $R' = \sqrt{z^2 + (\boldsymbol{\varrho} - \boldsymbol{\varrho}')^2}$, $R'' = \sqrt{z^2 + (\boldsymbol{\varrho} - \boldsymbol{\varrho}'')^2}$. Let us now introduce new integration variables $\boldsymbol{\xi} = \boldsymbol{\varrho}' - \boldsymbol{\varrho}''$, $\boldsymbol{\eta} = (\boldsymbol{\varrho}' + \boldsymbol{\varrho}'')/2$ and expand R' and R'' into a Taylor series in $\boldsymbol{\xi}$. Since the integrand in (5.14) includes the covariance $\psi_\zeta(\boldsymbol{\xi}) = \psi_\zeta(\boldsymbol{\varrho}' - \boldsymbol{\varrho}'')$ which promptly falls off to zero when $\xi \gg l_\xi$, we may keep only the first terms in these expansions. So, for $kl_\zeta^3/z^2 \ll 1$[1] we can substitute $R_1 = |\mathbf{r} - \boldsymbol{\eta}| = \sqrt{z^2 + (\boldsymbol{\varrho} - \boldsymbol{\eta})^2}$ for R' and R'' in the denominator of the integrand, and $-\mathbf{n}^s_\perp \boldsymbol{\xi}$ for $R' - R''$. Here, $\mathbf{n}_s = \mathbf{R}_1/R_1$ is the unit vector in the direction from the scattering point $\boldsymbol{\eta}$ in the plane $z = 0$ to the observation point $\mathbf{r} = (\boldsymbol{\varrho}, z)$, and $\mathbf{n}^s_\perp$ is the transverse component of the vector

---

[1] Note that the inequality is weaker than the condition $kl_\zeta^2/z \ll 1$, which corresponds to placing the observation point in the Fraunhofer zone of an individual inhomogeneity.

$$n_s = (n^s_\perp, n^s_z) = \left(\frac{\varrho - \eta}{R_1}, \frac{z}{R_1} \right) \quad .$$

Further, by placing the limits of integration with respect to ξ at infinity (this is possible even if the patch Σ is finite, but large as compared to the correlation radius l_ζ), we find

$$\int\limits_{-\infty}^{\infty} \psi_\zeta(\xi) \exp\left[ik(n^i_\perp - n^s_\perp)\cdot\xi\right]d^2\xi = 4\pi^2 F_\zeta(q_\perp) \quad , \quad \text{where}$$

$$F_\zeta(\kappa) = \frac{1}{4\pi^2} \int\limits_{-\infty}^{\infty} \psi_\zeta(\varrho) \exp(-i\kappa\cdot\varrho)d^2\varrho \tag{5.15}$$

is the Fourier transform of the covariance, i.e., the two-dimensional spatial spectrum of the inhomogeneities: $q_\perp = k(n^s_\perp - n^i_\perp)$ is the transvesrse component of the scattering vector $q = k(n_s - n_i)$ (Fig. 5.2). As a result, (5.14) now contains only the integral with respect to η

$$\overline{I}_1 = 4k^4(n^i_z)^2 z^2|A|^2 \int\limits_{-\infty}^{\infty} F_\zeta(q_\perp)\frac{d^2\eta}{R_1^4} \tag{5.16}$$

which covers the entire plane $z = 0$, or only some part of it, Σ, if the rough surface is finite.

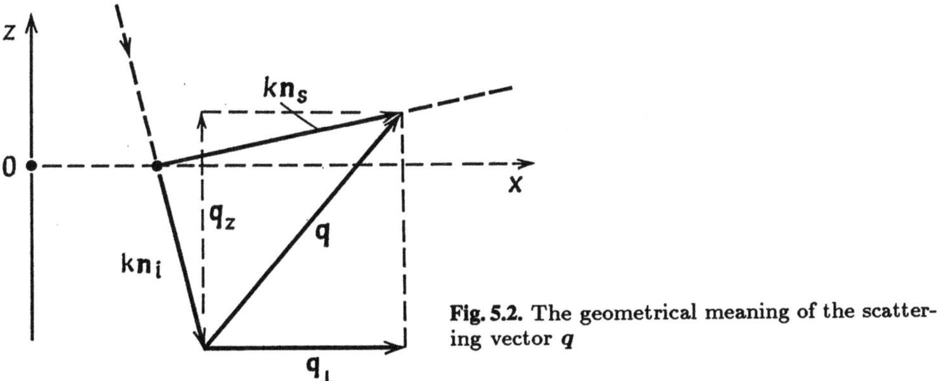

Fig. 5.2. The geometrical meaning of the scattering vector q

According to (5.16), each element of the surface $d^2\eta = d\Sigma$ with the center at η makes a contribution to the total intensity $\overline{I}_1$

$$\begin{aligned}
d\overline{I}_1 &= 4k^4(n^i_z)^2|A|^2 \frac{z^2}{R_1^2} F_\zeta(q_\perp)\frac{d\Sigma}{R_1^2} \\
&= 4k^4(n^i_z n^s_z)^2 F_\zeta(q_\perp)|A|^2 \frac{d\Sigma}{R_1^2} \quad ,
\end{aligned} \tag{5.17}$$

where $n^s_z = z/R_1 = \cos\theta$ is the cosine of the angle between $n_s = R_1/R_1$ and the z-axis (Fig. 5.1). Equation (5.16) thus reflects the selective nature of the scattering:

when scattered in a given direction, n_s is only a certain "harmonic" of the irregularities corresponding to the scattering vector $q_\perp = k(n_\perp^s - n_\perp^i)$, which is completely analogous to the selectivity of scattering from volume irregularities (Sect. III.4.2). In particular, when the primary wave is incident normally, i.e., when $n_z^i = -1$, $n_\perp^i = 0$ and $q_\perp = kn_\perp^s$, the wavelength of the "active" harmonic of the irregularities $\Lambda_q = 2\pi/q_\perp$ is equal to $\lambda/n_\perp^s = \lambda/\sin\theta$. With specular scattering, when $n_\perp^s = n_\perp^i$ and $q_\perp = 0$, Λ_q becomes infinite. Lastly, with backscattering ($n_\perp^s = -n_\perp^i$, $q_\perp = -2kn_\perp^i$)

$$\Lambda_q = \frac{2\pi}{2kn_\perp^i} = \frac{\lambda}{2\sin\theta_0} \quad .$$

The coefficient at $|A|^2 d\Sigma/R_1^2$ in (5.17)

$$\sigma_{\text{soft}}(q_\perp) = 4k^4(n_z^i n_z^s)^2 F_\zeta(q_\perp) \tag{5.18}$$

is the *scattering cross-section* of a unit area of absolutely soft surface in the n^s-direction. Unlike volume scattering, (5.18) is dimensionless. Equation (5.16), written as

$$\overline{I}_1 = |A|^2 \int \frac{\sigma_{\text{soft}}(q_\perp)d\Sigma}{R_1^2} \quad , \tag{5.19}$$

expresses explicitly the *incoherence* of waves scattered by individual elements of the rough surface and the *summation of their intensities*. For the absolutely soft surface, the cross-section (5.18), containing the factor $(n_z^i n_z^s)^2 = \cos^2\theta_0 \cos^2\theta$, vanishes at grazing angles of incidence and reflection, i.e., when θ or $\theta_0 \to \pi/2$, whereby (5.19) converges even when integrating over the entire plane $z = 0$.

If the surface has a finite area Σ and within the confines of this area $\sigma_{\text{soft}}(q_\perp)$ and R_1 are essentially constant, then

$$\overline{I}_1 = \frac{|A|^2 \sigma_{\text{soft}}(q_\perp)}{R_1^2} \Sigma \quad . \tag{5.20}$$

It is easily verified that the condition that $\sigma_{\text{soft}}(q_\perp)$ and R_1 be constant within Σ, with maximum size L, is written as:

$$R_1 \gg L \quad , \quad \text{if} \quad l_\zeta \ll \lambda \quad \text{(small-scale irregularities)} \quad ,$$

$$R_1 \gg kl_\zeta L \quad , \quad \text{if} \quad l_\zeta \gg \lambda \quad \text{(large-scale irregularities)} \quad . \tag{5.21}$$

The case of the *absolutely rigid* surface differs in that, even in a first approximation, the scattered field depends not only on ζ, but also on $\nabla_\perp \zeta$. It follows from (5.2, 3) that at the boundary,

$$\left[(\nabla_\perp \zeta \nabla_\perp u) - \frac{\partial u}{\partial z} \right]_{z=\zeta(\varrho)} = 0 \quad .$$

Substituting (5.8) gives

$$[\ldots]_{z=\zeta(\varrho)} = -\left(\frac{\partial u}{\partial z}\right)_0 - \left(\frac{\partial^2 u}{\partial z^2}\right)_0 \zeta + (\nabla_\perp \zeta \nabla_\perp u)_0 + \ldots = 0 \quad .$$

Unlike (5.9), we thus obtain for the successive approximations of the field the boundary conditions containing

$$\left(\frac{\partial u^{(0)}}{\partial z}\right)_0 + \left(\frac{\partial U}{\partial z}\right)_0 = 0 \quad ,$$

$$\left(\frac{\partial u^{(1)}}{\partial z}\right)_0 - \left(\nabla_\perp \zeta \nabla_\perp (u^{(0)} + U)\right)_0 + \left[\frac{\partial^2 (u^{(0)} + U)}{\partial z^2}\right]_0 \zeta = 0 \quad . \tag{5.22}$$

In order to find the fields $u^{(n)}$ from the values of the normal derivative $(\partial u^{(n)}/\partial z)_0$ specified in the plane $z = 0$, we can make use of Green's function (III.2.18)

$$u^{(0)}(\boldsymbol{r}) = \frac{1}{2\pi} \int \left(\frac{\partial U}{\partial z}\right)_0 \frac{e^{ikR}}{R} d^2 \varrho' \quad , \tag{5.23}$$

$$u^{(1)}(\boldsymbol{r}) = \frac{1}{2\pi} \int \left\{ (\nabla_\perp \zeta \nabla_\perp (u^{(0)} + U))_0 - \left[\frac{\partial^2 (u^{(0)} + U)}{\partial z^2}\right]_0 \zeta \right\} \frac{e^{ikR}}{R} d^2 \varrho' \quad . \tag{5.24}$$

Let the plane wave (5.11) be incident on a rough surface. Clearly, in the immediate vicinity of the area, the specular wave $u^{(0)}$, meeting the first of the boundary conditions (5.22), obeys the laws of specular reflection and is written in the form

$$u^{(0)} = A \exp\left[ik(\boldsymbol{n}_\perp^{\mathrm{i}} \cdot \varrho - n_z^{\mathrm{i}} z)\right] \quad .$$

The expression in braces in (5.24) will then be

$$\{\ldots\} \equiv f(\varrho') = 2A[ik(\boldsymbol{n}_\perp^{\mathrm{i}} \cdot \nabla_\perp)\zeta + k^2 (n_z^{\mathrm{i}})^2 \zeta] \exp(ik\boldsymbol{n}_\perp^{\mathrm{i}} \cdot \varrho') \quad .$$

The mean of this is zero, and the covariance $\psi_f(\boldsymbol{\xi}) = \langle f(\varrho') f^*(\varrho'') \rangle$ is expressed in terms of $\psi_\zeta(\boldsymbol{\xi})$ in the following way:

$$\psi_f(\boldsymbol{\xi}) = 4|A|^2 \exp(ik\boldsymbol{n}_\perp^{\mathrm{i}} \cdot \boldsymbol{\xi})[ik(\boldsymbol{n}_\perp^{\mathrm{i}} \cdot \nabla_\perp) + k^2 (n_z^{\mathrm{i}})^2]^2 \psi_\zeta(\boldsymbol{\xi}) \tag{5.25}$$

(the differentiation $\nabla_\perp$ here is with respect to $\boldsymbol{\xi} = \varrho' - \varrho''$).

Let us now calculate the mean intensity of the scattered field

$$\bar{I}_1 = \frac{1}{4\pi^2} \int \int \psi_f(\varrho' - \varrho'') \frac{\exp[ik(R' - R'')]}{R' R''} d^2 \varrho' d^2 \varrho'' \quad . \tag{5.26}$$

As before, it is convenient to go over to the new integration variables $\boldsymbol{\xi} = \varrho' - \varrho''$ and $\eta = (\varrho' + \varrho'')/2$ and substitute approximately $\exp(-ik\boldsymbol{n}_\perp^{\mathrm{s}} \cdot \boldsymbol{\xi})/R_1^2$ for $\exp[ik(R' - R'')]/R' R''$. Using (5.25), we now write

$$\overline{I}_1 = \frac{|A|^2}{\pi^2} \int \frac{d^2\eta}{R_1^2} \int \exp\left(-iq_\perp \cdot \xi\right)[ik(n_\perp^i \nabla_\perp) + k^2(n_z^i)^2]^2 \psi_\zeta(\xi) d^2\xi \quad .$$

The internal integral is expressed in terms of the two-dimensional spectral density of the irregularities, $F_\zeta(q_\perp)$:

$$\int (\ldots)d^2\xi = 4\pi^2[-k(n_\perp^i, q_\perp) + k^2(n_z^i)^2]^2 F_\zeta(q_\perp) \quad .$$

But, $q_\perp = k(n_\perp^s - n_\perp^i)$, and therefore

$$[-k(n_\perp^i \cdot q_\perp) + k^2(n_z^i)^2]^2 = k^4[(n_z^i)^2 + (n_\perp^i)^2 - n_\perp^i \cdot n_\perp^s]^2$$
$$= k^4[1 - n_\perp^i \cdot n_\perp^s]^2 \quad .$$

We thus have

$$\overline{I}_1 = 4|A|^2 k^4 \int \frac{\left[1 - n_\perp^i \cdot n_\perp^s\right]^2 F_\zeta(q)}{R_1^2} d^2\eta \quad . \tag{5.27}$$

We can represent this expression in the form (5.19) if we introduce the cross-section of the scattering from an absolutely rigid unit area

$$\sigma_{\text{rig}}(q_\perp) = 4k^4[1 - n_\perp^i \cdot n_\perp^s]^2 F_\zeta(q_\perp) \quad , \tag{5.28}$$

which differs from the scattering cross-section from the soft surface (5.18) by the factor at $F_\zeta(q_\perp)$. This factor $[1 - (n_\perp^i n_\perp^s)]^2$, and hence the cross-section σ_{rig}, does not tend to zero for grazing angles of incidence and reflection, so that the integral (5.27), taken over the entire plane $z = 0$, diverges. The intensity of a real scattered field cannot, of course, become infinite. The divergence is only associated with the technique used, namely with the use of a first approximation, and leads to divergence in higher-order approximations, the intensity of the total field $u = \Sigma_n u^{(n)}$ being finite.

The divergence can be remedied using improved forms of perturbation theory which take account of shadowing and multiple scattering, even in a zeroth approximation. Without going into detail, we will note that if we include the above-mentioned effects (to be discussed briefly in Sect. 5.3), the scattering cross-section σ_{rig}, unlike (5.28), vanishes at grazing angles of incidence and scattering, i.e., as $\theta_0 \to \pi/2$ or $\theta \to \pi/2$ (when $n_z^i \to 0$ or $n_z^s \to 0$), and at intermediate values of θ_0 and θ, the cross-section coincides with (5.28). It is thus only legitimate to compute the intensity of the scattered field by (5.27) for finite areas. If we pass over to infinite limits, we will need a more exact expression for $\sigma_{\text{rig}}(q_\perp)$.

Let us consider the angular dependence of the scattering cross-section σ for the soft and rigid boundaries referring to the rough surface with the isotropic Gaussian covariance for the irregularities

$$\psi_\zeta(\varrho) = \sigma_\zeta^2 \exp\left(-\varrho^2/2l_\zeta^2\right) \quad , \quad \sigma_\zeta^2 = \langle \zeta^2 \rangle \quad ,$$

with the spectral density

$$F_\zeta(q) = \frac{\sigma_\zeta^2 l_\zeta^2}{2\pi} \exp\left(-q^2 l_\zeta^2/2\right) \quad . \tag{5.29}$$

If we put $n_i = (\sin\theta_0, 0, -\cos\theta_0)$, $n_s = (\sin\theta\cos\varphi, \sin\theta\sin\varphi, \cos\theta)$, then, from (5.18) and (5.28), we find

$$\left.\begin{array}{c}\sigma_{\text{soft}}\\[4pt]\sigma_{\text{rig}}\end{array}\right\} = \left.\begin{array}{c}\cos^2\theta_0\,\cos^2\theta\\[4pt](1-\sin\theta_0\,\sin\theta\,\cos\varphi)^2\end{array}\right\}$$

$$\times \frac{2}{\pi}k^4\sigma_\zeta^2 l_\zeta^2 \exp\left[-\frac{k^2 l_\zeta^2}{2}(\sin^2\theta_0 + \sin^2\theta - 2\sin\theta_0\,\sin\theta\,\cos\varphi)\right] \quad .$$

For small kl_ζ (small-scale irregularities), the angular dependence is determined by the factor at the exponential function, scattering occurring within a wide angular opening of about 90°. The variation of σ_{soft} and σ_{rig} with θ (the angle of incidence θ_0 and φ being fixed) is given in Fig. 5.3a,b. The dashed line in Fig. 5.3 shows the true variation of the indicatrix $\sigma_{\text{rig}}(\theta)$ when multiple scattering is taken into account.

For $kl_\zeta \gg 1$, the form of the scattering diagram is mostly determined by the exponential factor, as for σ_{soft} and σ_{rig}. In that case, the maximum of σ is in the direction of specular reflection $\theta = \theta_0$, $\varphi = 0$, and the intensity decreases e-fold at an angular deviation from the maximum by about $\Delta\theta \sim 1/kl_\zeta$ (Fig. 5.4).

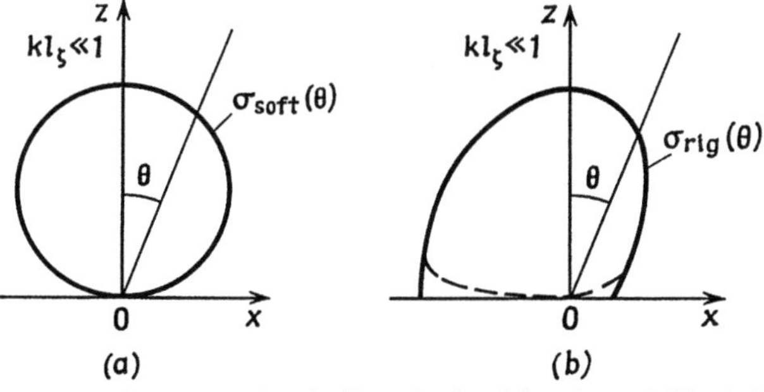

Fig. 5.3. The scattering indicatrix for **(a)** soft and **(b)** rigid rough surface with small-scale irregularities

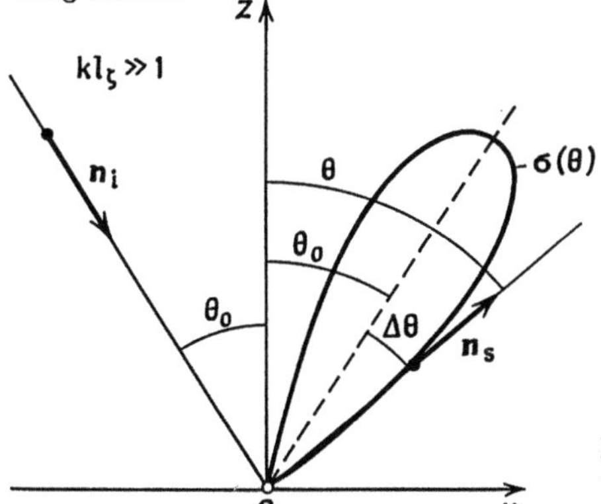

Fig. 5.4. The scattering indicatrix for a surface with large-scale irregularities

We can easily calculate that for $kl_\zeta \gg 1$ and incidence angles not very close to the grazing angle ($\cos \theta_0 \gg 1/kl_\zeta$), the ratio of the total intensity of the field scattered by a unit area

$$I_{\text{tot}} = |A|^2 \int\limits_0^{2\pi} d\varphi \int\limits_0^{\pi/2} \sigma \sin \theta \, d\theta$$

to the total intensity of the incident radiation, $I_0 = |A|^2 \cos \theta_0$, is approximately

$$\frac{I_{\text{tot}}}{I_0} \approx 4k^2 \sigma_\zeta^2 (\cos^2 \theta_0 + 2/(kl_\zeta)^2 \cos^2 \theta_0) \quad .$$

We see that as the height of roughness σ_ζ increases, the condition $I_{\text{tot}} \ll I_0$, that allows us to confine ourselves to a first approximation of perturbation theory, is violated sooner or later.

In actual practice, we more often encounter not the above formulation of the problem concerned with scattering from rough surfaces (a plane primary wave and on average plane scattering area), but the case where a large (essentially infinite) rough surface is illuminated by a wave beam, collimated or divergent. This is precisely the case in radar and sonar, and also in laboratory experiments on light scattering. By slightly modifying the treatment, we can easily generalize the expression (5.29) to this case, considering also a possible curvature of the reference surface.

Suppose that a surface is illuminated by a quasi-plane wave $U = A \exp(ik\varphi)$ whose amplitude A and local wave vector $k_{\rm i} = k\nabla\varphi$ are, for all practical purposes, constant in the scale of the correlation radius for roughness, l_ζ. Suppose also that the equation of the surface is given in the parametric form

$$r(\alpha, \beta) = \overline{r}(\alpha, \beta) + \zeta(\alpha, \beta) N_0(\alpha, \beta) \quad , \tag{5.30}$$

where $r = \overline{r}(\alpha, \beta)$ is the equation of the unperturbed (reference) surface Σ_0, and $\zeta(\alpha, \beta) N_0(\alpha, \beta)$ are random displacements along the normal to Σ_0 (Fig. 5.5). If the curvature radius of the reference surface is large as compared to the wavelength λ and correlation radius l_ζ, then the small element of the surface, $d\Sigma$, illuminated by the quasi-plane wave $A \exp(ik\varphi)$, scatters in the same way as an element of the plane surface contiguous to Σ_0. We will denote by $n_{\rm i}$ the gradient of the eikonal of the incident wave at the scattering point P lying in $\Sigma_0 (n_{\rm i} = k_{\rm i}/k = \nabla\varphi)$, and by $n_{\rm s}$ the unit vector in the direction from P to the observation point Q (Fig. 5.5). If, by $n_\perp^{\rm i}$ and $n_\perp^{\rm s}$, we denote the components of $n_{\rm i}$ and $n_{\rm s}$ tangent to Σ_0, and by R_1 the separation between P and Q, then the mean intensity of the scattered field will be

$$\overline{I}_1 = \int \frac{|A|^2 \sigma(q_\perp)}{R_1^2} d\Sigma_0 \quad , \tag{5.31}$$

where the cross-section $\sigma(q_\perp)$ is determined by the spectrum of the irregularities F_ζ near the scattering point P. The integration limits are determined either by the size of the scattering area, or by the size of the illuminated area, i.e., an area where the amplitude of the primary wave differs markedly from zero. Like (5.19), equation

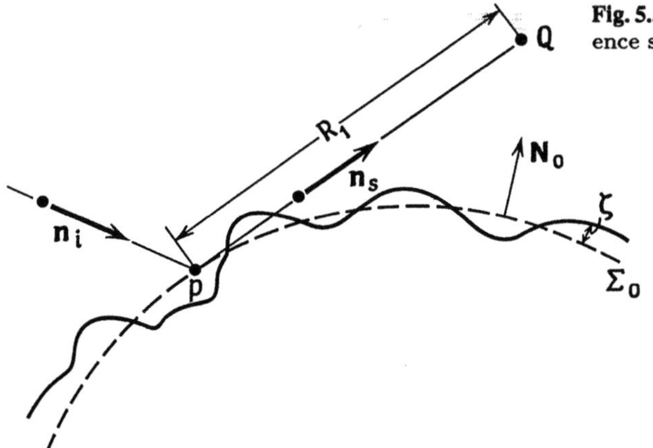

Fig. 5.5. Scattering for a curved reference surface

(5.31) is applicable subject to the condition $R_1 \gg \sqrt{kl_\zeta^3}$. For $kl_\zeta \gg 1$, this condition is weaker than the condition $R_1 \gg kl_\zeta^2$, which means that the observation point is placed in the Fraunhofer zone of an individual irregularity. But, if $kl_\zeta \ll 1$, equations (5.19, 31) hold even at $R_1 \sim \lambda$.

5.2 Scattering from Large-Scale Roughness. Kirchhoff's Technique

In acoustics, optics and radiophysics we are often interested in *large*, rather than small, inhomogeneities, where there is essentially no specular reflection. An example is the scattering of decimeter and centimeter radio waves from the surface of a rough sea, or the reflection of sound waves from the sea bed. This problem was first treated fairly completely by *Isakovich* [5.4], using Kirchhoff's approximation. For the most part, we will follow the treatment given in [5.4], and also in [5.5], where some of the constraints inherent in earlier publications were removed.

We will approach the problem using Green's function (III.2.18). Let a surface be illuminated by the locally plane scalar wave

$$U = A(r)e^{ik\varphi(r)} \quad , \tag{5.32}$$

which we will describe in the geometrical optics approximation. In particular, it may be plane or spherical (directional or omnidirectional). We assume that there is *no shadowing*, either for the incident or for the scattered waves. It is quite obvious that, however smooth the surface, this condition excludes grazing angles both for the incident wave and for the direction of observation.

It is well known that when the plane wave $U(r)$ is incident on the plane interface, the reflected field $u(r)$ and its normal derivative $\partial u/\partial N$ are related to U and $\partial U/\partial N$ on this boundary by the exact equations

$$u(r) = \mathcal{R}U(r), \quad \frac{\partial u}{\partial N} = -\mathcal{R}\frac{\partial U}{\partial N} \quad , \tag{5.33}$$

where $\mathcal{R}$ is the reflection coefficient that depends on the angle of incidence. According to the Kirchhoff principle, we assume that the boundary conditions (5.2) are *approximately* valid for the locally plane wave (5.1) that is incident on the locally plane surface Σ, i.e., on the surface with smooth irregularities. Of course, $\mathcal{R}$ here should be treated as a local reflection coefficient.

Substituting (5.2) into (III.2.18), we obtain for the secondary (scattered) field $u_{\text{sec}} \equiv u(\boldsymbol{r})$

$$u(\boldsymbol{r}) = \frac{1}{4\pi} \int \mathcal{R} \frac{\partial}{\partial N} \left[U(\boldsymbol{r}) \frac{e^{ikR}}{R} \right] d\Sigma \quad , \tag{5.34}$$

where $R = |\boldsymbol{r} - \boldsymbol{r}'|$ is the distance from the observation point $\boldsymbol{r} = (\varrho, z)$ to $\boldsymbol{r}' = (\varrho', z')$ lying in the rough surface $z' = \zeta(\varrho')$. We will now transform this expression in order to make it suitable for statistical averaging. For simplicity, we will consider the case where the reference surface is the plane $\overline{z}' = \overline{\zeta} = 0$.

First, in (5.3) we will go over to the integration over the reference plane $z' = 0$: if $\alpha = 1/\sqrt{1 + (\Delta_\perp \zeta)^2}$, then $d\Sigma = d\Sigma_0/\alpha = d^2\varrho'/\alpha$. Introducing the vector $\boldsymbol{N}_1 = \boldsymbol{N}/\alpha$ with components $(-\Delta_\perp \zeta, 1)$ normal to the surface Σ, we arrive at

$$\frac{\partial}{\partial N} \left(U \frac{e^{ikR}}{R} \right) d\Sigma = (\boldsymbol{N}\nabla') \left(\frac{U e^{ikR}}{R} \right) \frac{d^2\varrho'}{\alpha}$$

$$= (\boldsymbol{N}_1 \nabla') \left(A \frac{e^{ik(\varphi+R)}}{R} \right) d^2\varrho' \tag{5.35}$$

[the differential operator ∇' acts on the coordinates of the point $\boldsymbol{r}' = (\varrho', z')$].

Further, let the observation point $\boldsymbol{r}$ be separated from Σ by a distance that is larger than both the wavelength λ and the surface height: $kz \gg 1$ and $z \gg \sigma_\zeta$. We can then differentiate in (5.4) the fast-oscillating exponential function $\exp[ik(R+\varphi)]$ only, ignoring the derivatives of the slow-varying functions $A(\boldsymbol{r})$ and $1/R$

$$(\boldsymbol{N}_1 \nabla') \left(A \frac{e^{ik(R+\varphi)}}{R} \right) \approx ik \frac{A}{R} e^{ik(R+\varphi)} (\boldsymbol{N}_1 \nabla')(R + \varphi)$$

$$= -i \frac{A}{R} e^{ik(R+\varphi)} (\boldsymbol{N}_1 \cdot \boldsymbol{q})$$

$$= -i \frac{A}{R} e^{ik(R+\varphi)} [-(\boldsymbol{q}_\perp \nabla_\perp \zeta) + q_z] \quad , \tag{5.36}$$

where

$$\boldsymbol{q} = -k\nabla'(R + \varphi) = k(\boldsymbol{n}_{\text{s}} - \boldsymbol{n}_{\text{i}}) \tag{5.37}$$

is the scattering vector. Here $\boldsymbol{n}_{\text{i}} = \nabla'\varphi$ is the normal to the phase front of the incident wave, and the unit vector

$$\boldsymbol{n}_{\text{s}} = -\nabla'R = \nabla R = \frac{\boldsymbol{r} - \boldsymbol{r}'}{|\boldsymbol{r} - \boldsymbol{r}'|}$$

indicates the direction from $\boldsymbol{r}'$ to $\boldsymbol{r}$.

170

Considering that for gently sloping roughness $\nabla_\perp \zeta \ll 1$, and the scattering from such roughness occurs in the directions close to the direction of specular reflection (in this direction $n^i_\perp = n^s_\perp$ and $q_\perp = 0$), we will simply replace the scalar product $(N_1 q) = -(q_\perp \nabla_\perp \zeta) + q_z$ by q_z. As a result, we will have

$$u = -\frac{i}{4\pi} \int \mathcal{R} q_z \frac{A}{R} e^{ik(R+\varphi)} d^2 \varrho' \quad . \tag{5.38}$$

All the quantities in the integrand refer to the point $r' = \{\varrho', \zeta(\varrho')\}$ in the rough surface. In order to write explicitly the variation with the perturbation $\zeta(\varrho')$, we will expand the functions in the integrand into a Taylor series in ζ. In the slow-varying functions $\mathcal{R}$, q_z, A and $1/R$ we will keep only the zero-order term, and in the exponent we will retain in addition the term linear in $\zeta(\varrho')$

$$\mathcal{R} q_z \frac{A}{R} \approx \mathcal{R} q_z \frac{A}{R}\bigg|_{\zeta=0} \quad ,$$

$$\exp\left[ik(R+\varphi)\right] \approx \exp\left[ik(R+\varphi)|_{\zeta=0} - iq_z \zeta(\varrho')\right] \quad . \tag{5.39}$$

The local reflection coefficient $\mathcal{R}|_{\zeta=0}$ corresponds here to the direction of specular reflection, for which $n^s_\perp = n^i_\perp$ and $q_\perp = 0$. It is already a deterministic quantity. For the Fresnel reflection coefficients, which (if we exclude the case of total reflection) varies slowly with the incidence angle, this approximation is quite justified since, in this direction n_s, only those parts of the surface Σ (called specular points) scatter that are inclined at the same definite angle (Fig. 5.6).

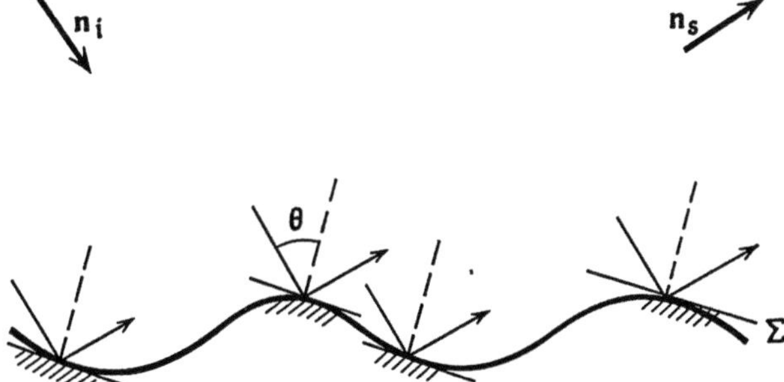

Fig. 5.6. Specular points on a surface with large-scale irregularities

By discarding the linear terms in the slow-varying amplitude functions and the quadratic ones in the exponent, we introduce a relative error that is not larger than $k\sigma_\zeta^2/D$, where D is either the curvature radius of the phase front of the incident wave, i.e., essentially the distance to the source, or the distance to the observation point, R. Next, we require that the condition

$$k\sigma_\zeta^2/D \ll 1 \tag{5.40}$$

be obeyed, which means that the source and observation point must lie within the Fraunhofer zone as regards the scale of the irregularities σ_ζ. Substituting (5.39) into (5.38) gives

$$u = -\frac{i}{4\pi} \int \mathcal{R} q_z \frac{A}{R} \exp\left(ik(R + \varphi) - iq_z\zeta\right) d^2\varrho' \quad , \tag{5.41}$$

where we take the values of q_z, A, R, φ, and $\mathcal{R}$ in the plane $z = 0$.

A more accurate treatment, in which the term $(q\nabla_\perp\zeta)$ in (5.36) is not discarded, but transformed by integrating by parts [5.4], leads to

$$u = -\frac{i}{4\pi} \int \mathcal{R} \frac{q^2}{q_z} \frac{A}{R} \exp\left(ik(R + \varphi) - iq_z\zeta\right) d^2\varrho' \quad , \tag{5.42}$$

which differs from (5.41) in that q^2/q_z is substituted for q_z. In (5.41, 42), as in (5.38), all the quantities are functions of $r' = \{\varrho', 0\}$ in the *reference* plane $z = 0$, rather than $r' = \{\varrho', \zeta(\varrho')\}$ in the rough surface Σ. Due to the presence of the perturbation $\zeta(\varrho')$ in the exponent, this problem bears a resemblance to the problem concerning the passage of the wave through a phase screen [5.6], but it should be remembered that in rough surface scattering, the random phase progression $-q_z\zeta$ is dependent (through q_z) on the directions of the primary and reflected waves.

Turning to the statistical part of the problem, we will concern ourselves with the case of *nonsmall* $q_z\zeta$. The nonlinear variation of the field u with ζ means that in order to find the moments of $u(r)$, it is now necessary to know not the moments of the random field $\zeta(\varrho')$ of the same order, but its distribution functions.

According to (5.42), the mean $\overline{u}$ is

$$\overline{u}(r) = -\frac{i}{4\pi} \int \mathcal{R} \frac{q^2}{q_z} \frac{A}{R} e^{ik(R+\varphi)} \langle e^{-iq_z\zeta}\rangle d^2\varrho' \quad .$$

The quantity

$$\langle e^{iq_z\zeta}\rangle = \int_{-\infty}^{\infty} e^{iq_z\zeta} w_{1\zeta}(\zeta) d\zeta \equiv f_{1\zeta}(q_z) \quad ,$$

where $w_{1\zeta}(\zeta)$ is the probability density of the deviations, represents the characteristic function of the field $\zeta(\varrho)$. If the field is homogeneous, $f_{1\zeta}(q_z)$ implicitly depends only on ϱ' — through the local value of the z-component of the scattering vector $q_z(\varrho')$. Thus,

$$\overline{u}(r) = -\frac{i}{4\pi} \int \mathcal{R} \frac{q^2}{q_z} \frac{A}{R} e^{ik(R+\varphi)} f_{1\zeta}^*(q_z) d^2\varrho' \quad . \tag{5.43}$$

Since $\overline{u}$ is determined by the characteristic function of ζ, (5.43) can easily be extended to include the case where surface roughness is a superposition of *independent* perturbations $\zeta = \sum_\nu \zeta_\nu$. then, $f_{1\zeta}(q_z)$ is the product of the appropriate characteristic functions.

We now apply the stationary phase technique [5.5] to calculate (5.43) and show that

$$\overline{u}(r) = \mathcal{R}_{st} f_{1\zeta}^*(q_{zst}) u^{(0)}(r) \quad , \tag{5.44}$$

where the subscript "st" implies that $\mathcal{R}$ and q_z are taken at the stationary point ϱ_{st} corresponding to the specularly reflected ray coming to the observation point r (Fig. 5.7); $u^{(0)}(r)$ is the value of the integral (5.42) at $\zeta = 0$, i.e., the field of the wave reflected from the reference plane $z = 0$, according to the principles of geometrical optics. For example, for the plane wave (5.11) incident on an absolutely soft surface ($\mathcal{R} = -1$), $u^{(0)}$ is given by (5.12), and for the spherical primary wave, this is the field of a mirror-image (relative to the plane $z = 0$) point source.

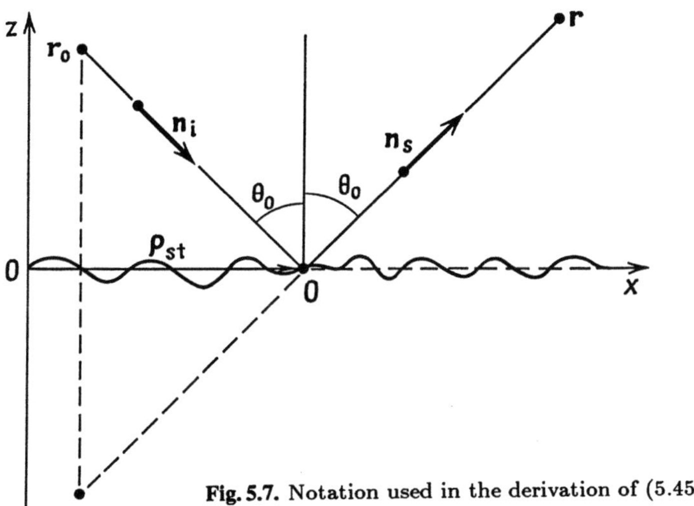

Fig. 5.7. Notation used in the derivation of (5.45)

It is seen from (5.44) that we can assign to the rough surface the *effective* coefficient of reflection of the *mean* field

$$\mathcal{R}_{ef} = \mathcal{R}_{st} f_{1\zeta}^*(q_{zst}) \ . \tag{5.45}$$

Clearly, for the perfectly reflective surface

$$|\mathcal{R}_{ef}| = |f_{1\zeta}(q_{zst})| \ .$$

For the slightly rough surface ($q_z \sigma_\zeta \ll 1$), we have approximately $f_{1\zeta} \approx 1 + iq_z\overline{\zeta} = 1$, since $\overline{\zeta} = 0$. This means that in a first approximation in ζ, the roughness is of no consequence for the mean field. As $q_z \sigma_\zeta$ increases, the mean field falls off quickly, because $f_{1\zeta}(q_z)$ decreases.

Let us now proceed to find the mean intensity of the fluctuational field

$$\overline{I}_1 = \langle|\tilde{u}|^2\rangle = \langle|u|^2\rangle - |\overline{u}|^2 \ . \tag{5.46}$$

According to (5.42),

$$\langle|u|^2\rangle = \frac{1}{16\pi^2} \int \left(\frac{\mathcal{R}Aq^2}{Rq_z}\right)' \left(\frac{\mathcal{R}^*A^*q^2}{Rq_z}\right)'' \exp\left[ik(R' + \varphi' - R'' - \varphi'')\right]$$
$$\times \langle\exp\left[iq_z'\zeta(\varrho') - iq_z''\zeta(\varrho'')\right]\rangle d^2\varrho' d^2\varrho'' \ , \tag{5.47}$$

173

where one and two primes mark the quantities referring to the points ϱ' and ϱ'' in the reference plane $z = 0$. The expression

$$\langle \exp(iq_z'\zeta' + iq_z''\zeta'') \rangle = \int \exp(iq_z'\zeta' + iq_z''\zeta'')w_{2\zeta}(\zeta',\zeta'')d^2\zeta'd^2\zeta''$$

$$= f_{2\zeta}(q_z', q_z''; \varrho', \varrho'') \tag{5.48}$$

is a two-dimensional characteristic function of the field that depends on ϱ' and ϱ'' as on the parameters.

For the intensity of the scattered wave, we find, from (5.43) and (5.47)

$$\begin{aligned}
\bar{I}_1 = \frac{1}{16\pi^2} \int &\left(\frac{\mathcal{R}Aq^2}{Rq_z}\right)' \left(\frac{\mathcal{R}^*A^*q^2}{Rq_z}\right)'' \exp\left[ik(R' + \varphi' - R'' - \varphi'')\right] \\
&\times \mathcal{P}(q_z', q_z''; \varrho', \varrho'')d^2\varrho'd^2\varrho'' \quad,
\end{aligned} \tag{5.49}$$

where $\mathcal{P}$ is the difference of (5.48) and the product of the unidimensional characteristic functions

$$\mathcal{P}(q_z', q_z''; \varrho', \varrho'') = f_{2\zeta}^*(q_z', -q_z''; \varrho', \varrho'') - f_{1\zeta}^*(q_z')f_{1\zeta}(q_z'') \quad. \tag{5.50}$$

This difference vanishes at large (as compared to the correlation radius of roughness, l_ζ) separations between ϱ' and ϱ'', because when $|\varrho' - \varrho''| \gg l_\zeta$ the values of $\zeta(\varrho')$ and $\zeta(\varrho'')$ become uncorrelated, and the two-dimensional characteristic function $f_{2\zeta}$ is represented as the product of the unidimensional characteristic functions. We shall see later in the chapter, by referring to an example, that a region where $\mathcal{P}$ is noticeably distinct from zero is actually even smaller than the area $|\varrho' - \varrho''| \lesssim l_\zeta$.

We will make use of the above property of $\mathcal{P}$ to compute (5.48) approximately. Let us go over from ϱ' and ϱ'' to the variables $\xi = \varrho' - \varrho''$ and $\eta = (\varrho' + \varrho'')/2$. We will expand the exponent of $k(R' + \varphi') - k(R'' + \varphi'')$ in a Taylor series expansion in ξ, keeping only the linear term,

$$k(R' + \varphi') - k(R'' + \varphi'')|_{\zeta=0} \approx k(\nabla_\perp R + \nabla_\perp\varphi)\xi = -q_\perp \cdot \xi \quad. \tag{5.51}$$

Next, we put $\xi = 0$ in the factor at the exponential function (i.e., $\varrho' = \varrho'' = \eta$), and also replace q_z' and q_z'' in $\mathcal{P}$ by the value of q_z at the "center of mass" $\eta = (\varrho' + \varrho'')/2$, i.e., we will set $q_z' \approx q_z'' \approx q_z(\eta)$. Then

$$\left(\frac{\mathcal{R}Aq^2}{Rq_z}\right)' \left(\frac{\mathcal{R}^*A^*q^2}{Rq_z}\right)'' \mathcal{P}(q_z', q_z''; \varrho', \varrho'') \approx \frac{|\mathcal{R}A|^2q^4}{q_z^2 R_1^2}\mathcal{P}(\xi, \eta) \quad, \quad \text{where}$$

$$R_1 = \sqrt{z^2 + (\varrho - \eta)^2} \quad, \quad \mathcal{P}(\xi, \eta) \equiv \mathcal{P}(q_z(\eta), q_z(\eta); \varrho', \varrho'') \quad.$$

We will now place the limits of integration in ξ in (5.49) at infinity in order to arrive at the equation for incoherent scattering (addition of intensities)

$$\bar{I}_1 = \int \frac{|A|^2}{R_1^2}\sigma d^2\eta \quad, \quad \text{where} \tag{5.52}$$

$$\sigma = \frac{|\mathcal{R}|^2 q^4}{16\pi^2 q_z^2} \int \exp\left(-iq_\perp \cdot \xi\right) \mathcal{P}(\xi, \eta) d^2\xi \tag{5.53}$$

has the sense of the scattering cross-section for a unit area.

The region of applicability of (5.52) is determined by those approximations that were made in its derivation. The most rigid condition is

$$R \gg \sqrt{kl_p^3} \quad . \tag{5.54}$$

If this holds, we can discard the cubic term in (5.51) (there are no quadratic terms, nor even power terms in general, in this expansion). Here, R is the separation between the point η in the plane $z = 0$ and the observation point or the source, and l_p is the characteristic space scale of $\mathcal{P}(\xi)$. For the slightly rough surface, when

$$\mathcal{P}(\xi, \eta) \approx \left\langle \left[1 - iq_z(\zeta' - \zeta'') - \frac{q_z^2}{2}(\zeta' - \zeta'')^2 \right] \right\rangle - \left\langle \left[1 - iq_z\zeta' - \frac{q_z^2}{2}\zeta'^2 \right] \right\rangle$$
$$\times \left\langle \left[1 + iq_z\zeta'' - \frac{q_z^2}{2}\zeta''^2 \right] \right\rangle \approx q_z^2 \psi_\zeta(\xi) \quad , \tag{5.55}$$

the characteristic scale l_p seems to coincide with the correlation radius of the inhomogeneities, l_ζ. In the opposite limiting case of $q_z\sigma_\zeta \gg 1$, the scale l_p has the order of magnitude of $l_\zeta/q_z\sigma_\zeta$. This can be verified, for example, for the two-dimensional normal distribution

$$w_{2\zeta}(\zeta', \zeta'') = \frac{1}{2\pi\sigma_\zeta^2\sqrt{1 - K_\zeta^2(\xi)}} \exp\left[-\frac{\zeta'^2 + \zeta''^2 - 2K_\zeta(\xi)\zeta'\zeta''}{2\sigma_\zeta^2(1 - K_\zeta^2(\xi))} \right] \quad , \tag{5.56}$$

for which

$$\mathcal{P}(\xi, \eta) = \exp\left(-q_z^2\sigma_\zeta^2\right)\left[\exp\left(q_z^2\sigma_\zeta^2 K_\zeta(\xi)\right) - 1\right] \quad . \tag{5.57}$$

The result $l_p \sim l_\zeta/q_z\sigma_\zeta$ follows from this, as does the similar result $l_u \sim l_S/\sigma_S$ for the phase screen (Sect. III.2.3), the role of the phase shift $\sigma_S = \sqrt{\langle S^2\rangle}$ being played here by the "phase height" of the rough surface, $q_z\sigma_\zeta$, [5.6].

In both limiting cases of $q_z\sigma_\zeta \ll 1$ and $q_z\sigma_\zeta \gg 1$, the quantity $\sqrt{kl_p^3}$ is far smaller than the distance kl_ζ^2, beyond which the observation point lies in the Fraunhofer zone of an individual irregularity. Thus, the equation for incoherent scattering (5.52) becomes valid even before the observation point and the source are placed within this Fraunhofer zone [5.5].

The scattering cross-section σ is of importance in calculating the energy-related characteristics of the field both for scattering from a finite area, and for the infinite rough surface. The integral in (5.53) can only be calculated exactly in a few cases; normally, it is taken approximately. We now consider some special cases.

1) In the case of a *slightly* rough surface ($q_z\sigma_\zeta \ll 1$), where, by (5.55), $\mathcal{P} \approx q_z^2\psi_\zeta(\xi)$, the cross-section σ is expressed in terms of the Fourier transform of the covariance $\psi_\zeta(\xi)$, i.e., proportional to the spectral density (5.15)

$$\sigma = \frac{|\mathcal{R}|^2}{4} q^4 F_\zeta(\boldsymbol{q}_\perp) \quad . \tag{5.58}$$

For the absolutely rigid and absolutely soft surfaces ($|\mathcal{R}| = 1$), this equation is equivalent to the results of perturbation theory (see problem 1), since in the directions close to that of specular reflection, we have $q \approx |q_z| = 2k|n_z^{\mathrm{i}}|$.

2) In the opposite case of surface inhomogeneities, *large* as compared to the wavelength ($q_z \sigma_\zeta \gg 1$), we can make use of the fact that the major contribution to (5.53) comes from the region of small ξ, and that the second term in (5.50), for $q_z \sigma_\zeta \gg 1$, is negligible in comparison with the first. Hence,

$$\mathcal{P}(\boldsymbol{\xi}, \boldsymbol{\eta}) \approx f_{2\zeta}^*(q_z, -q_z, \boldsymbol{\varrho}', \boldsymbol{\varrho}'') = \langle \exp\left[-iq_z(\zeta' - \zeta'')\right] \quad . \tag{5.59}$$

By way of example, we will calculate the scattering cross-section for irregularities that have the normal distribution (5.56) and are isotropic in the plane $z = 0$. The isotropy of the field ξ means that the correlation coefficient is only dependent on the magnitude of $\boldsymbol{\xi}$: $K_\zeta = K_\zeta(\xi)$. Expanding K_ζ in (5.57) into a Taylor series and ignoring the unity as compared to the exponential function in brackets, we obtain

$$\mathcal{P} \approx \exp\left[\frac{q_z^2 \sigma_\zeta^2}{2} K_\zeta''(0)\xi^2\right] \quad . \tag{5.60}$$

Substituting (5.60) into (5.53) and integrating, we get

$$\sigma = \frac{|\mathcal{R}|^2 q^4}{q_z^4} \frac{1}{8\pi\sigma_\zeta^2[-K_\zeta''(0)]} \exp\left[-\frac{q_\perp^2}{2q_z^2\sigma_\zeta^2[-K_\zeta''(0)]}\right] \quad . \tag{5.61}$$

This expression can be used in the case of roughness with the only space scale (correlation radius) $l_\zeta \sim [-K_\zeta''(0)]^{-1/2}$. If the correlation coefficient does not fall off in a monotonic manner, but oscillates (as in the case of high seas), then, apart from the vicinity of the point $\boldsymbol{\xi} = 0$, we should also take into account contributions of other points at which $\mathcal{P}$ has local maxima (an example of the calculation is given in [5.4]).

3) For high surface irregularities ($q_z \sigma_\zeta \gg 1$) with the only correlation radius, the scattering cross-section can also be worked out without assuming a normal distribution [5.7, 8]. Again, supposing that the major contribution to (5.53) comes from the vicinity of $\boldsymbol{\xi} = 0$, we will expand the difference $\zeta' - \zeta''$ in (5.59) into a Taylor expansion in $\boldsymbol{\xi} = \boldsymbol{\varrho}' - \boldsymbol{\varrho}''$

$$\zeta' - \zeta'' = \zeta(\boldsymbol{\varrho}') - \zeta(\boldsymbol{\varrho}'') = \zeta(\boldsymbol{\eta} + \boldsymbol{\xi}/2) - \zeta(\boldsymbol{\eta} - \boldsymbol{\xi}/2) \approx \boldsymbol{\xi}\nabla_\perp\zeta(\boldsymbol{\eta}) \quad .$$

Then

$$\mathcal{P} \approx \langle \exp(-iq_z\boldsymbol{\xi}\cdot\nabla_\perp\zeta)\rangle \quad \text{and}$$

$$\sigma = \frac{|\mathcal{R}|^2 q^4}{16\pi^2 q_z^2} \int\limits_{-\infty}^{\infty} \exp(-i\boldsymbol{q}_\perp\boldsymbol{\xi})\langle\exp(-iq_z\boldsymbol{\xi}\cdot\boldsymbol{\nu})\rangle d^2\xi \quad , \tag{5.62}$$

where $\boldsymbol{\nu} = \nabla_\perp \zeta$ characterizes the slope of the rough surface. But, $\langle e^{i\boldsymbol{a}\cdot\boldsymbol{\nu}}\rangle = f_\nu(\boldsymbol{a})$ is the characteristic function of $\boldsymbol{\nu}$ related to the distribution function of the slopes $w_\nu(\boldsymbol{\nu})$ by the Fourier transformation

$$f_\nu(\boldsymbol{a}) = \int\limits_{-\infty}^{\infty} \exp\left(i\boldsymbol{a}\cdot\boldsymbol{\nu}\right)w_\nu(\boldsymbol{\nu})d^2\nu \quad ,$$

so that the inverse transformation gives

$$w_\nu(\boldsymbol{\nu}) = \frac{1}{4\pi^2} \int\limits_{-\infty}^{\infty} f_\nu(\boldsymbol{a})\exp\left(-i\boldsymbol{a}\cdot\boldsymbol{\nu}\right)d^2a \quad . \tag{5.63}$$

If, in (5.62), we introduce a new integration variable $\boldsymbol{a} = q_z\boldsymbol{\xi}$, then, from (5.63), the cross-section σ can be represented in terms of $w_\nu(\boldsymbol{q}_\perp/q_z)$

$$\sigma = \frac{|\mathcal{R}|^2 q^4}{4q_z^4}\frac{1}{4\pi^2} \int\limits_{-\infty}^{\infty} \exp\left(-i\boldsymbol{q}_\perp\cdot\boldsymbol{a}/q_z\right)f_\nu(\boldsymbol{a})d^2a = \frac{|\mathcal{R}|^2 q^4}{4q_z^4}w_\nu\left(\frac{\boldsymbol{q}_\perp}{q_z}\right) \quad . \tag{5.64}$$

The cross-section σ is thus proportional to the probability of the slope at which a specular reflection occurs. It is maximal at $\boldsymbol{q}_\perp = 0$, as $w_\nu(\boldsymbol{\nu})$ has a maximum at $\boldsymbol{\nu} = \nabla_\perp \zeta = 0$ (the most likely orientation of the elements of the rough surface is parallel to the plane $z = 0$), and as $\boldsymbol{q}_\perp$ increases, σ decreases, because steep slopes are less probable.

If the roughness obeys the normal distribution law and is isotropic in the plane $z = 0$, we can easily see that (5.63) changes into (5.61). Note also that (5.61) and (5.64) are the geometrical optics approximations of the scattering cross-section (5.53). This follows already from the fact that both expressions do not contain the wavelength, since the ratios $q_\perp/q_z$ and q/q_z are independent of λ.

4) For the isotropic field ζ having a normal distribution with the Gaussian correlation coefficient $K_\zeta(\varrho) = \exp\left(-\varrho^2/2l_\zeta^2\right)$, an *exact* solution for σ in the direction of specular reflection ($\boldsymbol{q}_\perp = 0$) was obtained in [5.9]

$$\sigma_{\mathrm{sp}} = \frac{|\mathcal{R}|^2 q_z^2 l_\zeta^2}{8\pi}e^{-q_z^2\sigma_\zeta^2}[\mathrm{Ei}(q_z^2\sigma_\zeta^2) - C - 2\ln q_z\sigma_\zeta] \quad , \tag{5.65}$$

where $C = 0.577$ is the Euler constant, and $\mathrm{Ei}(z)$ the integral exponent (σ_{sp} at $\boldsymbol{q}_\perp \neq 0$ is represented as an infinite expansion). The behavior of the normalized cross-section $8\pi\sigma_{\mathrm{sp}}/|\mathcal{R}|^2 q_z^2 l_\zeta^2$ that depends on $q_z\sigma_\zeta$ is shown in Fig. 5.8. At first, σ_{sp} grows as σ_ζ^2, according to a first approximation of the method of small perturbation. At $q_z\sigma_\zeta \approx 1$, the cross-section attains a maximum, and at larger $q_z\sigma_\zeta$ decays following the law $1/\sigma_\zeta^2$. Of course, the total (integrated over all the angles) scattering cross-section grows further as $q_z\sigma_\zeta \to \infty$,but in the process the scattered radiation is spatially redistributed, with the result that the scattering indicatrix broadens and σ_{sp} reduces.

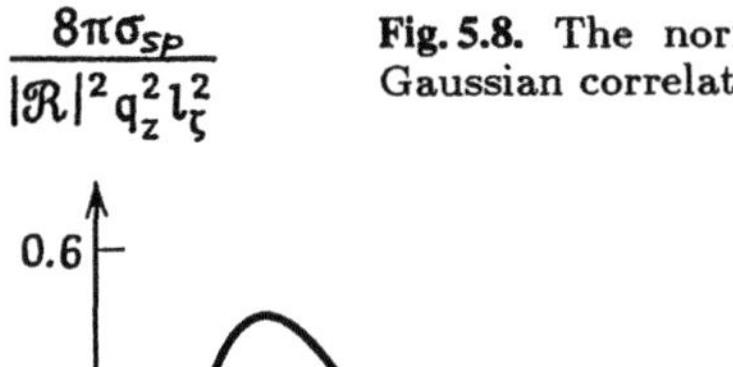

Fig. 5.8. The normalized scattering cross-section for a surface with a Gaussian correlation coefficient for irregularities

5.3 Additional Remarks. Other Approaches

As has already been said, most works on rough surface scattering use the perturbation technique and the Kirchhoff approximation. We will give here some of the results obtained using these techniques, and will also indicate other ways of treating the problem, proceeding primarily from the monograph [5.1] which gives a detailed discussion of the issues in question, and from the review article [5.2].

1) Electromagnetic scattering. In contrast to the scalar case electromagnetic scattering also takes polarization into account. For the primary wave in the geometrical optics approximation, the dynamic relationships are in principle derived in the same way as in the scalar problem, except that the treatment becomes more unwieldy since, instead of (III.2.18), we must use the vector form of Green's function. Moreover, in using the Kirchhoff approximation, we should take into account the difference in the local Fresnel coefficients of reflection for two orthogonal polarizations of the field of the incident wave.

The simplest formulas are obtained for scattering from a perfectly conducting surface. Specifically, the expression for the electric vector of the scattered wave in the Kirchhoff approximation can be obtained from (5.42) by putting into it $\mathcal{R} = 1$, and substituting $e_0 q^2 - 2q(e_0 q)$ for q^2, where e_0 is the unit vector of polarization of the primary wave. This replacement enables us to derive the mean strengths and bilinear characteristics of the scattered electromagnetic field (mean intensity, mean Poynting vector, elements of polarization matrix). At distances $R \gg \sqrt{k l_p^3}$ from the rough surface, the bilinear characteristics will be given by the equations for incoherent scattering of the form (5.52), where the intensity I_0 is clearly replaced by a more complex expression dependent on e_0. Significantly, however, the integrands in the expressions for bilinear quantities will now contain the cross-section (5.53) obtained by the scalar formulation of the problem, so that the results of the scalar theory can be directly applied to electromagnetic scattering theory. For example, the variance $\overline{I_1} = \langle |\tilde{E}|^2 \rangle$ of the electromagnetic field scattered by the rough surface is given by (5.52), if σ is understood to be $\sigma = \gamma\sigma_{sc}$, where σ_{sc} is given by (5.53), and $\gamma = 1 - |qe_0|^2/q^2$ is the polarization factor.

2) Kirchhoff's Approximation with Shadowing. We have already mentioned that, as the height of the surface increases and the grazing angle decreases, individual surface elements will sooner or later be shadowed, with the result that part of the rough surface will not be illuminated (Fig. 5.9), and part of the illuminated surface will not be seen from the observation point. We can easily estimate the range of the incidence angles ψ in which we may still ignore shadowing. Clearly, if l_ζ is the length of an irregularity and σ_ζ its root-mean-square height, we can ignore shadowing if $\psi \gg \sigma_\zeta / l_\zeta$.

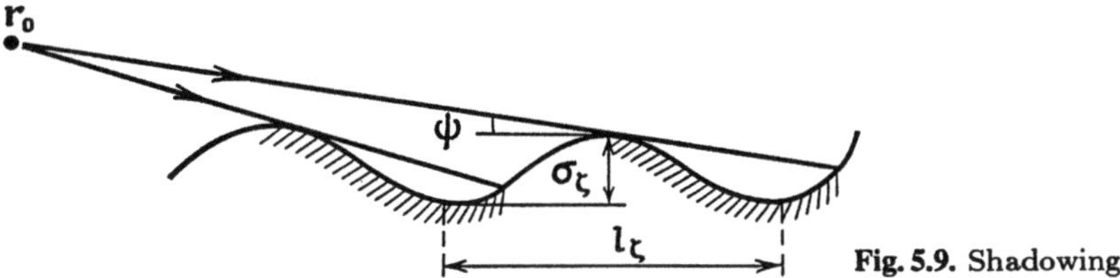

Fig. 5.9. Shadowing

Shadowing reduces the scattering cross-section σ in comparison with the result given by (5.53). The extinction factor here is the ratio of the illuminated part of the surface to the total surface, where the illuminated part can be worked out from simple geometrical arguments: the incident ray strikes the rough surface once in the illuminated parts, and three, five, etc., times in the shadowed parts. The task thus reduces to one of finding the probability of the ray striking a given random surface once only. Despite the seeming simplicity of the formulation, the problem appears to be fairly complex. A treatment of this issue and a number of other aspects of the problem are summarized in [5.1].

3) Scattering from Small- and Large-Scale Inhomogeneities (Combined Approach). Real surfaces often include both fine ($kl_\zeta \ll 1$) and coarse ($kl_\zeta \gg 1$) irregularities. Such surfaces can be regarded as large-scale roughness, superimposed by small-scale ripples ("two-scale surface"). Whereas with large-scale roughness the scattering diagram is relatively narrow, small-scale roughness essentially scatters omnidirectionally, and its influence in the direction of specular reflection is negligible. But, for small incidence angles, scattering is due precisely to the small-scale component. It also defines the spectrum of the scattered field in directions other than that of specular reflection. These, together with some other arguments, permit of a qualitative description of some of the experimental evidence, specifically of scattering from ocean waves. However, the theoretical treatment of scattering from "two-scale" surfaces is associated with certain difficulties. Here, we can use neither the perturbation technique (large-scale roughness), nor Kirchhoff's approximation (small-size roughness).

Kur'yanov [5.10] proposed a combined method in which a zeroth approximation is the Kirchhoffian solution of the form (5.42), which takes account of smooth large-scale roughness, whereas the ripples are taken into account by the first-order perturbation, assuming both types of roughness to be statistically independent. The method was improved in [5.11, 12]. A somewhat different approach is given in [5.13, 14], where a formula is used to sum up the intensities of the fields scattered by small-scale roughness.

The combined technique has two constraints: first, the final results must be independent of the way in which the deviations $\zeta(\varrho)$ are broken down into the independent parts $\mu(\varrho)$ and $\eta(\varrho)$; second, the conditions must be met for the perturbation technique to be applied to treat scattering from the small-scale component. It appears that these constraints are not obeyed for all types of waves [5.1].

4) Multiple Scattering. Both Kirchhoff's technique and the method of small perturbations (if we limit ourselves to a first approximation) only consider *singly* scattered (or *singly* reflected) fields. This is permissible as long as the irregularities are gentle enough, and relatively low. For larger σ_ζ and/or σ_ζ/l_ζ, multiple scattering comes into play.

Multiple scattering can be conveniently taken into account using the integral equation for Green's function [5.1, 2]. If we linearize the equation in perturbation ζ, we will be able to derive from it the Dayson equation for the mean Green's function $\langle G \rangle$, and the Bethe-Salpeter equation for the coherence function $\langle G(r_1, r_0)G^*(r_2, r_0) \rangle$. Further, we can solve both equations approximately, the one in the Bourret approximation, the other in the ladder approximation. As with volume scattering, these solutions of the above equations are equivalent to the summation of some subsequences of an infinite expansion in perturbations.

This approach has proved to be fairly effective in dealing with a number of problems, and in particular, waveguides having rough walls. It is possible to deduce here the attenuation coefficients for normal waves and the mode transformation ratios, and derive the RTE for the waveguide, including the mutual transformation of waves [5.1]. In addition, the fact that multiple scattering is taken into account makes it possible to give a consistent and improved treatment of the so-called "nonlocal" boundary conditions for the mean field, which was derived earlier using a different technique [5.1]. Similar ideas underlie the techniques which use integral equations for the field [5.15].

5) Steep Roughness. Despite the significant strides made by the theory of wave scattering from smooth rough surfaces, the difficult problem still remains of scattering from smooth, but *steep,* roughness, which is not amenable to the available approximation techniques. It is only natural, therefore, to attempt to give a *model* description of such roughness – either in the form of hemispheres, half-cylinders, etc., chaotically arranged over a plane (a typical example of such model is discussed, for example in [5.16]), or in the form of flat facets with randomly distributed slopes. The second model is widely used, in particular, in optical treatments of light reflection, using both the GOM [5.17, 18], and corrections for diffraction, which are taken into account by means of scattering indicatrices for elementary facets [5.19].

The model approach has at least two disadvantages. First, the applicability of specific model results is severely limited. Second, uncertainties due to the approximations are not easily estimated. Nevertheless, for want of more general techniques, the model description of steep roughness is rather common. What is more, the *scattering law*, i.e., $\sigma(\theta,\varphi)$, is sometimes modelled, rather than the roughness geometry. The best known model of this type is *Lambert's law*, according to which $\sigma(\theta) = \text{const} \cdot \cos\theta$. Both theoretically and experimentally, however, this simple law lacks a firm foundation.

In conclusion, the authors would like to refer the interested reader to the splendid overview [5.20]. There is no denying the opinion voiced in [5.20] that despite the remarkable advances in theory and experiment an amazing variety of problems still await solution. The authors would be happy if the book would appear of help in tackling some of these problems, many of which have both fundamental and applied value.

5.4 Exercises

5.4.1 Show that for large-scale roughness ($kl_\zeta \gg 1$), the scattering cross-sections obtained by the perturbation technique for the absolutely soft and absolutely rigid surface are essentially equal.

Solution. For $kl_\zeta \gg 1$, the spectrum $F_\zeta(q_\perp)$ differs markedly from zero only in the narrow interval $|q_\perp| \lesssim 1/l_\zeta$, i.e., for near-specular reflection, for which $q_\parallel = k(n_\perp^s - n_\perp^i) = 0$. In this interval we have $n_\perp^i \approx n_\perp^s$; therefore, the factors $(n_z^i n_z^s)^2$ in (5.18) and $(1 - n_\perp^i n_\perp^s)^2$ in (5.28) are approximately equal to $(n_z^i)^4 = \cos^4\theta_0$. Thus,

$$\sigma_{\text{soft}} \approx \sigma_{\text{rig}} \approx 4k^4(n_z^i)^4 F_\zeta(q_\perp) = 4(k\cos\theta_0)^4 F_\zeta(q_\perp) \quad .$$

5.4.2 Estimate the transverse correlation radius for the field scattered from a surface with *small-scale* roughness ($kl_\zeta \ll 1$) in the two cases where: (1) the observation point is separated from the rough surface by a distance R_1 larger than the diameter of the illuminated area, and (2) the diameter of the illuminated area L is determined by the radiation pattern width $\gamma \sim \lambda/d$ (d is the antenna diameter).

Solution. In the first case, by the van Cittert-Zernike theorem, the transverse correlation radius $l_\perp$ is dictated by the angle $\Delta\theta \sim L/R_1$ at which the scattering area is seen from the distance R_1: $l_\perp \sim \lambda/\Delta\theta \sim \lambda R_1/L$. Specifically, if $\Delta\theta \sim 1$, then $l_\perp \sim \lambda$.

In the second case, $L \sim \gamma R_0 \sim \lambda R_0/d$, where R_0 is the distance between the radiator and the center of the illuminated area on the surface. At the distance R_1 from the center of this area, the transverse correlation radius will therefore be $l_\perp \sim \lambda R_1/L \sim dR_1/R_0$. For backscattering, when $R_1 \sim R_0$, the correlation radius coincides with the antenna diameter ($l_\perp \sim d$), as was the case with scattering from volume irregularities (see Exercise III.4.9.3).

5.4.3 Estimate the transverse (as regards the specular ray) correlation radius of the scattered field for *large-scale* roughness.

Solution. The correlation radius can conveniently be estimated by using the results derived for the phase screen [5.6]. If a plane wave is *normally* incident on an infinite surface, with the roughness correlation radius $l_\zeta \gg \lambda$ and root-mean-square deviation σ_ζ, then the reflected wave appears to be phase modulated, with the variance $\sigma_S^2 = (2k\sigma_\zeta)^2$. According to item 2 in Sect. III.2.3, we have here $l_u \approx l_\zeta$ for $2k\sigma_\zeta \ll 1$ (low roughness) and $l_u \approx l_\zeta/2k\sigma_\zeta$ for $2k\sigma_\zeta \gg 1$ (high roughness).

If a rough surface is irradiated by a nonplane wave, the above estimates are only valid at small distances from the plane $z = 0$. At large distances, l_u will increase (or decrease) with the cross-section of the ray tubes corresponding to the specularly reflected (from the plane $z = 0$) rays. For example, if the rough surface is illuminated by a spherical wave, then

$$l_u \approx \left(1 + \frac{R_1}{R_0}\right) l_\zeta \quad \text{for} \quad 2k\sigma_\zeta \ll 1 \quad ,$$

$$l_u \approx \left(1 + \frac{R_1}{R_0}\right) \frac{l_\zeta}{2k\sigma_\zeta} \quad \text{for} \quad 2k\sigma_\zeta \gg 1 \quad ,$$

where R_1 is the distance from the point of specular reflection to the observation point, and R_0 is the distance from the point of specular reflection to the source.

References[1]

Preface

1. S.A. Akhmanov, A.S. Chirkin: *Statistical Phenomena in Nonlinear Optics* (Moscow University Press, Moscow 1971) (in Russian);
    S.A. Akhmanov, Y.Y. Dyakov, A.S. Chirkin: *Introduction to Statistical Radiophysics and Optics*, Vols. I, II (Springer, Berlin, Heidelberg, New York) in preparation
2. A.M. Yaglom: *An Introduction to the Theory of Stationary Random Functions* (Prentice Hall, Englewood Cliffs, New York 1962)
3. A. Blanc-Lapierre, R. Fortet: *Théorie des fonctions aléatoires* (Masson, Paris 1953)
4. J.L. Doob: *Stochastic Processes* (John Wiley, New York 1953)
5. M.S. Bartlett: *An Introduction to Stochastic Processes* (Cambridge University Press, Cambridge 1956)
6. Yu.A. Rozanov: *Stationary Random Processes* (Holden-Day, San Francisco 1967)
7. A.A. Sveshnikov: *Applied Methods of the Theory of Random Functions* (Pergamon, Oxford 1966)
8. H. Cramér, M.R. Leadbetter: *Stationary and Related Stochastic Processes* (John Wiley, New York 1967)
9. S. Karlin: *A First Course in Stochastic Processes* (Academic, New York 1966)
10. V.L. Lebedev: *Random Processes in Electrical and Mechanical Systems* (Gostekhizdat, Moscow 1958) (in Russian)
11. A. van der Ziel: *Noise* (Prentice-Hall, Englewood Cliffs, New York 1954)
12. B.R. Levin: *Fondaments théorique de la radiotechnique statistique* (MIR, Moscow) Vols. I, II (1973); Vol. III (1979)
13. W.B. Davenport, Jr., W.L. Root: *An Introduction to the Theory of Random Signals and Noise* (McGraw-Hill, New York 1958)
14. A. van der Ziel: *Fluctuation Phenomena in Semi-Conductors* (Butterworths Scientific Publications, London 1959)
15. R.L. Stratonovich: *Random Noise*, Vols. I and II (Gordon and Breach, New York 1963)
16. D.A. Middleton: *An Introduction to Statistical Communication Theory* (McGraw-Hill, New York 1960)
17. V.I. Tikhonov: *Statistical Radio Engineering* (Sovradio, Moscow 1966) (in Russian)
18. A.N. Malakhov: *Fluctuations in Self-Oscillatory Systems* (Nauka, Moscow 1968) (in Russian)
19. B.R. Levin: *Theoretical Foundations of Statistical Radio Engineering* (Sovradio, Moscow 1974) (in Russian)

Chapter 1

1.1 S.M. Rytov, Yu.A. Kravtsov, V.I. Tatarskii: *Principles of Statistical Radiophysics 3 – Elements of Random Fields* (Springer, Berlin, Heidelberg, New York 1989)
1.2 V.A. Krasilnikov: Dokl. Akad. Nauk SSSR **47**, 486 (1945)
1.3 P.G. Bergman: Phys. Rev. **70**, 486 (1946)
1.4 V.I. Tatarskii: *The Effects of the Turbulent Atmosphere on Wave Propagation* (IPST, Jerusalem 1971)

---

[1] Some Russian Journals are translated into English on a cover-to-cover basis: e.g., Akust. Zh. [Sov. Phys.-Acoust.]; Dokl. Akad. Nauk SSSR [Sov. Phys.-Dokl.]: Izv. Vyssh. Uchebn. Zaved. Radiofiz. [Radiophys. & Quantum Electron.]; Radiotekh. Elektron. [Radio Eng. Electron. (USSR)]

1.5   L.A. Chernov: *Propagation of Waves in a Medium with Random Irregularities* (Dover, N.Y. 1967)
1.6   Yu.N. Barabanenkov, Yu.A. Kravtsov, S.M. Rytov, V.I. Tatarskii: Sov. Phys.-Uspekhi, **13**, 3 (1971)
1.7   Yu.A. Kravtsov, Yu.I. Orlov: *Geometrical Optics of Inhomogeneous Media* (Nauka, Moscow 1980) (in Russian);
       Yu.A. Kravtsov, Yu.I. Orlov: *Geometrical Optics of Inhomogeneous Media,* Springer Series Wavephenomena Vol. 6 (Springer, Berlin, Heidelberg, New York) in preparation
1.8   S.M. Rytov: Dokl. Akad. Nauk SSSR **18**, 283 (1938)
1.9   Yu.A. Kravtsov, Yu.I. Orlov: Sov. Phys.Uspekhi, **23**, 750 (1980); **26**, 1038 (1983)
1.10   Yu.A. Kravtsov, Yu.I. Orlov: Radio Sci. **16**, 975 (1981)
1.11   Yu.A. Kravtsov: *Rays and Caustics as Physical Objects* in Progress in Optics, Vol. XXVI, 227–348, ed. E. Wolf (North Holland, Amsterdam, New York, Oxford 1988)
1.12   Yu.A. Kravtsov: Sov. Phys. ZhETP, **28**, 413 (1969)
1.13   Yu.A. Kravtsov: Akust. Zh. **14**, 1 (1968)
1.14   V.I. Klyatskin, V.I. Tatarskii: Izv. Vyssh. Uchebn. Zaved. Radiofiz. **14**, 706 (1971)
1.15   V.I. Klyatskin: *Statistical Description of Dynamic Systems with Fluctuating Parameters* (Nauka, Moscow 1975)
       Ondes et équations stochastique dans les milieux aléatoirement non homogénes. Les editions de physique. Les Vlis Cedex, France, 1985
1.16   N.G.Denisov: Izv. Vyssh. Uchebn. Zaved. Radiofiz. **1**, 34 (1958)
1.17   S.M. Golynsky, V.D. Gusev: Radiotekh. Elektron. **21**, 630 (1976)
1.18   N.G. Golynsky, V.G. Gusev: Radiotekh. Elektron. **21**, 1303 (1976)
1.19   N.G. Denisov, L.M. Erukhimov: Geomagn. Aeron. **6**, 695 (1966)
1.20   Yu.A. Kravtsov, A.I. Saitchev: Sov. Phys.-Uspekhi, **25**, 494 (1982)

Chapter 2

2.1   E.L. Feinberg: *Propagation of Radio Waves along the Earth's Surface* (USSR Acad. Sci., Moscow 1961) (in Russian)
2.2   V.E. Ostashev, V.I. Tatarskii: Izv. Vyssh. Uchebn. Zaved. Radiofiz. **21**, 714 (1978)
2.3   S.M. Flatté, R. Dashen, W.H. Munk, K.M. Watson, F. Zachaziasen: *Sound Transmission through a Fluctuating Ocean* (Cambridge University Press, London, New York, Melbourne 1979)
2.4   S.M. Rytov: Izv. Akad. Nauk SSSR, Ser. Fiz. **2**, 223 (1937)
2.5   A.M. Obukhov: Izv. Akad. Nauk SSSR, Ser. Geofiz. **2**, 155 (1953);
       G.S. Gorelik: *Oscillations and Waves,* 2nd edn. (Fizmatgiz, Moscow 1959) (in Russian)
2.6   V.I. Tatarskii: *The Effects of the Turbulent Atmosphere on Wave Propagation* (IPST, Jerusalem 1971)
2.7   V.I. Tatarskii: *A Theory of Fluctuational Phenomena in Wave Propagation Through a Turbulent Atmosphere* (USSR Acad. Sci., Moscow 1959) (in Russian)
       Wave Propagation in a Turbulent Medium. McGraw-Hill Book Comp., 1961, Dover Publ., N.Y. 1967
2.8   I.A. Ibragimov; Yu.V. Linnik: *Independent and Stationarily Related Variables* (Nauka, Moscow 1965) (in Russian)
2.9   A.S.Gurvich, A.I. Kon, V.L. Mironov, S.S. Khmelevtsov: *Laser Radiation in a Turbulent Atmosphere* (Nauka, Moscow 1976) (in Russian)
2.10   V.I. Tatarskii: Izv. Vyssh. Uchebn. Zaved. Radiofiz. **5**, 490 (1962)
2.11   V.V. Pisareva: Akust. Zh. **6**, 87 (1960)

Chapter 3

3.1   Rytov et al.: *Principles of Statistical Radiophysics* Vol 1
3.2   Rytov et al.: *Principles of Statistical Radiophysics* Vol. 2
3.3   V.I. Klyatskin, V.I. Tatarskii: Izv. Vyssh. Uchebn. Zaved. Radiofiz. **20**, 1040 (1977)
3.4   V.I. Klyatskin: JETP Lett. **57**, 952 (1969)
3.5   V.I. Tatarskii: *The Effects of the Turbulent Atmosphere on Wave Propagation* (IPST, Jerusalem 1971), Ch. 5B
3.6   L.S. Dolin: Izv. Vyssh. Uchebn. Zaved. Radiofiz. **7**, 380 (1964)
3.7   H. Bremmer: J. Res. Natl. Bur. Stand. **68**, 967 (1964)
3.8   V.I. Klyatskin, V.I. Tatarskii: Izv. Vyssh. Uchebn. Zaved. Radiofiz. **13**, 1061 (1970)

3.9   L.S. Dolin: Izv. Vyssh. Uchebn. Zaved. Radiofiz. **11**, 840 (1968)
3.10   I.M. Dagkesmanskaya, V.I. Shishov: Izv. Vyssh. Uchebn. Zaved. Radiofiz. **13**, 16 (1970)
3.11   W.P. Brown: J. Opt. Soc. Am. **62**, 966 (1972)
3.12   B.S. Elepov. G.A. Mikhailov: Zh. vych. met. i mat. fiz. **16**, 1264 (1976);
       A.S. Gurvich, A.I. Kon, V.L. Mironov, S.S. Khmelevtsov: *Laser Radiation in a Turbulent Atmosphere* (Nauka, Moscow 1976) (in Russian)
3.13   A.S. Gurvich, B.S. Elepov, V.V. Pokasov, K.K. Sabelfeld, V.I. Tatarskii: Izv. Vyssh. Uchebn. Zaved. Radiofiz. **22**, Iss. 2 (1979)
3.14   K.S. Gochelashvili, V.I. Shishov: Opt. Acta **18**, 767 (1971)
3.15   K.S. Gochelashvili, V.I. Shishov: Zh. Eksp. Teor. Fiz. **66**, 1237 (1974)
3.16   I.G. Yakushkin: Izv. Vyssh. Uchebn. Zaved. Radiofiz. **18**, 1660 (1975)
3.17   V.U. Zavorotny, V.I. Klyatskin, V.I. Tatarskii: Zh. Eksp. Teor. Fiz. **73**, 481 (1977);
       V.I. Klyatskin, V.I. Tatarskii: Zh. Eksp. Teor. Fiz. **58**, 624 (1970)
3.18   E.S. Fradkin: Tr. Fiz. Inst. Akad. Nauk SSSR **29**, 7 (1965)
3.19   E.S. Fradkin: Nucl. Phys. **76**, 588 (1966)
       R. Feynman, A. Hibbes: *Quantum Mechanics and Path Integrals* (McGraw-Hill, New York 1965)
3.20   V.I. Klyatskin, V.I. Tatarskii: Izv. Vyssh. Uchebn. Zaved. Radiofiz. **14**, 1400 (1971)
3.22   V.I. Tatarskii: Zh. Eksp. Teor. Fiz. **56**, 2106 (1969)
3.23   A.S. Gurvich, M.A. Kallistratova, F.E. Martvel: Izv. Vyssh. Uchebn. Zaved. Radiofiz. **20**, 1020 (1977)
3.24   V.I. Tatarskii, V.U. Zavorotnyi: *Strong Fluctuations in Light Propagation in a Randomly Inhomogeneous Medium* in Progress in Optics, Vol. XVII, 207–256, ed. E. Wolf (North-Holland, Amsterdam, New York, Oxford 1980)
3.25   J.W. Strohbehn (Ed.): *Laser Beam Propagation in the Atmosphere* (Springer, Berlin, Heidelberg, New York 1978)
3.26   A. Ishimaru: *Wave Propagation and Scattering in Random Media* (Academic, New York, San Francisco, London 1978)
3.27   S.M. Flatte (Ed.): *Sound Transmission Through a Fluctuating Ocean* (Cambridge University Press, Cambridge, London, New York, Melbourne 1979)
3.28   JOSA A: Feature Issue on Wave Propagation and Scattering in Random Media **2**, No. 12 (1985)
3.29   Appl. Optics, Feature Issue on Wave Propagation and Scattering in the Atmosphere **27**, No. 11 (1988)

Chapter 4

4.1   Yu.N. Barabanenkov: Sov. Phys. Usp., **18**, 673 (1975)
4.3   V.I. Tatarskii: *The Effects of the Turbulent Atmosphere on Wave Propagation* (IPST, Jerusalem 1971)
4.4   F.G. Bass, I.M. Fuks: *Wave Scattering at Statistically Rough Surface* (Nauka, Moscow 1972) (in Russian)
4.5   V.I. Klyatskin, V.I. Tatarskii: Usp. Fiz. Nauk **110**, 499 (1973)
4.6   V.I. Klyatskin: *Statistical Description of Dynamic Systems with Fluctuating Parameters* (Nauka, Moscow 1975) (in Russian)
4.7   L.A. Apresyan: Izv. Vyssh. Uchebn. Zaved. Radiofiz. **17**, 165 (1974)
4.8   V.N. Alekseev, V.M. Komissarov: Tr. Akust. Inst. **4**, 27 (1968)
4.9   O.G. Nalbandyan, V.I. Tatarskii: Izv. Vyssh. Uchebn. Zaved. Radiofiz. **20**, 549 (1977)
4.10   V.I. Tatarskii: Izv. Vyssh. Uchebn. Zaved. Radiofiz. **17**, 570 (1974)
4.11   Yu.N. Barabanenkov, Yu.A. Kravtsov, S.M. Rytov, V.I. Tatarskii: Sov. Phys. Uspekhi., **13**, 551 (1971)
4.12   Yu. A. Ryzhov, V.V. Tamoikin: Izv. Vyssh. Uchebn. Zaved. Radiofiz. **13**, 356 (1970)
4.13   Yu.N. Barabanenkov, V.M. Finkelberg: Izv. Vyssh. Uchebn. Zaved. Radiofiz. **11**, 719 (1968)
4.14   Yu.N. Barabanenkov, A.G. Vinigradov, Yu.A. Kravtsov, V.I. Tatarskii: Izv. Vyssh. Uchebn. Zaved. Radiofiz. **15**, 1852 (1972)
4.15   S. Chandrasekhar: *Radiative Transfer* (Oxford Univ. Press 1950)
4.16   Yu.N. Barabanenkov: Izv. Vyssh. Uchebn. Zaved. Radiofiz. **16**, 88 (1973)
4.17   A.G. Vinogradov, Yu.A. Kravtsov, V.I. Tatarskii: Izv. Vyssh. Uchebn. Zaved. Radiofiz. **16**, 1064 (1973)
4.18   L.A. Apresyan: Izv. Vyssh. Uchebn. Zaved. Radiofiz. **16**, 461 (1973)

4.19   G.I. Ovchinnikov, V.I. Tatarskii: Izv. Vyssh. Uchebn. Zaved. Radiofiz. **15**, 1419 (1972)
4.20   L.A. Apresyan: Izv. Vyssh. Uchebn. Zaved. Radiofiz. **18**, 1870 (1975)
4.21   L.A. Apresyan, Yu.A. Kravtsov: Sov. Phys. Uspekhi. **27**, 301 (1984)
4.22   V.M. Finkelberg: Zh. Eksp. Teor. Fiz. **53**, 401 (1967);
       G.V. Rozenberg: Usp. Fiz. Nauk **56**, 77 (1955)
4.23   Yu.N. Barabanenkov, V.M. Finkelberg: Zh. Eksp. Teor. Fiz. **53**, 978 (1967)
4.24   Yu.A. Kravtsov, S.M. Rytov, V.I. Tatarskii: Sov. Phys. Uspekhi., **18**, 118 (1975)
4.25   L.A. Apresyan, Yu.A. Kravtsov: *Theory of Radiative Transfer: Statistical and Wave Aspects* (Nauka, Moscow 1983) (in Russian)
4.26   Yu.A. Kravtsov, A.I. Saichev: Sov. Phys. Uspekhi. **25**, 494 (1982)
4.27   Yu.A. Kravtsov, A.I. Saichev: JOSA-A **2**, 2100 (1985)

Chapter 5

5.1   F.G. Bass, I.M. Fuks: *Wave Scattering from Statistically Rough Surface* (Pergamon Press, Oxford 1979)
5.2   A.B. Shmelev: Usp. Fiz. Nauk **106**, 459 (1972)
5.3   Beckman, P.A. Spizzichino: *The Scattering of Electro-Magnetic Waves from Rough Surfaces* (Pergamon, Oxford 1963)
5.4   M.A. Isakovich: Zh. Eksp. Teor. Fiz. **23**, 305 (1952); and Tr. Akust. Inst. **5**, 152 (1969)
5.5   Yu.A. Kravtsov, I.M. Fuks, A.B. Shmelev: Izv. Vyssh. Uchebn. Zaved. Radiofiz. **14**, 854 (1971)
5.6   V.V. Tamoikin, A.A. Fraiman: Izv. Vyssh. Uchebn. Zaved. Radiofiz. **11**, 56 (1968)
5.7   E.V. Chaevsky: Izv. Vyssh. Uchebn. Zaved. Radiofiz. **8**, 1128 (1965)
5.8   D.E. Barrick: Proc. IEEE **56**, 1728 (1968)
5.9   E.V. Chaevsky: Izv. Vyssh. Uchebn. Zaved. Radiofiz. **9**, 400 (1966)
5.10   B.F. Kuryanov: Akust. Zh. **8**, 325 (1962)
5.11   A.I. Kalmykov, I.E. Ostrovsky, L.D. Rozenberg, I.M. Fuks: Izv. Vyssh. Uchebn. Zaved. Radiofiz. **8**, 1117 (1965)
5.12   I.M. Fuks: Izv. Vyssh. Uchebn. Zaved. Radiofiz. **9**, 876 (1966)
5.13   B.I. Semenov: Radiotekh. Elektron. **11**, 1351 (1966)
5.14   B.I. Semenov: Radiotekh. Elektron. **15**, 595 (1970)
5.15   A.N. Teokharov, A.B.Shmelev: Izv. Vyssh. Uchebn. Zaved. Radiofiz. **25**, 40 (1982)
5.16   J.E. Burke, V. Twersky: J. Acoust. Soc. Am. **40**, 883 (1966)
5.17   W.A. Rense: J. Opt. Soc. Am. **40**, 55 (1950)
5.18   A.A. Gershun, O.I. Popov: Svetotekhnika **1**, 3 (1955)
5.19   V.K. Polyanskii, V.P. Rvachov: Opt. Spektrosk. **22**, 279 (1967)
5.20   J.A. Ogilvy: *Wave Scattering from Rough Surface*, in Reports on Progress in Physics **50**, 1553 (1987)

Subject Index

MIX
Papier aus verantwortungsvollen Quellen
Paper from responsible sources
FSC® C105338

If you have any concerns about our products,
you can contact us on
ProductSafety@springernature.com

In case Publisher is established outside the EU,
the EU authorized representative is:
Springer Nature Customer Service Center GmbH
Europaplatz 3, 69115 Heidelberg, Germany

Printed by Libri Plureos GmbH
in Hamburg, Germany